Soil Fungicides

Volume II

Authors

A. P. Sinha, Ph.D., F.P.S.I.
Senior Research Officer
Plant Pathology
G.B. Pant University of
Agriculture and Technology
Pantnagar
India

Kishan Singh, Ph.D., F.P.S.I., F.N.A. Sc.
Director
Indian Institute of
Sugarcane Research
Lucknow
India

A. N. Mukhopadhyay, Ph.D., F.P.S.I.
Professor
Plant Pathology
G.B. Pant University of
Agriculture and Technology
Pantnagar
India

CRC Press, Inc.
Boca Raton, Florida

Library of Congress Cataloging-in-Publication Data

Sinha, A. P.
  Soil fungicides.

  Bibliography: p.
  Includes index.
  1. Soil fungicides. I. Singh, Kishan.
II. Mukhopadhyay, A. N. (Amar Nath), 1940-
III. Title.
SB951.3.S57   1988   632'.952   87-24232
ISBN 0-8493-4548-0 (v. 1)
ISBN 0-8493-4549-9 (v. 2)

This book represents information obtained from authentic and highly regarded sources. Reprinted material is quoted with permission, and sources are indicated. A wide variety of references are listed. Every reasonable effort has been made to give reliable data and information, but the author and the publisher cannot assume responsibility for the validity of all materials or for the consequences of their use.

Direct all inquiries to CRC Press, Inc., 2000 Corporate Blvd., N.W., Boca Raton, Florida, 33431.

© 1988 by CRC Press, Inc.

International Standard Book Number 0-8493-4548-0 (v. 1)
International Standard Book Number 0-8493-4549-9 (v. 2)

Library of Congress Number 87-24232
Printed in the United States

DEDICATED
TO
OUR WIVES

Kiran, Sheela, and Sumitra

PREFACE

The need for fungicides is created by fungal activities detrimental to the welfare of mankind. As the bases for such destructive activities are identified, compounds are sought to control the fungi involved. Soil fungicides have been used for a long time to control soil-inhabiting plant pathogens successfully, and within the last two decades, a striking increase has occurred in their use. Moreover, a great number of compounds have since been discovered, new techniques of application have been developed, and new concepts on modes of action have evolved. Tremendous achievements have also been made in their use with reference to plant disease control. At no time in the past has interest in fungicides been greater than during the current period of transition from the protective to systemic fungicides.

The success of soil fungicides depends on the complex nature of soil, which modifies the action of these compounds. Moreover, in field tests, a number of factors are operative, i.e., formulations, intrinsic weakness in the compound, mode of action, changes in physical and chemical properties of soil, and mechanical pretreatment of the soil. Also the amount of chemical residues and economic considerations have to be taken into account for the effective and efficient utilization of the soil fungicides. An understanding of the basic scientific principles involved would clear the way for future success. There has been a rising awareness of the toxicological hazards and the environmental impact of fungicides. With the advent of more selective fungicides, fungal resistance has emerged as a major problem.

Of the many aspects of pesticide behavior in the soil environment which have currently received scientific attention, there has been a notable acceleration in the microbiological field in research institutions globally. The vast amount of information generated is widely dispersed in diverse scientific journals, in several languages. This set represents an attempt to bring this information together in two volumes, to summarize and evaluate recent developments, to integrate them with significant developments of the past, and to attempt some projections for the future.

Our intention has been to obtain a set which, apart from serving as a reference source for research workers or scientists already in the field, might also interest those who work in allied areas such as pesticide microbiology, toxicology, and microbial ecology. We have also considered the growing number of postgraduate students who are interested in soil fungicides. For these reasons we have included an "introductory" chapter to give basic information on the subject. Overlap has been minimized, but retained where necessary for clarification or emphasis.

Soil Fungicides contains 13 chapters. In Volume 1, the first chapter deals with introduction, definitions, history, classification, formulations, and methods of application. Chapters 2 through 8 deal with individual groups of fungicides and their members, with regard to their manufacturers, chemical name, toxicological information, mechanism of action, and disease control developed for commercial use of experimental work. In Volume 2, Chapters 9 through 12 deal with the various aspects of soil fungicides, such as factors affecting their efficacy, microbial interactions, nontarget effects of pesticides on soil borne plant pathogens, and development of resistance to fungicides. In the last chapter, evaluations of fungicides have been included because of their significance in the fungicidal recommendation. Literature pertaining to each chapter has been cited in the last, so that the reader can obtain more detailed information on the topic(s).

We are indebted to Mr. B. G. Starkoff, President, CRC Press for his help and guidance at the initial stage of developing the outline for this book, to Ms. Sandy Pearlman, Managing Editor, Ms. Jan Boyle, Administrative Manager, Ms. Renee Taub, Coordinating Editor, and other skillful staff of CRC Press for their excellent cooperation, patience, and helpfulness in resolving problems encounted during the writing of the manuscript. Our thanks are also due to the authors and publishers who have allowed us to reproduce much published information as acknowledged at the appropriate places.

Our sincere thanks are due to Drs. J. P. Upadhyay, U. S. Singh, and H. S. Verma for critically going through the manuscript. We should like to thank Mr. J. C. Papnai who has typed the entire manuscript with patience and cheerfulness. Finally, we sincerely thank our families for their great forbearance over the past years during which we were preoccupied with the manuscript of the book. In particular, we express our appreciation and thanks to our wives, Kiran, Sheela, and Sumitra, for their constant inspiration, encouragement, and cheerfulness during the entire period of preparation of the manuscript. It would be unrealistic to suppose that the text is free from errors and notification of any errors or omissions will be appreciated.

A. P. SINHA
KISHAN SINGH
A. N. MUKHOPADHYAY

THE AUTHORS

A. P. Sinha, Ph.D. is Senior Research Officer and Project Leader at G.B. Pant University of Agriculture and Technology, Pantnagar, India. Dr. Sinha graduated in 1968 from the University of Gorakhpur with a B.Sc. degree in agriculture. He obtained his M.Sc. (Agric.) in mycology and plant pathology from Banaras Hindu University and the Ph.D. degree in plant pathology from Kanpur University. He worked on various aspects of soil fungicides for his Ph.D. degree. Currently Dr. Sinha is Principal Investigator of a research project on "Soil Fungicides" financed by the Indian Council of Agricultural Research.

Dr. Sinha is a Fellow of the Indian Phytopathological Society. Dr. Sinha has much experience working on several crops, such as wheat, soybean, triticale, sorghum, and sugar cane, with special reference to the use of fungicides for disease control. He has been actively engaged in teaching and guiding research theses of graduate students in plant pathology. Dr. Sinha has authored nearly seventy scientific publications, mostly on the interactions between pesticides and soil microorganisms. He has also contributed chapters to three books.

Kishan Singh, Ph.D. is Director of the Indian Institute of Sugarcane Research. Dr. Singh received his M.Sc. (Agric.) in 1952 from Agra University and the Ph.D. in 1955 from Louisiana State University, U.S. He was appointed Assistant Plant Pathologist at the Indian Agricultural Research Institute, New Delhi, in 1956 and Mycologist and Head of the Division in 1958 at the Indian Institute of Sugarcane Research, Lucknow. In 1968, he was appointed Director of the Institute.

Dr. Singh is a Fellow of the Indian National Academy of Sciences and the Indian Phytopathological Society. Dr. Singh has been a member of the editorial boards of several learned societies in the country, Vice President of the Association of Plant Pathologists of India, and President of the Association of Sugarcane Technologists of India. He has also been a member of expert panels on Plant Pathology and Entomology/Nematology of the Indian Council of Agricultural Research (ICAR), Advisor to the Industrial Finance Corporation, and served on the Governing body of ICAR. He has traveled widely as a member or leader of teams to look into the feasibility of the sugar industry in several places in India and also in Malaysia, the Sudan, Ethiopia, Tanzania, and Cuba.

In the course of his varied and distinguished career, he has been awarded the Shanti Swaroop Bhatnagar Award for Biological Science (1976), the B.K. Srivastava Memorial Award (1980), and the Andhra Pradesh Farmer's Award (1982). Dr. Singh has published more than 200 papers contributed chapters, and edited several books.

A. N. Mukhopadhyay, Ph.D. is Professor of Plant Pathology at G.B. Pant University of Agriculture and Technology, Pantnager, India, Dr. Mukhopadhyay received his B.Sc. (Agric.), M.Sc. (Agric.), and Ph.D. degrees from Banaras Hindu University in 1960, 1962, and 1968, respectively. He has been on the staff of the G.B. Pant University of Agriculture and Technology since 1967. He was appointed an Assistant Professor in 1967, Associate Professor in 1971, and Professor and Head in 1984. He also worked as a Commonwealth Academic Staff Fellow at the University of Newcastle upon Tyne, U.K., during the year 1977 to 1978. Dr. Mukhopadhyay also served as Professor and Head of Plant Pathology at Gujarat Agricultural University, India for two years from 1982 to 1984.

Dr. Mukhopadhyay has been engaged in sugar beet disease research since 1967 as a leader of the sugar beet pathology research project. Thus far, 16 students have received their M.Sc. (Agric.) and Ph.D. degrees under his guidance. Dr. Mukhopadhyay's work has been highly acclaimed through the Professor M. J. Narasimhan Academic Award of the Indian Phytopathological Society for outstanding research contributions in the field of plant pathology. Nationally, he has served on numerous top level committees. Currently, Dr. Mukhopadhyay is President of the Indian Society of Mycology and Plant Pathology, Member of the Scientific Panel of the Indian Council of Agricultural Research on plant pathology, Technical Advisor to the journal *Pesticides,* and Editor-in-Chief of the *Indian Journal of Mycology and Plant Pathology* (1984 to 1986), and several others. He is a Fellow of the Indian Phytopathological Society and Chairman of the Working Group on Soilborne Diseases. Dr. Mukhopadhyay has authored more than seventy research publications and has written or coedited four books. His latest publication *Handbook on Diseases of Sugar Beet* has been published by CRC Press.

TABLE OF CONTENTS

Volume I

Volume II

Chapter 9

FACTORS AFFECTING EFFICACY OF FUNGICIDES IN SOIL

I. INTRODUCTION

The only criterion by which to judge an agricultural fungicide is its success in controlling plant diseases, but this is composed of many things. Besides by direct application, fungicides may also enter into the soil from a variety of treatments. When chemicals are used as spray applications to the aerial parts of plants, excess solution/suspension may drop onto the soil surface or may reach the soil surface directly. In preplanting, planting, or other treatments, toxicants are usually applied either in solution/suspension, as dusts, or in granular formulations. The ultimate behavior of a fungicide in soil basically depends on the physical, chemical, and biological properties of both the soil and chemical. Sorption, diffusion, volatilization, leaching, runoff, microbial and chemical degradation, photodegradation, and plant uptake are all significant processes affecting fungicide behavior in soil. Interaction of these various parameters determines the effectiveness of the chemical applied to soil and its residual life in soil. Therefore, it should not be surprising that out of many candidate compounds only a few are successful. Yet from each failure there is a lesson to be learned if we look closely enough. Failure may also be due to a defect in formulation or may arise from an intrinsic weakness in the chemical itself that can be remedied only by the synthesis of a new derivative with improved properties. Often the case is hopeless. Whatever the cause, an understanding of the basic scientific principles involved will help clear the way for future successes.

According to Munnecke,[1] much of the early research on soil fungicides was empirical. A change from empirical to experimental research began about 1950, as emphasized by Fuchs,[2] when he stated that the research on chemical disinfestation was just at the beginning of its development. This view, however, was not entirely shared by McNew,[3] who made the rather pithy comment that studies on fungicides in general had a great weakness, because scientists were working in a vacuum. It is interesting to note that the vacuum has been gradually overcome and significant research contributions on fungicides in soils have been made to further advance our overall understanding of the behavior of all pesticides in soil. Kaufman[4] rightly pointed out that since the publication of Carson's[5] *Silent Spring,* increased awareness of our environment dictates that we not only understand the efficacy of chemicals released into the environment, but that we also aim for a total concept of the ultimate fate of the chemical. The fate and behavior of insecticides and herbicides are relatively well understood and, with few exceptions, very little attention by comparison has been devoted to fungicide research.

Several review articles have been published on the environmental behavior of fungicides.[6-12]

II. BEHAVIOR AND FATE OF FUNGICIDES IN SOIL

A. Research Methods Used in Studying the Fate of Fungicides in Soil

Munnecke[1] summarized techniques[13-22] used to assess fungicide behavior in soil, including chemical analysis response of spores or mycelium of test fungi buried in treated soil,[23-29] response of spores in agar media to diffusates directly from soil[30] or from extracts from fungicide-treated soil,[36] response of fungi to various vapors from treated soil,[19,24,32] disease reaction of susceptible plants in infested soil treated with fungicides,[33-36] infestation of agar media in treated soils,[37] and combination of one or more of the above.[19,36-40]

Chemical analyses are mostly qualitative and quantitative in terms of actual persistence of the active ingredient and characterization of degradation rates and products. However, these should be used in conjunction with other methods to determine the various processes occurring in the soil. No single method is suited for all purposes. Bioassay closely measures actual fungitoxicity. Reproducible results can be obtained if the assay is rigidly standardized, particularly with respect to the quantity and genetic stability of the test material used. Viability, number, and age of the assay organism are important aspects of the bioassay method. Demerits of bioassays are that they do not measure effects that may occur after the fungicide is no longer detectable.[1] This technique is also dependent upon water or gas diffusion and a chemical may possibly be advantageous or active in the soil without having these properties. However, this has little usefulness for determining plant disease response to assay fungicide under field conditions. At the same time, the method is time consuming, requires considerable space, and is difficult to make quantitative. Regarding chemical assays which indicate an actual extractable chemical, they are not necessarily indicative of actual residual toxicity. According to Munnecke,[1] inherent complexities of soil itself must be considered in studying soil fungicides.

B. Sorption

A survey concerning the sorption of fungicides in the soil has been given by different authors (Burchfield,[6] Goring,[8,41,42] Hartley,[43] Munnecke,[1] Kaufman,[4] Knight et al.,[44] and Osgerby[45]). They have discussed the relevance of adsorption to fungicide and/or fumigant behavior in soil. When a toxicant which is introduced into the soil is retained or held by the soil, we consider that the chemical has been sorbed. The concentration of fungicide in soil water or air and the length of time it is maintained decide the extent to which the organism is controlled and how long control will be maintained. This depends on the physical, chemical, and biological properties of the chemical and the soil and their interaction. Adsorption to soil constituents will affect the rate of volatilization, diffusion, or leaching as well as availability of the chemical to microbial or chemical degradation, or uptake by plants or other organisms. Sorption is physical or chemical. Physical sorption occurs with nonionic compounds, while the chemical sorption occurs with anionic or cationic compounds. Hydrogen bonding is intermediate between physical and chemical sorption. Physical sorption generally involves several layers and low binding strengths. Chemical sorption involves high binding strength; although several layers may be present, only the first layer is chemically bonded to the surface. Hartley[46] has proposed that solution of organic chemicals in oily constituents of soil organic matter also occurs. Since the particulate matter provides sorptive surfaces, it has been the basis of several studies. The Langmuir isotherm equations generally apply to compounds adsorbed in monomolecular layers. S-shaped isotherms have been obtained from fumigants in soils, and the Brunauer, Emmett, and Teller (BET) equations have been applied to these cases, based on the assumption that the divergence from linearity is due to multimolecular adsorption resulting from action of van der Waals' forces.[47] The BET equations have been widely used, although Burchfield[6] stated that the basic theory is unsound.

Clay minerals and organic matter are among the most important agents of sorption.[48,49] Components of soil influence sorption. The principal sorbing agent in the soil is the colloidal aluminoclay micella, which has a clay mineral core consisting of sheets of silica and alumina giving it its fundamental properties. Although the montmorillonite, kaolinite, or illite core forms the principal mass of the clay particle, it may also contain organosilicate gel complexes.[50] Jurinak[51] reported that the sorption of chemicals to soil micelles is influenced first by the nature of the mineral forming the core of the clay particle. Nonpolar chemicals such as ethylene dibromide (EDB) are less readily sorbed on montmorillonite than on kaolinite or illite.[47] Montmorillonite, because of its structure, seems to offer less sorbing surface-to-nonpolar chemicals than keolinite or illite does. Water bound into the porous montmorillonite lattice seems to block mechanically the sorption of nonpolar chemicals such as ethylene dibromote.[52] Soil adsorption is often dependent on intermolecular surface reactions: the sorption capacity of the soil increases with decreasing particle sizes. Soil particle size relationships, especially the amount of clay and silt, are of great importance. Stark[53] observed a direct correlation between soil particle size and adsorption, with regard to chloropicrin. Considerable effort has been made to determine the relative importance of clay and organic matter in sorption. The task is difficult because both clay and organic matter sorb physically and chemically. It is now generally accepted that organic matter is most important, except in dry soils where clay content and specific surfaces are the principal factors regulating sorptive capacity.[54] Hartley[43] pointed out that clay has "preoccupied" most researchers' minds, and that it was not important with large molecular particles, especially uncharged ones. This was not to say that the clays were of no importance; however, Goring[7] has come to the same brief conclusions. Increased organic matter, as a rule increasing sorption, was reported to be true for D-D mixture and 1,3-dichloropropene,[55-57] methyl bromide (MB),[27] EDB,[58] dibromochloropropane,[59] and methylisothiocyanate (MIT).[60] According to Stark,[53] however, this was not true for chloropicrin. Since organic materials increase the possibility of both covalent bonding and solution in solvents of an organic nature, we would not expect all biocides to respond alike.

Soil moisture also has great influence on sorption of some fungicides. It was demonstrated that EDB was rapidly sorbed by dry soil, but the amount sorbed dropped off sharply as moisture contents rose above approximately 15% and rose very slightly as a moisture content of 60% was reached.[61] Wade[62] later reported that sorption isotherms of soils held at less than 50% relative humidity were sigmoidal and corresponded to BET equations, but above 50% the curves did not get the equations. Jurinak[51] also fitted sorption isotherm data for EDB to the BET calculations and assumed that van der Waals' forces were operative after a monolayer was formed. Call[52] worked out that EDB was displaced from sorption by water in three stages. At low humidities water molecules more successfully competed for adsorption sites and displaced the EDB molecules. At higher humidities (10 to 20%) when the formation of a monolayer of water was complete, a greater number of water molecules of EDB was sorbed onto the sorbed water. Then, solution of EDB in soil water did not become important until the sorbed water films attained the properties of bulk water, which probably occurred at field capacity.[1] Call[52] examined EDB sorption at 20 different soils held at moisture contents corresponding to field capacity. The data were expressed as sorption coefficients (slope of the isotherms), and correlations with surface area, organic matter content, moisture content, and clay content were obtained. The last three properties also correlated with each other. Biocides which are insoluble in water and nonpolar are adsorbed less as the expansion of the clay fraction increases.[63-65] Dry clay and dry peat can adsorb higher quantities of MB than moist material. Sand has the lowest sorption index and shows only slight differences in respect to the water content.[66] Munnecke and Ferguson[27] confirmed these results with MB and have observed that the absolute sorption indices decline in the following sequence: peat, clay, and sand. Munnecke[1] mentioned that the sorption coefficient could be predicted with reasonable accuracy from moisture content alone, on the basis of moisture level at field capacity. As indicated by the correlation of sorption coefficients of the various soils, clay content and specific surface area were the principal factors regulating sorptive capacity of dry soils, whereas organic matter was critical at field capacity. Leistra[56] obtained similar results with *cis*- and *trans*-1,3-dichloropropene (1,3-D, D-D), but also noted effects of temperature and isomeric form. The amount adsorbed at 2°C was about three times greater than that adsorbed at 20°C. The *trans*-isomer was more strongly adsorbed than the *cis*-isomer.

The pH of soil as well as the charge of clays affects the dissociation of toxicants and thus is very important in soil processes. Frissel[67] pointed out that all compounds are adsorbed strongly at low pH, anionic substances are adsorbed negatively at slightly basic conditions, and nonionic compounds are moderately adsorbed. Craft[68] reported that cationic compounds in moist soil are immediately affected by the base-exchange complex in a manner similar to the rapid immobilization of ammonia by soil. A similar effect occurs with quaternary nitrogen compounds. Although they may be precipitated as insoluble compounds of soil basis, nonpolar and anionic compounds are not so affected by soil pH. The pH effect was shown strikingly by Harris and Warren[69] with 2-*sec*-butyl 4,6-dinitrophenol, dinoseb (DNBP), which was strongly adsorbed at low pH but only slightly adsorbed at high pH. Munnecke and Martin[70] obtained a similar effect in three soils treated with MIT. The amount of MIT is released linearly as soil pH is increased, although the slopes of the three vary.

C. Diffusion and Volatilization

Success and failure of soil fumigation depend on the optimum distribution and volatilization of the biocide in soil. The significance of these processes becomes most apparent when pesticides are found in air or on soil at sites far from those originally treated. Excellent reviews have been prepared on diffusion[1,4,8,9,43,71] and several papers are pertinent for this very abbreviated discussion.[72-75]

If a chemical is not mechanically mixed into the soil, there are only two ways through which it can be distributed: (1) dispersion in water solution or (2) diffusion as a vapor. There are two factors which influence diffusion: (1) the chemical and physical nature of the diffusing materials and (2) the nature of the medium (soil) in which the material diffuses. The toxicant and soil may interact, bringing about sorption and chemical reactions which affect the degree of diffusion. Basic properties of a soil biocide — i.e., tendency for sorption and chemical reactivity of vapor pressure and water solubility — influence its dispersion through the soil.[71] The characteristics of the diffusion medium. e.g., sorption capacity, pore volume, moisture content, temperature, and the application methods, are also of great importance.

Biocides, like allyl alcohol and formaldehyde, which are volatile but highly water soluble, and the nonvolatiles like sodium ethylenebisdithiocarbamate and sodium *N*-methyldithiocarbamate, must penetrate soil as water solutions to be effective. However, nonvolatile and water-insoluble fungicides such as maneb, zineb, and captan must be mechanically mixed with soil for effective results.[76]

Information on diffusion comes mainly from the work on biocides or soil fumigants, which move in the soil in the vapor phase. Potential diffusibility of a volatile biocide in soil is enhanced by high vapor pressure, low banding properties, and low water solubility. The degree of diffusion of a chemical through soil as a gas is reduced with decrease in vapor pressure, increase in bonding properties, and increase in water solubility. According to Kreutzer,[71] these are good rules to be remembered. The rate of the diffusion is important and it may explain the report by Youngson et al.,[73] who observed that factors that tend to increase diffusion of chloropicrin increased control, whereas factors decreasing diffusion of methyl bromide increased control. Also, incorporation of peat moss to soil increased depth of control by methyl bromide probably by decreasing the rate of diffusion. It has also been pointed out that diffusion is very important in transport of molecules over distances up to 1 mm from a surface of solid whose presence retards flow processes.[43] They were of the opinion that with chemicals, which are of low volatility, air diffusion may be important, because a substance partitioning even as much as 5000 times in favor of water from air would diffuse more rapidly in the air spaces of dry soil than in solution of water-logged soil. For a substance partitioning even as much as 50,000 in favor of water, air diffusion may be important.

According to Goring,[8] Fick's first law of diffusion was probably applicable to all diffusion processes in soil, including movement of pesticides through air, water, and organic matter:

$$\frac{dm}{dt} = -Do \frac{dc}{dx}$$

where Do is the diffusion coefficient of the vapor and dm/dt is the rate at which it diffuses through a plane of unit area against a concentration gradient of dc/dx. In simpler words, the rate of movement of a toxicant is directly proportional to the concentration of the toxicant and its diffusion coefficient. The diffusion coefficient is a quantitative way of expressing the difficulty diffusing molecules have in moving through different transfer media such as air, water, or organic matter. Goring[8] stressed the significance of water/air ratios and organic matter/water ratios. The ratio of concentrations of chemical in water and air will be constant for ideal gases at constant temperature and a wide range of concentrations.[7,46,73-75,77,78] These rates are estimated by dividing the weight of the chemcal per unit volume of air calculated from vapor pressure. Ratios have been calculated by Goring,[7] Hartley,[46] and Peachey and Chapman.[78] These can be used for estimating the distribution of toxicants in soil air and soil water with varying soil conditions, and the relative contribution of diffusion through soil air, and soil water to movement of the toxicant in soil. Toxicants with a water/air ratio under 10,000 should diffuse primarily through air, whereas those with ratios over 30,000 should diffuse primarily through water. Methyl bromide and carbon disulfide have water/air ratios under 10 and move so rapidly they are lost within hours to days when introduced into water. Compounds such as chloropicrin, EDB, 1,3-D, and MIT, with ratios of 10 and 100, disappear within a few weeks. 1-Chloro-2-nitropropane, 1,2-dibromo-3-chloropropane (DBCP), and 2-chloro-6-(trichloromethyl) pyridine with ratios of 100 to 2000 could remain in the soil for months if not decomposed by other mechanisms.

Some toxicants have very high ratios, varying from about 1 million to 70 million. It is unlikely that vapor transfer contributes significantly to movement of these chemicals through the soil. Significant losses from the soil surface have been demonstrated for urea herbicides[79,80] and triazines.[81] Munnecke et al.[19] demonstrated an unidentified volatile toxicant from soil treated with methylmercuric dicyanimide, where the ratio is about 19 million. Kimura and Miller[82] obtained evidence that the toxicant evolved was methylmercuric dicyanimide, which is unexpected considering its high water/air ratio.

Increased quantities of chemical introduced into soil increase the rate of diffusion.[83,84] This has been observed for carbon disulfide[85,86] and for dichloropropene-dichloropane mixture and 1,3-D;[57,84,87] chloropicrin, dazomet,[88] Vapam,[24] trapex,[89] DBCP,[90] and EDB.[86] However, the method of application and the formulation of the biocides in some cases greatly influence diffusion and lead to changes in the relationships of the above-mentioned factors as is reported for Vapam.[91] Furthermore, the sensitivity of pest organisms does not always correspond with the conditions for optimum diffusion. Quantity of fumigant used is one of the most important factors that determine the depth of fumigant penetration into soil. Baines et al.[87] conduced an experiment for the control of the citrus nematode *Tylenchulus semipenetrans* in sandy loam soil and observed that 45 gal of D-D mixture was needed to control nematodes to a depth of 3 ft. However, 80 gal/acre was required to obtain control of 6 ft. As has already been pointed out, the second overall factor in the diffusion of a biocidal fumigant is the nature of the medium in which the diffusion occurs: with increasing pore volume and temperature and decreasing moisture, the rate of diffusion in soil rises. It is expected that the degree of porosity is the most important soil factor in diffusion.[65,72] Call[72] proposes that the apparent diffusion coefficient of a compound in soil is related to its value in air by

$$D = (S - S') \frac{Do}{K}$$

where S is the total porosity of the soil, S' is the fraction of pores that is blocked, and K is the average tortuosity of the pores. The tortuosity factor takes into account the fact that the pores in soil through which vapors diffuse are unlikely to be straight, so that a gas must travel a longer distance to reach any given destination than it would in air. In general, with increasing pore volume, greater is the degree of gaseous diffusion. Several factors, including soil type, soil moisture content, and soil compaction influence the degree of porosity. Fumigants like carbon disulfide,[64] dichloropropene-dichloropane mixture,[87] and chloropicrin[53] diffuse best in sandy soils and least in clay soils. Kreutzer[71] suggested that this may not be entirely a matter of porosity and sorption may also be involved. The texture and structure of a soil may influence the movement of a gaseous material in the soil. The lighter the texture, the better the gaseous flow of a soil biocide.[64] Soil compaction also affects soil porosity. The greater the compaction, the less permeable the soil to the movement of volatile chemicals.[92] Maximum compaction of the soil is brought about by a combination of rain and the use of heavy machinery. The plow sole is a compacted layer commonly encountered 8 to 12 in. below the surface of the soil. It is formed by mechanical compaction and the deposition of mineral salts. Reports show that the hardened layer is impermeable to CS_2, EDB, D-D mixture, and dichlorobromopropene.[92-94]

A high soil water content decreases the porosity and in turn interferes with the distribution of nonpolar biocides by mechanical blockage. Kreutzer[71] cited that this is an old observation. Sebate, one of the pioneers in the use of carbon disulfide, is quoted by French[95] as stating, "never inject a solution into damp soil, because the diffusion of the poisonous gases has no effect beyond the sides of the hole made by the injector." High moisture impedes the diffusion of carbon disulfide,[96] and EDB,[58,72] D-D mixture,[97] and chloropicrin.[53]

Kreutzer[71] observed that movement of polar chemicals may be affected by even moderate soil moisture. At the soil moisture of 50% of the moisture equivalent, the diffusion of allyl alcohol and other volatile polar materials in the soils is reduced. Contrarily, the movement of nonpolar materials shows no interference until the soil moisture content equals or exceeds the soil moisture equivalent. According to Hoffman and Malkomes,[9] moderate soil moisture inhibits sorption to the soil complex and therefore also to the chemical reactions, thereby favoring the diffusion.

Soil temperature also affects the diffusion of volatile soil fumigants. Diffusion of EDB, chloropicrin, and dichloropropene is accelerated[65,98] with an increase in soil temperature. In accordance with the gas laws, this would be an expected response.

The dosage of the applied material also influences the depth of biocide penetration. MB has the highest rate of diffusion: D-D, chloropicrin, and MIT have middle to low rates. To determine the depth effect of soil fumigants, test organisms placed at different soil levels beforehand are used.[24,27,97,99-110] Drosihn[111] estimated the diffusion of MB in the soil with the aid of thermal conductivity. This method was first used by Monro et al.[112] when fumigating slips. It was observed that the concentration of MB was about the same throughout the top 20 cm of the soil. A decrease in concentration was observed at a depth of 30 cm. It has been proved that MB reaches a depth of 60 cm after 2 hr and 100 cm after 6 hr.[111] The concentration at such depths remains low for quite a while and is insufficient to destroy pathogens.[113] A quick diffusion of MB from the top 30 cm of soil into the atmosphere was performed after the removal of the sheet (48 hr after application). However, at a depth of 60 to 100 cm, MB continued to accumulate for a longer time. Hoffmann and Zinkernagel[103] mentioned that due to the concentration time effect, soil microbes including parasitic fungi are eliminated at these soil levels. It has been suggested that to obtain the optimum depth effect for MB on soil the latter should not be too moist: in case chloropicrin is added, even less moisture is needed.[114,115] Similarly, the humus and clay content should not be too high. Mixed preparations sometimes require different optimum conditions from the pure substances.

Kreutzer[71] rightly mentioned that a discussion would not be complete without some consideration of diffusion patterns of soil fumigants. Several methods of applying volatile

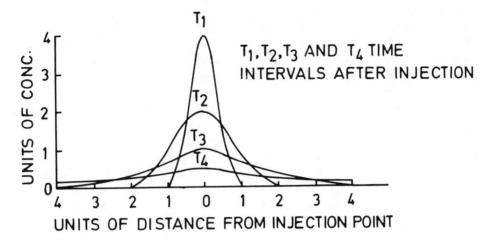

FIGURE 1.   Relation between distance from injection point and concentration of fumigant in the soil at various time intervals after infection. (From Goring, C. A. I., *Annu. Rev. Phytopathol.*, 5, 301, 1967. (C) 1967 by Annual Reviews. With permission.)

chemicals to soils have been very well reviewed.[7,8,78] If a chemical is applied in the soil at a single point, it diffuses as a gas outward from this point forming a pattern of definite size and shape. This size and shape of the pattern is usually determined by the limits of biocidal effectiveness. Figure 1 clearly demonstrates the basic pattern of diffusion once injected. The area under each curve represents the dose an organism would receive at that distance from injection point.

Diffusion pattern data, unless related to all the factors which influence diffusion, are difficult to interpret.[71] Kreutzer[71] further concluded that comparisons cannot be made between chemicals unless tests are conducted under identical conditions. He compared the diffusion pattern of allyl alcohol and chloropicrin, following injections in a sandy loam soil in optimum planting conditions. A spherical pattern with a maximum radius of 2 in. was obtained with allyl alcohol at a dose of 1.0 mℓ, while chloropicrin at a 0.5-mℓ rate gave an oblate pattern with a maximum radius of 10 in. The type of pattern formed is also influenced by the presence or absence of a soil "seal". A general rule needing to be mentioned here is that if a fumigation is not sealed into the soil by use of surface watering or a surface cover, a lethal concentration of fumigant fails to build up on the top of a 2-in. zone. Several workers have demonstrated this phenomenon with the use of CS_2,[116] chloropicrin,[53] D-D mixture,[55] and dibromochloropropane.[117]

Fumigants differ remarkably in the speed with which diffusion patterns form. Dibromochloropropane appears to move slowly, whereas MB and CP diffuse rapidly. Ichikawa et al.[117] observed that dibromochlorpropane attained its maximum diffusion radius of 15 in. 9 weeks after application as injection at the rate of 0.22 mℓ/ft in field.

Goring[7] stated that volatilization is dependent upon the inherent volatility of the compound, temperature dilution and biological materials, and general restraint in diffusion caused by soil. Chemicals with high vapor pressure penetrate field soil best, if the gas is not allowed to escape into the air. At 20°C the vapor pressure of MB is 1380 mm, but for chloropicrin it is only 20 mm.[73] Therefore, it is useless to apply MB without covering the soil. The effectiveness of biocides with lower vapor pressures such as MIT, CP, and CS_3 frequently enhances by trapping. According to Goring,[8] there are certain advantages with the use of semivolatile compounds. On one hand, thorough mixing with soil of these chemicals is not required. On the other hand, as compared to volatile chemicals, they are more efficient because of their greater tendency to stay in the treated zone. However, nonvolatile compounds required thorough mixing with the soil for effective control. According to Goring,[8] the partition coefficient is important in the process of volatilization. Generally, the quantity of the chemical applied is much less than that required to saturate the soil water phase. Adsorption of pesticides will be greater on dry soil surface. However, at normal soil moisture content, volatilization of toxicants takes place from dilute aqueous solution at the exposed soil surface. From the ratio of water solubility to vapor pressure, an estimate of potential volatility can be obtained. This will indicate what portion of the chemical is in vapor phase. This ratio is advantageous as a guide, but it is affected by adsorption, which decreases the amount of material present in vapor phase,[118,119] or by seed treatment. Munnecke et al.[19] were unable to show that soil treated with pentachloronitrobenzene (PCNB) contained fungitoxic vapors, but Richardson and Munnecke,[120] using a more sensitive bioassay method, demonstrated fungitoxic concentration of PCNB in the soil air. Redistribution of methylmercuric dicyanimide on seed because of its volatility is well known,[121] while it is questionable whether the actual chemical or some decomposition products might diffuse in the soil zone around the seed.[19,82] Whatever diffuses is highly toxic to *Pythium* but not to *Rhizocotonia*.[19]

Many ingenious cultural operations have been devised to help prevent loss by volatilization to the air.[1] Half of the rate is injected deep in the soil, permitting biocide to diffuse for a time. After that, the soil is plowed in order to place the surface soil at the bottom. Finally, the second half of the rate is injected to kill the remaining organisms. This technique has been modified for those chemicals, such as chloropicrin, which have the ability to sorb large quantities. For this, only one injection is made. A few days after, the surface soil is inverted by plowing and the surface is compacted by rollers to restrict gas emission. The principle involved here is that the gas which is sorbed deep in the soil releases sufficient gas after the plowing to kill these organisms which could survive initially on the surface.[1] There are several other operations designed to prevent loss by volatilization, like irrigation or packing the surface with a heavy roller after injection of the fumigant. However, these techniques have hardly been used since the advent of cheaper means of applying traps.

D. Leaching

The movement of a fungicide in the soil matrix with water, commonly known as leaching, is an important factor in relation to the biological activity of the fungicide. Pesticides are moved into soil by rainfall or irrigation. Three processes govern the movement of the toxicant: desorption, diffusion through water, and hydrodynamic dispersion.[8,122] Solubility of the pesticides in soil solution has also been considered important.[8,43] Toxicants highly or moderately soluble in water are leached from soils more quickly than those that are less soluble. Desorption is proportional to the amount of toxicant sorbed.[122,123] Diffusion through water obeys Fick's law. Hydrodynamic dispersion[123] is caused by water percolating downward through the centers of the pores more rapidly than along the sides. The net result of these three processes is increased spread of the toxicant throughout the soil. The spreading nature of the concentration-depth curves with increasing amounts of water is reminiscent of the curves for diffusion of fumigants. Fick's law actually underlies both processes, except that with the fumigant itself, whereas with toxicants being leached percolation of water through the soil pores is an additional driving force.[8]

Hydrodynamic dispersion is of considerable practical significance when applying pesticides in irrigation water. They will be distributed to greater depths if the soil is initially moist rather than dry, or if the water infiltrates rapidly rather than slowly.[59] The distribution patterns for solutes moved through by water have been discussed in detail by Thomas.[123] Figure 2 shows typical patterns for a toxicant placed on the soil surface and leached with increasing amounts of water. Curves obtained with increasing amounts of water have widening bases, decreasing peak heights, and are skewed upward. The amount of water required to move concentration peaks to selected depths is directly related to the organic matter/water

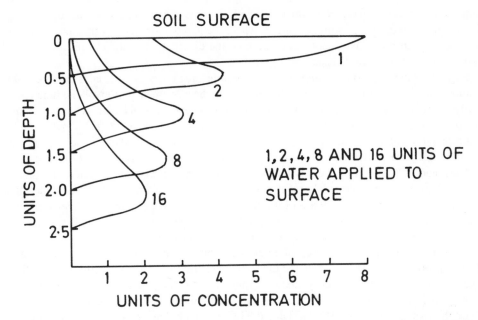

FIGURE 2.   Relation between amounts of water applied and concentration of toxicant at various depths, assuming application of the toxicant to the soil surface. (From Goring, C. A. I., *Annu. Rev. Phytopathol.*, 5, 304, 1967. (C) 1967 by Annual Reviews. With permission.)

ratios of the toxicants being leached.[8] Adsorption to soil colloids affects leaching. Soil organic matter is correlated more closely to retention against leaching than is clay for nearly all compounds, the organic cations diquat and paraquat being exceptions.[124] Toxicants are more readily leached in light than in heavy textured soils, reflecting the typically higher clay and organic matter contents and higher field capacity of heavy soils. Field capacity is often negatively correlated with pesticide mobility. A given quantity of water, from rainfall or irrigation, will penetrate further in soils of low-field moisture capacity. This term is also correlated with the adsorptive organic matter and clay components and, thus, the relationship is complex. In relatively acidic soils, the reduced pH may restrict leaching (by increased adsorption). The effect of initial soil moisture content and water flux is unclear. Mass transfer of toxicants occurs upward in drying soil, and laterally from wetter to drier soils, as well as downward. Movement of water in soil is upward as well as downward. This alternating movement tends to distribute the toxicant evenly through the profile, especially if it is neither too easy nor too difficult to leach.

According to Freed and Haque,[125] in a three-dimensional coordinate system leaching may occur in a downward, lateral, or upward direction. In practice, the important factors controlling leaching are adsorption, water solubility, soil type, and amounts of moisture and percolation velocity. The downward leaching of a chemical accompanies the movement of water. Water arriving at the soil surface first dissolves the chemical and carries the chemical with it as it passes through the soil. The situation in soil is analogous to column chromatography. Under rapid water percolation, the bulk movement of a chemical will be in the direction of water flow. During the leaching process, there is always a dynamic equilibrium between the free chemical and the chemical in the adsorbed state. Consequently, a tightly bound chemical should leach slowly and vice versa. Heat of solution can give a qualitative indication of the strength of binding. A highly soluble chemical, if not tightly adsorbed, will be leached more easily because of its tendency to go into solution. Temperature, which affects the solubility, will also play an important role. The intensity and frequency of moisture or rainfall will markedly affect the distribution of a chemical. Since the adsorption processes depend to a great extent on the nature of the soil, the type of soil will also influence leaching. Under static conditions of soil moisture or where the percolation rate is very slow, diffusion will become an important factor in determining the distribution of the chemical in the soil profile, at least over short distances.

Regarding the vapor-phase movement of volatile fungicides in soil, considerable information is available. However, very little is known of the movement of nonvolatile fungicides. Early efforts with some soil fungicides focused on soil drenches which carried the assumption of some fungicide transport. Munnecke[126] studied the movement of 13 fungicides in large-diameter soil columns and observed that water-soluble compounds such as nabam and Panogen are greatly affected by leaching. The less soluble chemicals such as thiram, captan, ferbam, and PCNB were less easily leached, and their movement was affected to a greater degree than is explainable solely on the basis of water solubility. He also concluded that penetration was greater for solution formulations than for suspensions, for finer fungicide particles than the coarse particles, for coarse-textured soils, and for initially dry soils.

The process of leaching may be considered to be analogous to chromatography, and consequently one should expect a maximum concentration of a chemical at some point of the soil profile. Helling[127] summarized the techniques used in pesticide-mobility evaluation. These include field and laboratory methods. In the former category are residue analyses with depth and lysimeter experiments. Among laboratory methods are soil columns and thin-layer chromatography (soil TLC). The soil TLC technique has proved useful in studies of relative mobility of toxicants.[128-130] Water moves in the soil layer by unsaturated flow,[128] resembling typical water flow under field conditions. Helling[131] demonstrated 33 fungicides in terms of leaching and diffusion characteristics with a fungal bioassay developed for use with soil TLC. Cycloheximide, cycloheximide oxime, Ceresan L, diazoben, and oxycarboxin were found to be relatively mobile compounds. Chloronil, Chloroneb, DCNA, dichlone, dodine, hexachlorophene, oxythioquinox, PCNB, TCNA, ETMT, and zineb were immobile. The behavior of dodine was similar to that of the herbicides paraquat and diquat. All three are organic cations and are strongly adsorbed to soil because of the cation exchange capacity of the soil. The movement of a number of chemicals on soil TLC plates was characterized by several inhibition zones. Few spots were presumably due to fungitoxic degradation products formed either on the soil TLC plate, or more likely in the original samples. Ceresan L contains two active ingredients, methylmercuric acetate and a methylmercuric mercaptide, whose leaching characteristics are completely different. It was reasoned that the methyl-mercury cation of the methylmercuric acetate would be strongly adsorbed by soil ion exchange and, therefore, be the immobile fraction. The dithiocarbamates do yield many fungitoxic by-products.[132] The mobility order Nabam — maneb — zineb was determined by both bioassay and autoradiography. Movement of these toxicants in five soils was inversely related to soil organic matter content.[124]

The relative mobility in soil TLC of the fungicides evaluated was similar to that demonstrated by other more cumbersome methods for zineb, PCNB, thiram, captan, nabam,[126,133] Panogen,[19] benomyl,[134] ACNQ,[135] dichlone,[8,29] chloranil,[8] diazoben,[136] chloroneb,[137] DCNA,[139] PCP,[29] and Dinocap.[28,29]

The theoretical treatment of leaching has led to the development of mathematical models of the process.[125] The basic equation in building all the models for leaching is well known. Leaching of chemicals was treated theoretically, and a model was developed by using the techniques of Lapidus and Amundson[139] and of others. For details readers are advised to consult the chapter of Freed and Haque[125] and Helling et al.[124]

III. PERSISTENCE AND DEGRADATION OF FUNGICIDES IN SOIL

A. Persistence of Fungicides

The ambiguous character of the term "persistence" has been pointed out by Frehse and Anderson.[140] It is used, on one hand, as an expression of the much desired duration of efficacy of a compound (positive effect); on the other hand, it is used as an expression of its undesirably long life in the environment and, in particular, in soil (negative image). At a symposium on this theme, which took place in the U.K. in 1976, Frehse[141] could not really solve this dilemma. A suggestion of the IUPAC Commission on Pesticide Chemistry deserves consideration:[142] "Persistence is the residence time of a chemical species in a specially defined compartment of the environment." In the context of this definition "chemical species" denotes a specific chemical which may be the parent compound or a derivative, but not both; "resistance time" is the period in which the chemical remains in one compartment; a "compartment" is one phase of the environment (i.e., soil, water, air, animal, or plant tissues) which must be described.[140] The dispersal of the chemical from its primary compartment will necessitate a further determination of persistence. Domsch[143] indicated two possibilities of testing fungicide persistence in soil, i.e., determination of remaining activity of the fungicides or determination of the remaining viability of pathogenic fungi. Further, if a biological indicator is used to measure persistence, it cannot be decided whether the chemical itself or an effective decomposition product is responsible for its activity. Results from chemical analysis are confined to the original chemical and can be interpreted in terms of biological activity only with reservation.

Persistence of fungicides may be advantageous from the point of view of pest control, and in principle a soil fungicide can persist with the ability to control disease or to control a fungus. Furthermore, it can remain stable as a chemical, or it can remain active in its biological effectiveness. Stability of a fungicide, from a chemical point of view, depends on its reactivity with active groups. Captan, for instance, has a high group specificity, in comparison to Dyrene (2,4-dichloro-6o-chloroaniline-s-triazine). Captan is about seven times more stable than Dyrene in a silt loam. Nabam loses "diffusible" activity rapidly after application. The detoxification of nabam as well as that of ferbam, which is more persistent than nabam, is abiotic.[144] Similarly, organic mercury is generally reduced to metallic mercury immediately after its contact with the soil. The distribution of metallic mercury is limited by precipitating substances by the sorption capacity of the soil against mercury. By the oxidation potential of soil[145] under the influence of light, p-dimethylamino-benzenedizo sodium sulfonate undergoes a photochemical reaction. The corresponding diazonium compound and, in a further reaction, p-dimethylamino-phenol are formed. The subsequent steps are oxidation and polymerization. The reaction products have lost the fungicidal properties of the original material.[146,147]

Since reactivity depends upon the availability of the two reaction members, it is greatly influenced by chemical soil conditions.[143] This may be one explanation for so many varying reports on the persistence of fungicides in soil.[148-152] Another reason for lack of accordance is the direct dependency of fungicide persistence upon concentration. The higher the application rate, the longer the fungicidal effect lasts, and the lower is the decomposition rate.[152,153] Edaphic, climatic, and pesticide structural characteristics together greatly determine the persistence of a toxicant in soil.[124] Adsorption strongly affects movement, whereas the depth of movement is dependent both on soil properties and rainfall received. The residue remaining at any depth reflects the conditions governing chemical or microbial degradation. The innate nature of the toxicant is superimposed on these factors. Regarding soil factors, higher soil moisture content, however, is more generally associated with increased microbial activity. For toxicants which are anaerobically metabolized, flooding the soil may cause very rapid loss. Similarly, higher temperature usually decreases persistence. Soil-type differences may outweigh temperature as a factor controlling persistence. There is a tendency for greater persistence of toxicants in soils containing relatively more organic matter and clay than in light-textured sandy soils. The phytotoxic persistence of soil fumigants is longer on loamy clay soils than a sandy loam, presumably because adsorption is greater in the fine-textured soils.

Edwards[154] summarized insecticide persistence as a function of formulation: granules, emulsions, miscible liquids, wettable powders, and dusts. According to him this relative order probably holds for all pesticides. Helling et al.[124] further mentioned the implications of persistence which include (1) the time span over which a pesticide will remain effective, (2) the danger of carryover to succeeding crops, and (3) the possibility of adverse residue accumulation in other segments of the ecosystems. External manipulation of pesticide persistence is a highly desirable goal, from both organic and environmental standpoints (if one can separate these two). Foy and Bingham[155] reviewed techniques for minimizing herbicide residues. These include the use of alternate control methods, increased efficiency, selectivity and removal, and inactivation or alternation of persistence. Helling et al.[124] cited that mass inoculation of soil microorganisms represents an interesting but apparently unsuccessful attempt to modify persistence. No commonly accepted method exists for describing pesticide persistence in soils. Some workers use the expression "half-life", assuming that the time necessary for dissipation of 50% of a pesticide residue is independent of concentration. Freed and Haque[125] pointed out that while the half-life is a useful parameter in estimating the disappearance of a substance from the soil, it does not necessarily tell when the concentration has been sufficiently reduced so as to have little or no further biological activity. Further, because of various uncertainties and assumptions that must be made in deriving this so-called half-life, it should be reorganized only as a useful first approximation and not as a precise figure.

Analysis of persistence data revealed that small applications disappeared proportionately faster than larger doses.[154] Hamaker[156] treated a mathematical prediction of pesticide residues in more detail, pointing out that a large range of concentrations must be studied to reliably determine the appropriate rate law. He calculated the ultimate accumulation of residue in soil for annual additions of one concentration unit and losses of one half in the 1st year. Residues are shown immediately after addition for various assumed reaction orders: zero order (00), half order (3.41), first order (2.00), and second order (1.62). Classification of rate seems quite feasible in pure systems in which the pesticide undergoes chemical degradation. Loss of pesticides from soil in the field would be far more complex, with volatilization, leaching, adsorption, and biological and/or chemical degradation occurring simultaneously.

Studies on persistence of fungicides have been reviewed by Alexander,[157] Helling et al.,[124] Domsch,[143] and Brown.[158] Agnihotrudu and Mithyantha[159] have reviewed the Indian work on pesticide residues. As compared to insecticides and herbicides, very little work has been done on the persistence of fungicides in soil. Soil texture, pH, and organic matter are the most important factors affecting the activity of soil fungicides.

1. Organosulfur Compounds

Richardson[31] demonstrated that the half-life of thiram was less than 1 day when mixed with compost soil and about 7 days in sandy soils; whereas Domsch[153] observed half-life for thiram 38 days, Nabam 16 days, ziram 45 days, and zineb 77 days applied in soil at the concentrations of 160, 100, 1800, and 800 ppm. respectively. Methan has an effective life of 2 days, its loss of activity being independent of microorganisms. Thiram, when mixed with soil, showed very low persistence, having a half-life of between 1 and 2 days. In contrast, when the fungicide was added to the soil on the surface of glass beads, the persistence of the chemical was considerably increased and even after 21 days it almost maintained its initial concetration in soil.[160] Persistence of thiram was directly related to the concentration applied initially.[161] The amount of thiram remaining after 64 days in soils supplied with 1000 ppm Thiram was very large and the rate of disappearance was very gradual, whereas the amount remaining in soils supplied with 250 ppm thiram was undetectable after 27 to 49 days. The breakdown was most rapid in the unsterilized soil, indicating that a part of the breakdown was due to microbial activity on thiram. Maneb and thiram at 100 ppm persisted for less than 4 weeks. Thiram showed no activity at 3 weeks. At 1000 ppm maneb was also least persistent. It persisted for 11 weeks whereas thiram persisted for more than 32 weeks. Mehrotra et al.[162] reported that thiram at 10 ppm persisted for less than 7 days, but 100 ppm thiram persisted for 28 days, and 1000 ppm Thiram persisted for over 105 days. Vyas[163] determined persistence of thiram, Dithane M-45, and Dithane Z-78 in sterilized and unsterilized soil. Each was applied to 100 g of soil at the rate of 2500 ppm (w/w). Thiram persisted for more than 77 days in both soils. Dithane M-45 was not detectable on the 56th day in nonsterile soil, but could be detected up to 63 days in sterile soil. Dithane Z-78 had a better life by one week in two soils as compared to Dithane M-45. In nonsterile soil, the rate of degradation of Dithane M-45 was relatively rapid; only 36% of the initial activity remained on the 28th day, while it was about 58% in sterile soil over the same period. Only 50% of the initial activity was evident on the 77th day in nonsterile soil mixed with thiram, whereas it was of the order of 63% in sterile soil.

2. Organomercurials

Indulkar and Grewal[164] observed that Panogen 15, Ceresan, and Agrosan 5 W persist in soil for 6 days. Fungicides were equally effective in sterilized and unsterilized soils, indicating that microorganisms present in unsterilized soil do not have any effect on degradation of fungicides. Methylmercury dicyandiamide (MMDD) at 10 ppm or lower persisted for less than 4 weeks,[165] whereas at 100 ppm it persisted for 15 weeks. Conversely, Munnecke and Moore[166] reported that MMDD could not be detected in nonsterile soil after 2 weeks. Saha et al.[167] noted that about $^1/_2$ of the MMDD applied at rates of 3.3 and 6.6 ppm to soil could be measured in its organic or inorganic forms after 5 months. According to Sinha et al.[168] the disappearance of aretan was closely related to the amount of fungicide applied initially. The presence of aretan could not be detected on the 3rd day at 5 ppm, on the 9th day at 10 ppm, on the 15th day at 20 ppm, and on the 18th day at 40 ppm (Figure 3).

3. Captan and Difolatan

The persistence of captan in soil has also been found to be low. It had a half-life of 3 to 4 days when mixed in moist soil and more than 50 days when the soil was dry.[14] It has extremely low persistence in a forest nursery soil, where it was fully degraded within a week when applied at or below 250 ppm concentration.[169] Similarly, at 125 ppm initial concentration of the chemical mixed with the soil showed very low persistence, having a half-life between 1 and 2 days. In contrast, when the fungicides were added to the soil on the surface of glass beads the persistence of the chemicals was considerably increased, and even after 21 days in soil they remained at approximately their initial concentration.[160] Higher persistence of captan has also been reported by some workers. A half-life up to 70 days for captan was observed when it was applied at the concentration of 250 ppm. The period of appreciable fungicidal activity in soil was determined to be at least 65 days for captan.[170]

Persistence of Difolatan (2500 ppm) may differ from crop to crop and is also affected by environmental conditions. Difolatan persisted for more than 21 days in soils cropped with potato and was not detectable on the 29th day.[163] Difolatan (10 ppm) remained biologically active for a period of less than 21 days, but 100 and 1000 ppm of it remained persistent for a period of 35 and 105 days, respectively.[162]

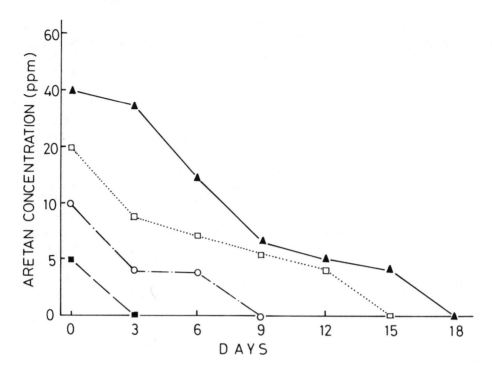

FIGURE 3.   Persistence of aretan in soil.

4. Dexon

Alconero and Hagedorn[171] found up to 563 ppm of Dexon in soil after 48 hr of treatment. There were indications that greater amounts of Dexon remained in soil when it was incorporated as a powder than as a surface drench. It almost disappeared from the soil after 60 days of application.[172] Dexon had extremely low persistence as soil drenches; when applied in the form of dry dressings on glass beads it had a half-life of up to 25 to 30 days.[173]

5. PCNB and DCNA

Loss of 1,2,4,5-tetrachloronitrobenzene in fine sandy loam soil was much faster, more than 50% of the fungicides disappearing within 2 months. PCNB, however, degrades slowly in soil and 20% of it may remain in soil even after 10 months.[174] Wang and Broadbent[175] reported that the half-life of PCNB was 4.7, 7.6, and 9.7 months and of DCNA 30, 16.2, and 13.6 months in sandy loam, clay, and peaty muck soils, respectively. Mehrotra et al.[162] stated that the half-life of PCNB in soils when the quantities added were from 93 to 465 μg/g soil or clay varied from 15 to 40 days in Karl soils under submergence, to 153 days in laterite soil, kept at 1 bar moisture tension. They further mentioned that 10 ppm PCNB persisted for less than 7 days, but 100 and 1000 ppm of it persisted for 7 and 21 days. Pimentel[170] observed that the period of appreciable fungicidal activity in soil exceeds 8 months for PCNB. However, Kher et al.[176] reported that PCNB applied @ 2500 ppm totally lost fungitoxicity after 28 and 35 days in unsterilized and sterilized soils, respectively.[176]

6. Benzimidazole Fungicides

The persistence of benomyl was studied by several workers including Hine and co-workers,[177] who were among the very first investigators who demonstrated the stability of benomyl in soil. They found that benomyl was detectable at least 22 weeks after final soil application in greenhouse studies, whereas Erwin et al.[178] detected fungitoxic activity in benomyl-treated soil even 6 months after treatment. They further demonstrated that after 19

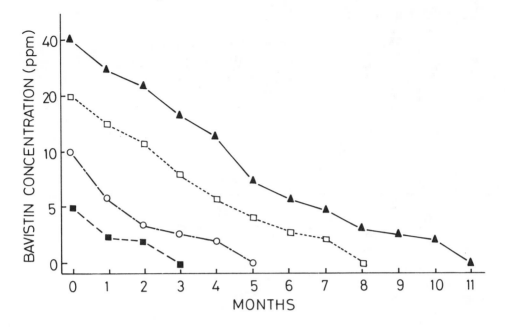

FIGURE 4.   Persistence of bavistin in soil.

weeks of incubation at 40°C benomyl at 100 ppm level had decreased to approximately 30 ppm, whereas at the lowest temperature (16°C) there was little or no inactivation. After 19 weeks, inhibition zones at 16°C were similar to 0-hr readings. Smith and Worthing[179] reported the decreased loss of benomyl and methyl-2-benzimidazole carbamate (MBC) in brick earth soil with an organic matter content of 1 to 2% carbon after steam treatment or soil disin-festation with MB. Raynal and Ferrari[180] studied the persistence of benomyl in soil by using *Penicillium* and *Cyclopium* as biological test material, and observed only 8 ppm of benomyl in soil 6 months after treatment with 100 ppm. Netzer and Dishon[181] also reported that this fungicide retains its biological activity for long periods. According to Baude et al.,[182] the half-life of the fungicide in different soils of the U.S. was between 6 and 12 months. Initially, thiabendazole was more uniformly distributed in the soil and subsequently was moved by leaching than was carbendazim.[183] Thiabendazole and carbendazim persisted for 22 and 10 weeks, respectively.[183] According to Sinha et al.,[184] Bavistin could not be detected after 3, 5, 8, and 11 months, when applied in sandy loam soil at the rate of 5, 10, 20, and 40 ppm, respectively (Figure 4). In sandy loam soil, 10 ppm Benlate persisted for 105 days, but 100 and 1000 ppm persisted for over 105 days.[162]

Different factors may affect the persistence of benomyl and MBC in soil. Persistence of thiabendazole (TBZ) and MBC in soil incubated at 25°C in small glass vials was examined by Aharonson.[185] Nine months after adsorption of these chemicals to the soil, 85 to 95% of the applied TBZ and 65 to 75% of the applied MBC were recovered from air-dried soils. However, in moist soil after 9 months of incubation, only 75 to 90% of the applied TBZ and 20 to 30% of the applied MBC were recovered. Van Wainbeke and Van Assche[186] conducted a detailed study where benomyl had been used in soil mixtures for investigating the influence of soil composition, steam treatment, moisture content, and temperature on its persistence. The soils used in the experiment were a sandy loam of 9 pH, 5.70 and 2.5% organic matter and a decomposed leaf soil with a pH of 6.85 and 28.4% organic matter. The mean moisture contents of both the soils were, respectively, 70 and 140, and 110 and 150% of the water-holding capacity. The temperatures were 10 and 23°C. Benomyl was applied in an equimolar dosage based on 50 mg MBC per liter of soil. Analysis of the MBC

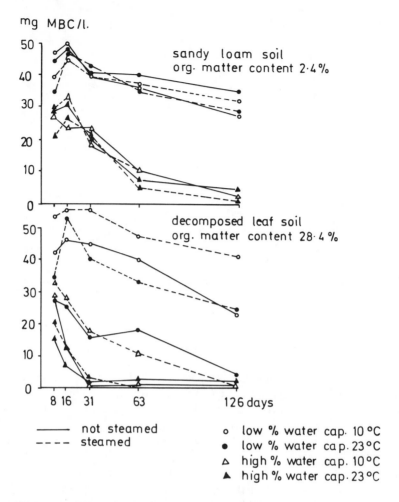

FIGURE 5.   MBC residues in soil under different conditions after treatment with benomyl. (From Van Wambeke, E., *Soil Disinfestation*, Mulder, D., Ed., Elsevier Scientific
Publishing Company, Amsterdam, 1979, 353. With permission.)

content in soil was made at several dates at increasing intervals during 4 months. The results
show the importance of the parameters studied and are summarized in Figure 5. Different
soil composition does affect the disappearance of the MBC residues. However, remarkable
effects were obtained when combined with other parameters. Steam treatment affected the
MBC content especially in decomposed leaf soil, where breakdown was strongly inhibited
as compared to the nonsteamed soil. Thus, the organic matter content of the leaf soil and
its microbial activity, which are presumably significant when comparing steamed to nonsteamed soil, acted as favorable circumstances for fungicide loss at high moisture content.
The higher temperature also enhanced the disappearance of MBC residues especially under
conditions of higher moisture content. The difference between two extreme cases is very
striking: 36.8 mg MBC per liter of steamed sandy loam soil at 70% of the water-holding
capacity at a temperature of 10°C and 2.4 mg MBC per liter of nonsteamed decomposed
leaf soil at 150% of water-holding capacity at a temperature of 23°C, both considered on
the 63rd day after incorporation of 50 mg MBC per liter of soil. They concluded that these
four factors — soil composition, steam treatment, moisture content, and temperature —
thus affect the persistence of benomyl and MBC.

Singh[187] also studied persistence of MBC in nine soil types which revealed that the fungicide could be detected up to 90 days in sandy loam. In loam, clay loam, and silty loam (alluvial) soils, the fungicide persisted for 120 days. In silty clay loam, fine sandy loam, loam (hill), silty loam (bhabhar), and silty (saline alkaline) soils the fungicide persisted up to 105 days of incubation. He concluded that the persistence of MBC in nine soils is different due to variations in their adsorptive capacities which depend on physicochemical properties of the soils.

7. Vitavax

Persistence of carboxin was related to the amount of fungicide added, and it was completely degraded between 20 and 24 days when applied at 17.5 kg/ha and between 32 and 36 days when applied at 35 kg/ha.[188]

8. Metalaxyl

In a study on persistence of metalaxyl in soil, Austin et al.[189] observed an initial rapid decline followed by a much slower decline in concentration of metalaxyl after soil application. A residue of higher than 1 µg/kg dry soil (sufficient to control *Phytophthora syringae*) was recorded at 2 to 4 months, depending upon dosage applied. In another study, metalaxyl applied to soil to control *P. syringae* rot of apple fruits was not only effective during the year in which it was applied, but it also reduced the disease incidence significantly in the subsequent year.[190] Soil drench of metalaxyl is effective for 8 to 24 weeks in *Rhododendron* spp. against *Phytophthora* species,[191,192] 53 to 60 days in tobacco against blue mold,[193] 2 weeks in turf grass against *Pythium* sp., and 63 days to entire crop season (depending on applied dose) in broccoli against downy mildew.[194] In these studies, however, one cannot be sure whether metalaxyl persisted in the plants or remained in the soil from where the plant gradually took it up during the growing season.

9. Fumigants

The fumigant fungicides quickly dissipate from soils as may be demonstrated by bioassaying the treated soil with *Pythium*. An application of allyl alcohol, like formalin, loses its fungitoxicity by the 2nd day of treatment. Metham and dazomet have an effective life of 2 days.[153] As cited above, fumigants normally disappear from the soil soon after treatment. Supplementary interpretation is needed on this concept. The effectiveness of soil fumigants is judged by killing action, but subtoxic concentration (for pathogens and plants) is not generally observed. The presence of subtoxic concentrations seems to be an important additional effect of every successful control by fumigants. Therefore, fumigants which do not completely disappear, but remain in low concentration in soil, help to stabilize the effect by preventing rapid recolonization.[195-199] In a rough approximation, fumigants which need a relatively long time to disappear from soil belong to this group of chemicals. According to Domsch,[143] typical representatives of this type are *N-N'*-dimethylthiuram disulfide and the Zn salt of *N*-methyldithoiocarbamate.

It can be concluded that the persistence of pesticides reflects the sum of all other processes modifying pesticides in soils. Factors influencing persistence include the pesticide itself, soil type, moisture status, temperature, application rate, depth of placement, formulation, soil pH, and microbial activity. Domsch[143] has rightly pointed out that on one hand we want to keep the concentration down for reasons of soil hygiene, and on the other, high persistence of a fungicide applied to the growing plant is wanted. To meet in a reasonable compromise, a ratio of 10:1 has been postulated between application rate (ppm) and half-life (days) of soil fungicides.[143] Materials which fit into this postulate with respect to a fungicidal (not chemical) half-life are, for instance, captan (half-life, 50 days), TMTD, and nabam (half-life, 50 days).[143] More and more work is needed on persistence of toxicants, considering all the factors singly or in all the possible combinations which may prevail in nature. Helling et al.[124] recommended that metabolic inhibitors may be added to pesticide formulations to modify their persistence. Addition of this prolonged the herbicidal activity which improved weed control.[200] The mechanism is inhibition of microbial hydrolysis of the herbicide.

B. Degradation of Fungicides

Upon application, the chemical distributes or partitions itself in the various compartments of the environment in accord with its physicochemical properties and interactions with existing conditions. For most toxicants applied in a liquid or solid form, the bulk will ultimately reach the soil surface. The chemical is subjected to a variety of forces and processes that tend to result in an alteration of the chemical. Physical processes like leaching, vaporization, or adsorption in themselves have but little effect on the chemical, though decomposition may occur during these processes. The three basic processes leading to decomposition of a chemical are chemical reaction, biochemical or biological reaction, and photochemical reaction.

Because of the problems of studying metabolic formation in field (particularly with labeled compounds) most of the detailed work on the pathway and the rates of fungicide degradation in soils is conducted in the laboratory with model systems. For the research of degradation and metabolism in soils, models used in the laboratory consist of a radioactive compound, a plant-free soil, and an incubation system which allows "full" recoveries of the applied radioactivity. The position of the radioactive label in each fungicide and the isotope used will depend on the objectives of the research. The incubation systems used must allow the rates of dissipation of the chemical through volatilization as well as degradation to be measured. The system described by Anderson[201] has been found to be useful. Recoveries of radioactivity with this system are good (e.g., for 794 cultures, recoveries were 97.5 ± 3.0% of the applied radioactivity) and the system can conveniently be used for aerobic as well as anaerobic incubation. The soil used for degradation experiments is also important. Generally, the main emphasis in studying fungicide dissipation has been on the chemical and physical aspects of the soil; biological aspects have been largely ignored. In the different guidelines of conducting degradation tests, it is mentioned that chemical and physical data for the description of the soil are always needed (biological data are rarely, if at all, mentioned). In the past, since several persistent fungicides were poor substrates for enzymatic attack, their behavior in soils appeared to follow simple physical and chemical rules. Therefore, data from laboratory studies with many of these study compounds could be interpreted with a reasonable degree of accuracy. Presently, most of the modern fungicides are non-persistent and many are "good" substrates for enzymatic attack. A closer look at soil microbiology is essential as it relates to the rates of metabolism of the fungicide. Frehse and Anderson[140] mentioned three major variables that determine the rates of microbial metabolism of fungicide (and for that matter other biodegradable organic chemicals) in soil.

1. The availability of chemical to the organisms or enzyme system responsible for metabolism
2. The quantity of microorganisms or enzyme systems which have capacity to degrade the chemical
3. The activity level or physiological state of the organisms

Availability is determined by adsorption. Its role in determining degradation rates is well documented by Frehse and Anderson.[140] The relationship between quantity of microbial biomass in soil and the rate of fungicide degradation has also been well illustrated.[140] It can be concluded that when fungicides are biodegradable, both the quantity of microbial biomass in a soil and the availability of the chemical to the biomass determine the rate of degradation and, thus, the resistance time of the chemical in the soil. The third important point is the influence of activity level of the microbial biomass on the rates of degradation of a fungicide. The influence of temperature on the rates of degradation of diallate in soil was examined and mentioned in detail by Frehse and Anderson.[140]

1. Chemical and Microbial Degradation

Fungicide degradation takes place by chemical reactions with soil constituents or as a result of biochemical interactions with soil microorganisms. The rate at which fungicides decompose in soil is characteristic of both the chemical and the soil. Whether chemical or biochemical processes are most important in degradation of a specific fungicide is probably most dependent upon the chemical characteristics of the fungicides, but may also be dependent upon specific soil characteristics. One must be very careful to accurately examine whether degradation is actually biological or chemical. Because of apparent similarities, chemical degradative mechanisms are sometimes misinterpreted as biological mechanisms. In environments such as soil or growing plants, the effects of air, light, water, and enzymes are often completely integrated, and it is only with the aid of complementary in vitro experiments that we can possibly separate contributions with any degree of certainty. Various types of reactions that may take place in the environments include hydrolysis of alkyl or aryl esters, amides, phosphates, and carbamates, not only in a fungicidal formulation, particularly during storage under favorable conditions, but also in soil. Oxidation also takes place in carbon, sulfur, and nitrogen atmospheres and isomerizations involving pentavalent phosphorus are known. However, almost any of the known organic constituents of soil are potential reducing agents, especially under oxygen-deficient conditions, and indeed because of its nature, which provides reactive surfaces supplied with oxygen and water in varying amounts, soil affords an almost ideal medium for the degradation of fungicides. The recently discovered free radicals in soil humic acid are potential reagents for a variety of reactions, and the same semiquinones also could serve as oxidizing agents.[202] Chemical reactions most likely to occur are decomposition in water or hydrolysis and nucleophilic substitution by active groups of organic matter.[4] The role of free radicals (high-energy hydroxyl groups) in decomposition of fungicides in soil has been investigated recently.[203,204] The dithicarbamates tend to hydrolyze or degrade readily in water.[205,206] Dichlone chloranil, anilazine, PCNB, captan, nitrohalobenzenes, and compounds with allylic halogens participated in substitution reactions.[4] Since the concentration of toxicant is generally less as compared to substrate with which it is reacting, first-order kinetics are believed applicable whether the reaction occurring is hydrolysis or uni- or bimolecular substitution.[8,207,208]

Microbial degradation of fungicides involved detoxification of the compound and its degradation by some soil microorganisms.[1] Algae, fungi, actinomycetes, and bacteria are the primary microorganisms present in soil. Most of these are dependent on organic compounds for energy and growth. When an organic toxicant is added to soil and reaches an equilibrium between the soil colloids and the soil solution, any molecules in the soil solution are immediately attacked as potential energy sources. If a toxicant is readily available in the soil solution, any organisms that can adapt to it as an energy source rapidly enhance in numbers until they have completely degraded it. Microbial inactivation takes place in the following stages: initial uptake of toxicant, metabolic breakdown, and finally possible utilization of suitable fragments for energy purposes.

Microbial degradation accounts for much of the loss of chemicals from soil.[8,11,209,210] The kinetics of decomposition typical for enrichment cultures were established in early studies with 2,4-D.[211] The work spawned much effort towards isolating microorganisms especially responsible for decomposition of selected toxicants. While many microorganisms capable of degrading specific toxicants have been isolated and characterized, few chemicals failed to support microbial growth or enrichment during the degradation process. Other chemicals appeared persistent or otherwise resistant to microbial degradation. Therefore, the concepts of microbial fallibility[212,213] and cometabolism[214,215] were introduced. Some organisms thrive and enrich on compounds that are decomposed easily. Some toxicants are initially unsuitable as energy sources for microorganisms. The more persistent toxicants are degraded inefficiently by the myriad of enzymatic reactions the microorganisms generate. Alexander[212] was perhaps first to challenge the then-present notion of "microbial infallibility" (i.e., all organic molecules would be subject to attack by some microbe in the soil), citing, among others, the persistence of DDT in the biological food chain. Before this time the persistence in soil of substances like DDT, dieldrin, and PCNB was looked upon as an exception to the rule of "microbial infallibility" and numerous attempts were made to develop other persistent compounds. Since the hazards of persistent compounds became apparent, research efforts switched to developing compounds with less persistent activity. Coincidently, researchers found that even DDT may be microbially degraded. Also, soil microflora may be "trained" more readily to alter a toxicant when successively higher applications of the compound are applied.[166] Munnecke[10] was of the opinion that microbial infallibility may remain as a reasonable concept that applies to most situations in soil, although a number of compounds obviously resist microbial decomposition under most conditions.

When concentration of a toxicant is low relative to the level of biological activity in the soil, first-order kinetics apply.[80,156,216,217] Michaelis-Menten kinetics seem to apply when the toxicant concentration increases and the rate of decomposition changes from being proportional to being independent of concentration.[156,218]

According to Freed and Haque,[125] in so complex a milieu as the soil, with its variable conditions, it would be unreasonable to expect a simple straightforward behavior in breakdown. Rather the process is influenced by the availability of moisture, the nutrient level in the soil, the pH, the oxygenation of the soil, the temperature, organic matter content, and such physical processes as diffusion, leaching, and adsorption.[219] Compounds, by virtue of their structure and physical properties, vary in rates at which they are decomposed.[220] Many of the more refractory compounds may show a considerable period of time between application and the first detectable initiation of decomposition.[125,220] The interval between application and initiation of breakdown is called "the lag period"; the rate at which one breaks down in the soil has been determined for many toxicants. Various factors influencing this rate have been elucidated and expressing the quantitative relationship to rate of breakdown has been attempted, at least to a first approximation. For some compounds, the rate appears to follow an order analogous to that of the first-order rate law of chemical kinetics.[125]

$$\frac{dc}{d+} = K \cdot C$$

Here C is the concentration of chemical and K is the rate constant. However, the observations on breakdown of toxicants in soil and calculations based on assumption of a first-order rate should never be mistaken as representing the actual detailed kinetics.[125] It rather measures the differences between the boundary conditions, without reflecting the intervening mechanisms. The application of this approximation to study the rates of breakdown of pesticides has proven useful. From such data, one derives estimates of an energy value that gives an index of the stability of the compound in the soil and, hence, its probable persistence under specific conditions.[125] It has, in certain cases, even been used to suggest the probable mechanism of breakdown. Another concept frequently used in relation to decomposition of toxicants in the soil is the half-life expressed in this equation:

$$t_{1/2} = \ln \frac{2}{K}$$

where K is the estimated reaction or decomposition constant.

Extensive literature has developed concerning biological breakdown of toxicants in soil. Goring[8] cited various factors that may influence decomposition of fungicides in soil. Decomposition by soil microorganisms is enhanced by increasing temperature and moisture. Organic matter content is a fairly good indicator of total biological activity, and decomposition also seems to be increased with increased organic matter. However, bioassay methods are often used to evaluate decomposition, and they may not measure the portion of the toxicant in the soil sorbed by the organic matter. Organic matter has been subject to rapid and complete remineralization by microorganisms after life had ceased. However, there may be exceptions, especially under strictly anaerobic conditions or at extreme pH values or very low temperatures of the environments. With such extreme conditions, products like oil, coal, and methane accumulated, and not only materials like wood, but even whole animals were conserved.[221] Heterotrophic microbes that contain a wide variety of enzymes capable of decomposing this natural organic matter are especially equipped for providing themselves with the energy and organic molecules necessary for growth and reproduction.

As mentioned earlier, chemicals can serve sufficiently as a source of carbon and energy for microbes. In the case of fungicides, this has thus far been demonstrated only in a few instances: the growth of a species of *Beyerinckia* on biphenyl, of a species of *Pseudomonas* on pentachlorophenol, and of a species of *Nocardia* on carboxin as sole source of carbon. During shortage of natural substrate, as is very common in soil, such old substrates will only be utilized in nature. Moreover, if a sufficient nitrogen source is available, microbial development takes place, thereby enhancing the rate of breakdown of the substrate involved. Another type of microbial degradation is called "cometabolism", which is defined as any oxidation of substances without utilization of the energy derived from the oxidation to support microbial growth. Transformations like the oxidation of a sulfide to sulfoxide or an amino group to a nitro group belong to this category. The phenomenon of cometabolism appears to occur widely in microbial metabolism.[221] As compared to former types of breakdown, where the fungicide serves as the sole source of carbon for growth, the rate of cometabolic type of decomposition is sped up by the addition of a natural oxidizable substrate. It was thought earlier that several foreign compounds by the principle of cometabolism were recalcitrant to microbial attack; these were later shown to be degradable to a certain extent. All the biological conversions that fungicides undergo can be roughly grouped as oxidations, reductions, hydrolysis, and conjugate formation. Examples of oxidation are the oxidative demethylation of chloroneb, hydroxylation reactions of mebenil and other compounds, the formation of sulfoxide and a sulfone from carboxin, as well as of an N-oxide from tridemorph. Reduction of aromatic NO_2 to NH_2 is well known, but the formation of benzene from phenylmercuric acetate is also regarded as a reduction process. Hydrolysis takes place with carboxin, pyrazophos, and other products. Conjugate formation is noticed in the formation of a glucoside from the demethylated product of chloroneb, the formation of an anisole from pentachlorophenol, of an alanine derivative and a glucoside from dimethyl dithiocarbamate, and of a thiomethyl derivative from PCNB under certain circumstances. In the case of conjugate formation, a reverse reaction may again give rise to the parent compound.[221]

Knowledge of the pathway of breakdown of fungicides along with the various intermediates formed, as well as the enzyme systems involved, is of much interest. However, the information is still very scarce.

a. Mercury Compounds

The breakdown of mercury-containing compounds illustrates the need for factual information in assessing the impact of pesticides on the environment. Several well-documented cases of mercury poisoning in certain Scandinavian countries, Japan, and North America have resulted in restricted use or banning of these materials. Out of about 20 various mercury-containing fungicides, 8 contain alkylmercury radicals that could be released into the environment during degradation. All 20 could eventually yield alkylated mercury through degradation to free mercury with subsequent aklylation of the free mercury. This later process would perhaps require more time than the direct release of alkylated mercury. The breakdown of both aliphatic and aromatic compounds of mercury in the soil was originally ascribed to chemical breakdown,[145] but microbial involvement has become increasingly apparent. A mercury-resistant strain of *Penicillium roquefortii* able to absorb a large amount of the metal from nutrient media containing phenylmercury acetate was reported.[222] Spanis et al.[223] found that semesan (2-chloro-4-[hydroxymercury] phenol) and Panodrench 4 (cy-ano[methylmercury] guanidine) were inactivated by soil microorganisms. Semesan was de-graded by isolates of *Penicillium* spp. and *Aspergillus* spp. Panodrench was inactivated by several *Bacillus* spp. The inactivating organisms of one fungicide were unable to inactivate the other fungicide. They postulated that inactivation occurred in a three-step process: (1) uptake by sorption by the fungus protoplast, or by sorption to a cell membrane complex, in the case of bacteria; (2) metabolism of the toxicant; and (3) possible utilization of the innocuous products of the decomposed toxicants for growth. Spanis[224] amplified the work on Panodrench breakdown and proposed that the *Bacillus* spp. act on the fungicide at the cytoplasmic membrane surface. Further binding to the membrane may be facilitated by a guanidine-membrane attachment, perhaps resulting in an innocuous mercury-sulfo complex. The report of the inactivation of organomercury fungicides by sulfur in groundwood made from salt water-stored pulpwood is significant. Pulpwood logs stored in fresh water or water of low salinity were protected from fungal attack by treatment with phenylmercuric acetate or phenylmercuric acetate-oxine, but logs stored in seawater were not.[225] In salt water storage, logs gradually accumulate mercury-inactivating organosulfur compounds, collectively re-ferred to as thiolignin, which arise from the reduction by sulfate-reducing bacteria of sulfates adsorbed from the sea. Hydrogen sulfide liberated by these bacteria combines chemically with the lignin in the tissues forming organosulfur compounds, which have great affinity for mercury fungicides and render them biologically inert. This report was mentioned by Munnecke[1] with the idea that the soil may act in a similar manner where organic materials may have been combined with active sulfur groups.

The chemical breakdown of mercury fungicides in soil led to apparently conflicting results. Booer[226] reported that organic mercury compounds reacted by base exchange with soil clays to form an intermediate compound. The intermediate subsequently gave a dialkylmercury or diphenylmercury and a mercury clay compound. The diphenylmercury would accumulate in soil while dialkylmercury would escape into the atmosphere. From degradation of the clay mercury complex, metallic mercury would form which eventually was rendered in-nocuous by formation of mercury sulfide. These results were not confirmed in later studies. Kimura and Miller[82] investigated the residual mercurials in soils and in the atmosphere above treated soil. The objective was to elucidate the degradation in soil and to estimate whether or not biological inactivation and mercury evolution occur together. Experiments were carried out in pots filled with Puyally sandy loam, the principal bulb-growing soil in Pierce Country, Wash. The soil was dried and intimately mixed with phenylmercury acetate (PMA), ethyl-mercury acetate (EMA), MMDD, or methylmercury chloride (MMC) to make a concentration of about 11 μg mercury per gram of soil. Autoclaved and nonautoclaved soils held at different moisture capacities were used. They estimated the concentration of the parent compounds in the soil and in air above the soil, ionic mercury in soil, and total mercury. Mercury vapor and trace amounts of PMA were present in the air surrounding PMA-treated soils. Much of the loss occurred from the upper few centimeters of a soil column. EMA was also degraded mostly as a vapor, but trace amounts of the parent compound were detected in the air. With the use of methylmercury compounds, methylmercury vapor was present with trace amounts of mercury vapor. MMC was about twice as volatile as MMD when added to soil. In other words, it can be said that MMD and MMC were not broken

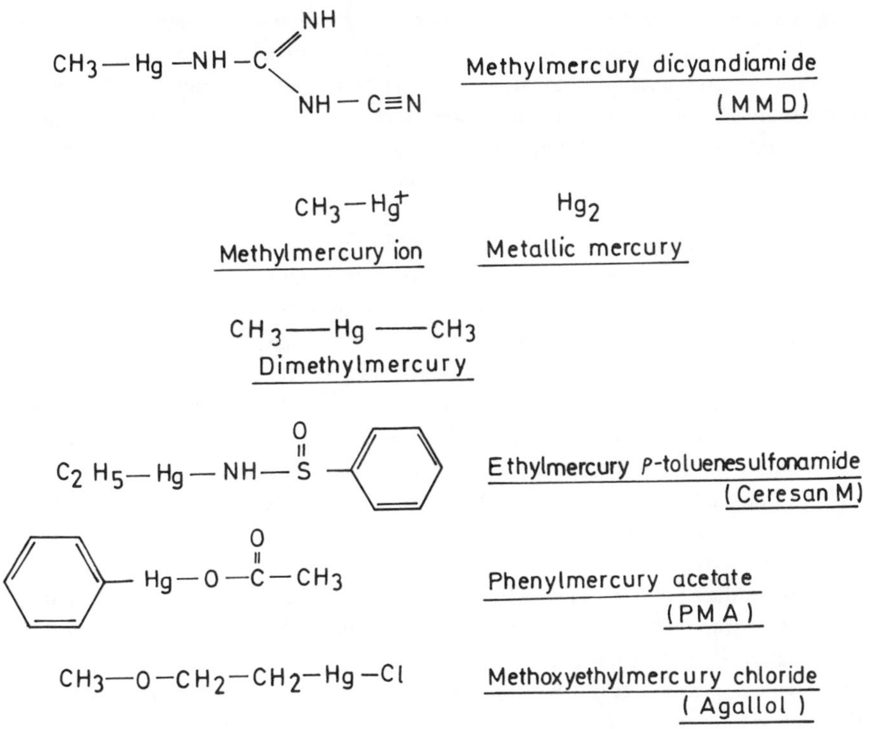

$$CH_3-Hg-NH-C\underset{NH-C\equiv N}{\overset{NH}{\diagup}}$$

Methylmercury dicyandiamide
(MMD)

$$CH_3-Hg^+$$ $$Hg_2$$

Methylmercury ion   Metallic mercury

$$CH_3----Hg----CH_3$$
Dimethylmercury

$$C_2H_5-Hg-NH-\overset{O}{\overset{\|}{S}}-\bigcirc$$

Ethylmercury *p*-toluenesulfonamide
(Ceresan M)

$$\bigcirc-Hg-O-\overset{O}{\overset{\|}{C}}-CH_3$$

Phenylmercury acetate
(PMA)

$$CH_3-O-CH_2-CH_2-Hg-Cl$$

Methoxyethylmercury chloride
(Agallol)

FIGURE 6. Interrelationship between organomercury fungicides, methyl mercury, and metallic mercury. (From Brown, A. W. A., *Ecology of Pesticides*, John Wiley & Sons, New York, 1978, 437. Copyright © 1978. Reprinted by permission of John Wiley & Sons, Inc.)

down to mercury vapor and, thus, loss from soil was due almost entirely to the volatility of the organomercury compound. After 30 to 50 days, a large portion of the compounds applied to soil was still in the organomercury form. The tendency for mercury vapor to escape from soils decreased at low soil moisture contents. From these findings they presumed that biological inactivation and mercury evolution do not necessarily occur together. These workers further observed a remarkable increase in amount of mercury vapor from nonautoclaved soil as compared to autoclaved soil with PMA and EMA, which were degraded in soil. After 20 days with PMA, approximately 2.9 times more mercury vapor was produced from nonautoclaved as compared to autoclaved soil, and after 40 days with EMA approximately 1.7 times more was produced in nonautoclaved soil. Yet, essentially the same product was recovered from both autoclaved and nonautoclaved soils. This would support the assumption that microbial as well as nonmicrobial breakdown might be the same.

A review of the mercury research in Sweden[227] shows that metallic mercury, inorganic divalent mercury, phenylmercury, and alkoxialkylmercury can all be converted to methylmercury. The latter compound undergoes biomagnification in the food chain, primarily in fish, and causes brain damage in humans consuming contaminated fish. The first step in the biomagnification process apparently occurs in the sediment layers of lakes and streams where metallic mercury is oxidized to divalent mercury. This is a chemical reaction. The subsequent conversion of divalent inorganic mercury to methylmercury and dimethylmercury (Figure 6) is a biological process.

Putrescent homogenates of *Xiphophorous helleri* and *Gadus colja* alkylate Hg^{2+} to dimethylmercury, presumably through the intermediate formation of methylmercury.[228] Microorganisms in river sediment can produce methyl- and dimethylmercury from divalent mercury under laboratory conditions. The reaction rate is enhanced by organic matter and increasing amounts of mercury in the sediment. A rapid rise in methylmercury content occurred when the concentration of inorganic mercury was enhanced from 1 to 10 ppm. Methyl- and dimethylmercury both arise under aerobic and anaerobic conditions. Under anaerobic conditions mercuric sulfide may be produced, precipitating without forming mon-omethylmercury. Wood et al.[229] demonstrated that the methyl group was transferred from methylcobalamine to Hg^{2+} by the action of an anaerobic bacterium isolated from sediment and provided direct evidence for a microbial methylation process.

Vice versa, a *Pseudomonas* was found that could remove the alkyl groups from salts of methylmercury, ethylmercury, and phenylmercury, producing metallic Hg and methane, ethane, and benzene, respectively.[230] Furukawa and Tonomura[231] used an enzyme preparation from *Pseudomonas* sp. which decomposed PMA to metal mercury. They demonstrated that the inducible enzyme, a reduced NAD(P)-generating system, glucose dehydrogenase or arabinose dehydrogenase, and cytochrome C-1 were required for the decomposition of PMA. Bacteria have been isolated which convert PMA into metallic mercury and methane.[232,233] Some bacterial isolates resistant to PMA have been shown to degrade PMA into metallic mercury and benzene[234] or to convert PMA into diphenlymercury.[235] They reported that PMA was quickly metabolized by soil and aquatic organisms and that one of the major products was diphenylmercury. Methylmercury was not found, although it had been sus-pected. Although they pointed out that PMA does not yield the highly toxic methylmercury, the toxicological implications of diphenylmercury remain obscure and need investigation. It was also demonstrated[82] that soil, provided it had not been sterilized, degraded PMA to elemental Hg vapor which could be collected and weighed, the loss due to volatiliz⸗ on from the soil amounting to 10 to 15% in 4 weeks. Subsequently, a bacterium with the characteristics of *Pseudomonas* was isolated from soil which produced Hg vapor from PMA,[236] whereas EMA was lost at about the same rate by volatilization both as elemental Hg and the unchanged organomercurial; MMD was volatilized from the soil as the unchanged organic mercurial at about half the rate.[82]

b. Organosulfur Compounds

Recent review of the chemistry and mode of action of dithiocarbamate fungicides suggests chemical degradation of these compounds into organic compounds, including ethylene dia-mine, carbon disulfide, ethylene thiourea (ETU), ethylenethiuram disulfide, ethylene thiuram monosulfide, isothiocyanates, metallic sulfides, elemental sulfur, and carbonyl sul-fide.[19,31,35,237-246] Limited microbial degradation has been observed.[4,247-251] Thiram degra-dation occurred more rapidly in nonsterile soil than in sterile soil. The degradation of two unidentified divalent sulfur degradation products of thiram also occurred more rapidly in nonsterile than in sterile soil. Thiram was shown to be degraded by a certain strain of *Pseudomonas* and dimethylamine was the important product.[252] Thiram rapidly disappeared from compost soil, but it persisted for over 2 months in sandy soil. The role of disappearance suggested that more than one factor was involved and microbial involvement appeared likely.[31] Ferbam, ziram, and thiram produce the dimethyldithiocarbomate ion (DMDC) in fungi and yeast[241] (Figure 7). In alkali, ferbam yields the DMDC ion, and in acid below pH 5 it yields dimethylamine and CS_2.[253] The soil fungi *Penicillium notatum, Glomerulla cingulata,* and *Fusarium roseum* produce CS_2 from thiram.[250] The formation of dimethy-lamine (DMA) in thiram-treated soils has been demonstrated.[252,254] Although CS_2 had been indicated as one of the degradation products, very little has been reported about its presence in soil. Gas chromatographic analysis of the head-space gas of flasks containing nonsterile thiram-treated soil revealed the presence of CS_2.[252] No CS_2 was isolated from sterile soil. Ayanaba et al.[254] reported the formation of dimethylnitrosamine (DMNA) in flooded thiram-treated soil in the presence of nitrate or nitrite. They also demonstrated the formation of DMNA with added DMA and nitrite. However, both DMA and DMNA disappeared with time from the soil.

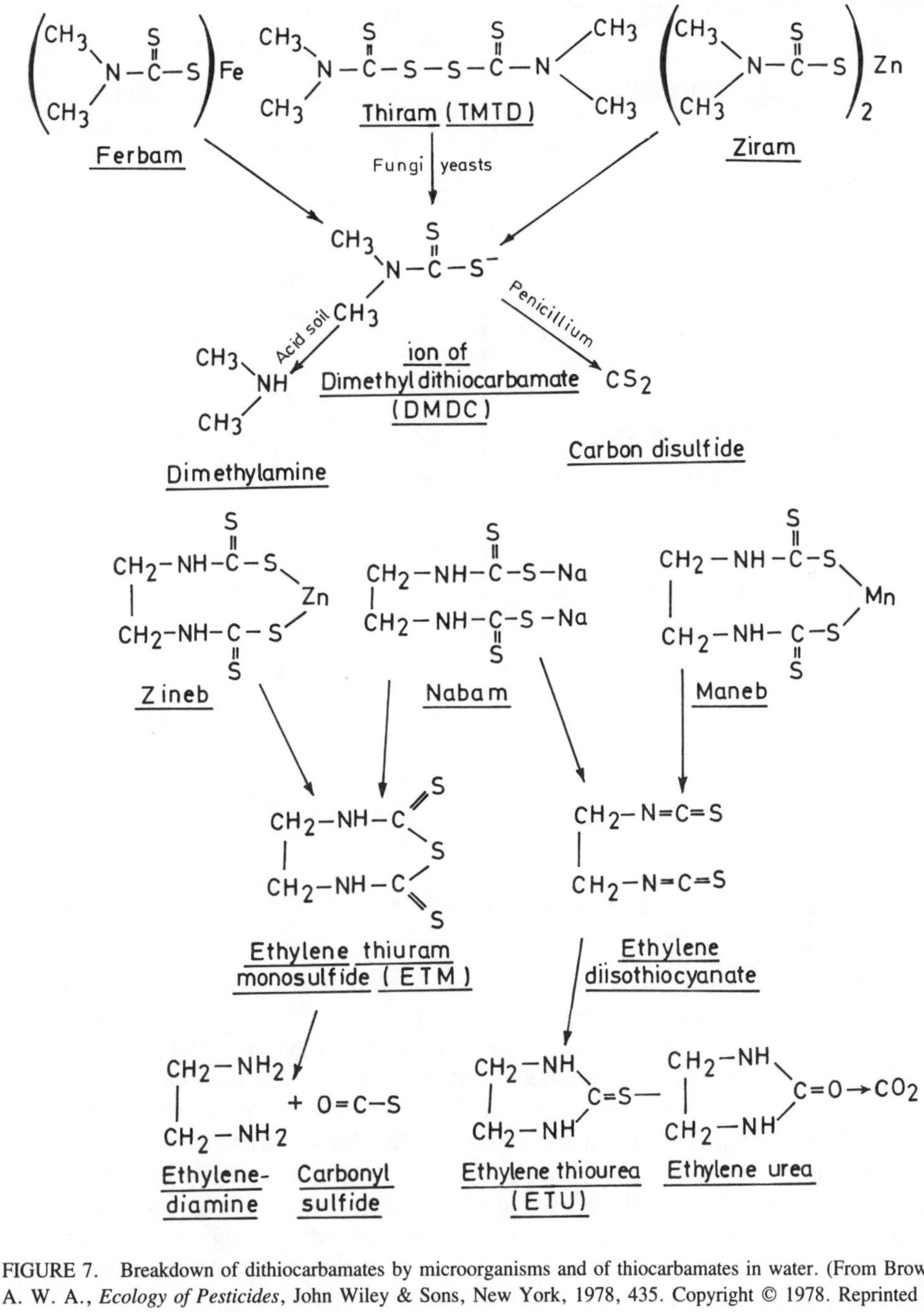

Ferbam

Thiram (TMTD)

Fungi | yeasts

Ziram

CH_3 $N-C-S^-$ CH_3

Acid soil

Penicillium

ion of
Dimethyl dithiocarbamate
(DMDC)

CS_2

Dimethylamine

Carbon disulfide

Zineb

Nabam

Maneb

Ethylene thiuram
monosulfide (ETM)

Ethylene
diisothiocyanate

Ethylene-
diamine

Carbonyl
sulfide

Ethylene thiourea
(ETU)

Ethylene urea

$+ O=C-S$

$C=O \rightarrow CO_2$

FIGURE 7. Breakdown of dithiocarbamates by microorganisms and of thiocarbamates in water. (From Brown, A. W. A., *Ecology of Pesticides*, John Wiley & Sons, New York, 1978, 435. Copyright © 1978. Reprinted by permission of John Wiley & Sons, Inc.)

A possible degradation pathway for thiram in soil was proposed by Raghu et al.[252] based on their results and those of others (Figure 8). Thiram is degraded to the dithiocarbamate ion which can readily form heavy metal complexes with Cu, Zn, Fe, etc. The microbial formation of the amino and ketobutyric acids has been indicated.[255] An isolate of *Pseudomonas,* capable of utilizing thiram as a source of carbon, nitrogen, and sulfur, was isolated. Dithiocarbamate (DTC), DMA, formaldehyde, elemental sulfur, and methionine were detected in the culture as degradation products of thiram.[162] *Aspergillus* spp. utilized thiram as a sole source of carbon. Sodium DMDC was converted to α-aminobutyric acid, not only

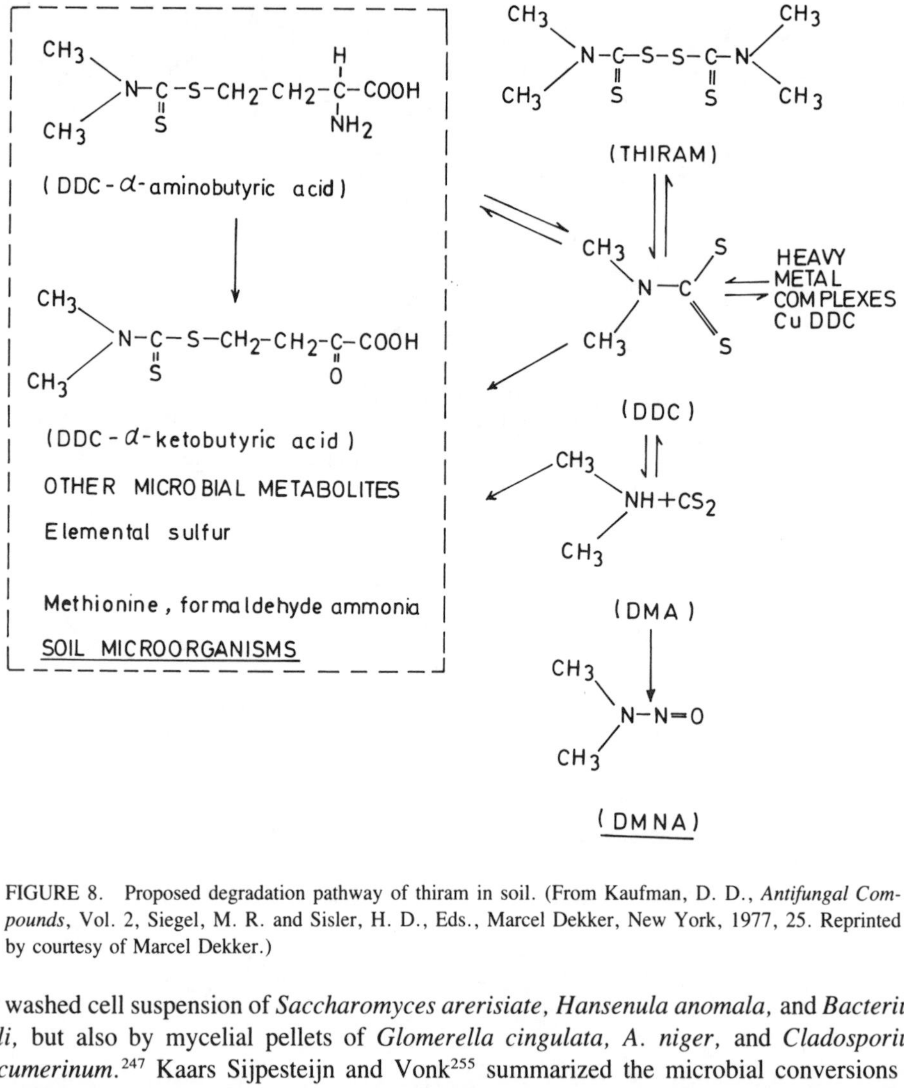

FIGURE 8. Proposed degradation pathway of thiram in soil. (From Kaufman, D. D., *Antifungal Compounds*, Vol. 2, Siegel, M. R. and Sisler, H. D., Eds., Marcel Dekker, New York, 1977, 25. Reprinted by courtesy of Marcel Dekker.)

by washed cell suspension of *Saccharomyces arerisiate, Hansenula anomala,* and *Bacterium coli,* but also by mycelial pellets of *Glomerella cingulata, A. niger,* and *Cladosporium cucumerinum.*[247] Kaars Sijpesteijn and Vonk[255] summarized the microbial conversions of some DTC fungicides. Thiram is reduced to DMDC by microbes or reducing agents. When soil pH is below 7, dimethyldithiocarbamic acid is formed; eventually this resolves to DMA and CS_2. The DMA may be volatile or be metabolized by soil bacteria. Nabam, maneb, zineb, NIA 9102, and the zinc ion-maneb complex have recently come under close scrutiny, since they are known to degrade to ETU, a known carcinogen. Kaufman[4] pointed out that in most soils ETU degrades by formation of ethylene urea, which is subsequently degraded to other products including CO_2. Degradation may occur by photolytic, chemical, or biological mechanisms. The ethylenebisdithiocarbamate, in contrast, has not been shown to be degraded microbiologically. However, it is unstable in aqueous solution and may rapidly form ethylene diamine and CS_2, which are both volatile. In aerated soil, however, nabam forms ethylene thiram monosulfide (ETM) and ETU or ethylene diamine and CS_2. ETU appears to be the only decomposition products of bisdithiocarbamate fungicides that persist in soil.[255] Moje et al.,[18] showed that when an acid soil was treated with nabam, carbonyl sulfide (COS) and H_2S were gaseous products in air above the soil (Figure 8); while H_2S was quite innocuous to test fungi, vapors of COS were quite toxic to *Pythium irregulare* in vitro. Vonk and Kaars Sijpesteijn[256] made similar remarks that among the ethylenebisdithiocarbamate fungicides, nabam, manab, and zineb break down in water to yield sulfur,

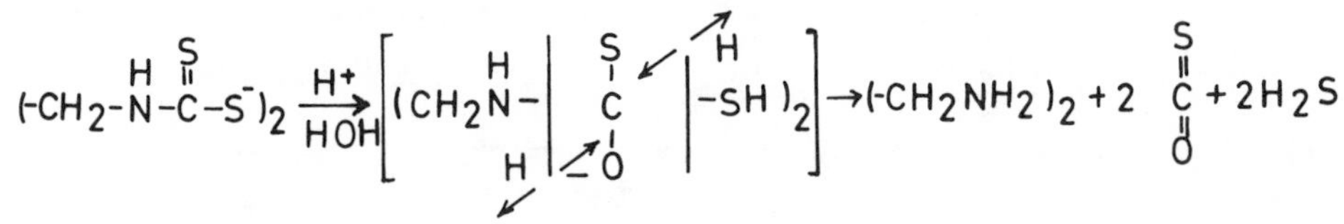

$$\text{(-CH}_2\text{-N-C-S}^-\text{)}_2 \xrightarrow[\text{HOH}]{\text{H}^+} \left[\text{(CH}_2\text{N-} \middle| \overset{\text{S}}{\underset{\text{O}}{\text{C}}} \middle| \text{-SH)}_2 \right] \rightarrow \text{(-CH}_2\text{NH}_2\text{)}_2 + 2 \overset{\text{S}}{\underset{\text{O}}{\text{C}}} + 2\text{H}_2\text{S}$$

FIGURE 9.   Gaseous products in air above the soil after treating acid soil with nabam. (Reprinted by permission from Moje, W., Munnecke, D. E., and Richardson, L. T., *Nature*, 202, 831, 1964. Copyright © 1964, Macmillan Journals Limited.)

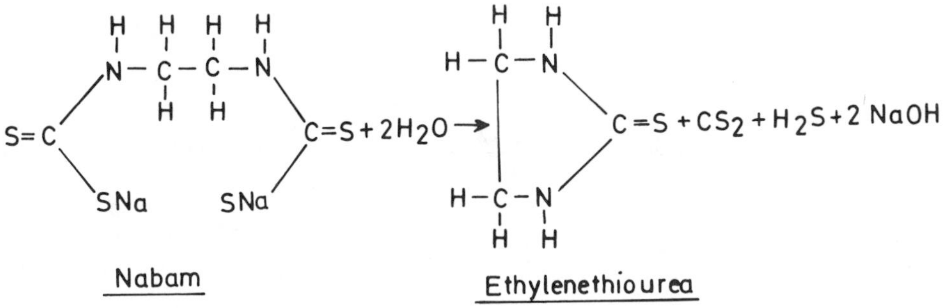

FIGURE 10.   Degradation of nabam in the presence of weak acid. (From Lopatecki, L. E. and Newtonk, W., *Can. J. Bot.*, 30, 131, 1952. With permission.)

ETM, and ETU. When aerated in aqueous solution, they yield several degradation products (Figure 9) including ETM, ethylenediisothiocyanate (EDI), ETU, ethylenediamine, and sulfur.[257] Sodium diethyldithiocarbamate, decomposed under slightly acidic conditions producing carbon disulfide and a salt of diethylamine.[242] Nabam (disodium ethylenebisdithiocarbamate) decomposed under slightly acid conditions to produce approximately equal volumes of hydrogen sulfide, and CS_2 and presumably left a residue of ETU. Apparently, therefore, nabam decomposes in the presence of weak acid in the following fashion[242] (Figure 10).

Kaufman and Fletcher[258] examined degradation of (ethylene-^{14}C) ETU in autoclaved and nonsterile soil of two types. At rates as high as 200 ppm, essentially all the compounds were converted to ethylene urea in 8 days and a similar slower, but steady, conversion occurred in sterile soil. Further degradation of ethylene urea with evolution of [14]CO$_2$ occurred in nonsterile soil only. Four degradation products of ETU were isolated and characterized. Vander Kerk[259] demonstrated that the compounds like TMTD, di-sodium ethylenebisdithiocarbamate, and sodium *N*-methyldithiocarbamate are converted into biocide isothiocyanates in the soil. This is probably the reason for their fungicidal potency. Vapam decomposes chemically in soil to MIT which is the primary toxicant.[19,21,205,260,261] Gray[205] demonstrated that in soil the main decomposition product of Vapam was MIT which was recovered in 78% yield without heating. When dilute solutions of Vapam in water were applied at the recommended rate to peat and heavy clay soils in pots, the Vapam disappeared completely (100%) in 1 hr. Under the same conditions, the amount of Vapam disappearing in 1 hr was 92 to 94% in lighter clay soils, 71 to 74% in sandy loam soils, 28 to 43% in lighter sandy soils, and 10% in builders' sand. Vapam decomposed (95 to 100% in 1 hr) in pots of moist Montmorillonite clay, vermiculite, and sphagnum peat moss, but it decomposed very slowly (6 to 14%) in calcite, talc, sand, and cellulose paper powder. Turner and Corden[21] conducted a detailed study on some of the chemical reactions of Vapam and its degradation products under varying conditions of the soil environment. They reported that Vapam decomposed in dilute aqueous solution at pH 9.5 to MIT and elemental sulfur. In acid solution, carbon disulfide (CS_2), hydrogen sulfide (H_2S), *N,N'*-dimethylthiuram disulfide (DMTD), methyl-

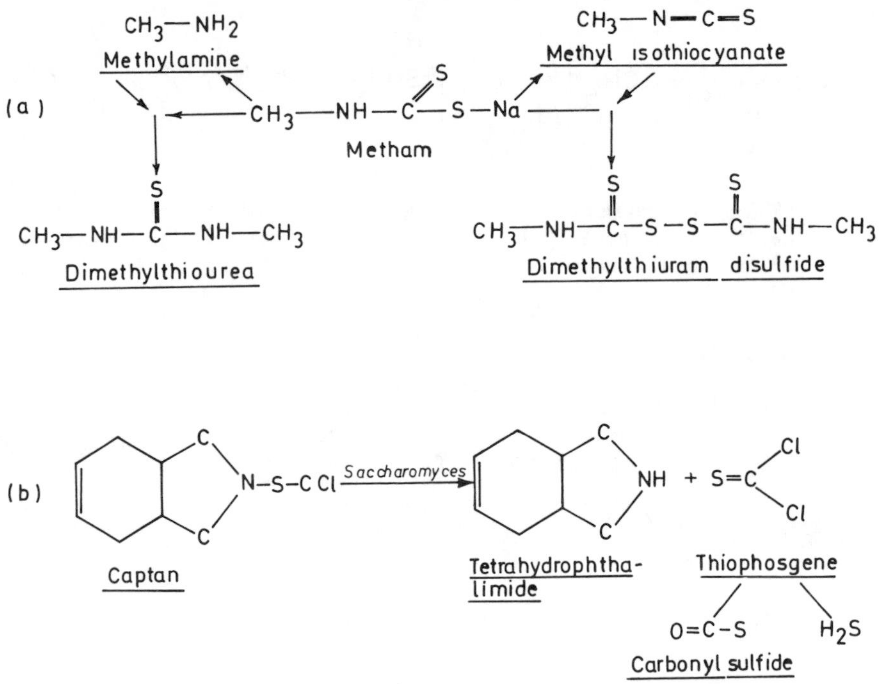

FIGURE 11. Metabolic fate of metham in soil and of captan in yeast. (From Brown, A. W. A., *Ecology of Pesticides,* John Wiley & Sons, New York, 1978, 434. Copyright © 1978. Reprinted by permission of John Wiley & Sons, Inc.)

amine, and MIT are formed. Methylamine, MIT, and N,N'-dimethylthiourea (DMTU) were identified in commercial samples of Vapam. Methylamine and CS_2 can react to form MIT, which in turn can react with Vapam to yield DMTD, and with methylamine or H_2S to give DMTU (Figure 11); they further observed that in a sandy loam soil, over 87% of the Vapam applied was accounted for as the highly fungitoxic MIT. In addition, more DMTU was found than was present in Vapam before it was added to the soil. High temperature and low moisture increased the rate of Vapam decomposition. Increasing the soil temperature from 10 to 40°C decreased the time required for maximum Vapam decomposition from 7 to 1.5 hr; decreasing the moisture content from 20 to 6% decreased the time required from 7 to 2.5 hr.

Production of MIT from Vapam is primarily an oxidative process and is therefore inhibited by a nitrogen atmosphere. Environmental conditions that favor aeration of Vapam in soil, e.g., low soil moisture and an increased liquid-air interface associated with the greater surface area of smaller soil particles, increase decomposition of Vapam to MIT. Hughes[15] suggests that decomposition is associated with direct action of soil causing only a slow decomposition similar to that obtained in distilled water. Decomposition of metham[205] and MIT[262] is accelerated by added organic matter. Organic matter content is generally considered a good indicator of microbiological activity, and decomposition also seems to be increased with increased organic matter.[8] Soil pH is known to affect microbial populations. Metham decomposition is accelerated by metallic cations such as copper and iron.[13]

Effectiveness of Vapam as a soil fumigant is largely dependent on its conversion to MIT; production of DMTU and DMTD reduces its effectiveness. These relatively nonvolatile decomposition products represent a potential hazard of contamination of soil, water, and plants with undesirable residues.[21] Since MIT persists in sandy soils longer than in the heavier soils, phytotoxicity is most likely a problem.[262] Gray and Streim[263] were of the

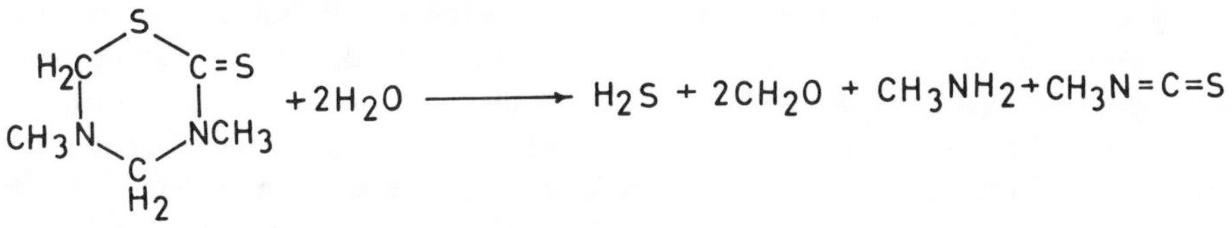

FIGURE 12.   Degradation of dazomet in soil by hydrolysis.

opinion that DMTU present in commercial Vapam is associated with phytotoxicity following Vapam treatment, particularly in sandy soils.

Within a short period, dazomet disintegrates in the soil by hydrolysis, forming MIT, hydrogen sulfide, formaldehyde, and methylamine[70,262,264-267] (Figure 12). In acid solutions, dazomet produces carbon disulfide, methylamine, and formaldehyde, even at low temperatures. Under alkaline and neutral conditions MIT rather than carbon disulfide results from hydrolysis. This process happens within a few minutes in humus and moist soils. Ammonium humates accelerate the release of MIT from dazomet. The clay fraction of the soil favors this reaction by buffering the pH value and by increasing the surface of the water active in hydrolysis. Together with methylamine, MIT is transformed into the symmetrical DMTU. At the same time, a small amount of monomethylthiourea is produced in the soil originating from MIT and ammonium. Monomethylthiourea is presumably oxidized to sulfide, carbon dioxide, and water under the influence of microorganisms.[266] Similar to metham degradation products, many of the dazomet products may react with one another and form additional products. Methylamine and hydrogen sulfide react with formaldehyde and form (methylamino)methanol, trimethylamine, and 1,3,5-trithiocyclohexane which eventually degrades to CO_2, NH_3, SO_2, and H_2O. MIT reacts with ammonia, amines, and sulfhydryl groups. MIT also will react with water and CO_2, H_2S, and Methylamine.

The decomposition of dazomet increases with increased moisture,[70] perhaps because it is less water soluble and more strongly sorbed by soil organic matter than metham sodium. Sorbed chemical is more or less unavailable for degradation. Increasing moisture releases more of the chemical from the organic matter, thus tending to hasten decomposition.

c. Polychlorinated Benzenes

PCNB is reduced to pentachloroaniline (PCA) by a large number of soil microorganisms. Chacko et al.[268] used an enrichment technique to develop and isolate organisms capable of degrading PCNB. Eight species of fungi (*Aspergillus niger, Fusarium solani* f. sp. *Phaseoli, Glomerella cingulata, Helminthosporium victoriae, Mucor ramannianus, Myrothecium verrucaria, Penicillium frequentans,* and *Trichoderma viride*) degraded PCNB, and also eight actinomycetes could reduce quintozene to PCA and MTPCB. Degradation occurred only during active growth phases of the organisms. It was proposed that growth in soil is not very active, hence chlorinated hydrocarbons persist for a long time in natural soils. The reduction of PCNB to PCA obviously is an unspecific biological process in which PCNB acts as a hydrogen acceptor. This process may be reversible under less reducing conditions leading again to PCNB. The mechanism of the formation of MTPCB is not yet clear. Possibly the $-SCH_3$ moiety is derived from methionine, the only $-SCH_3$ compound abundantly present in biological material; but experimental proof is still lacking. *Streptomyces aureofaciens* grown in liquid nutrient medium with quintozene (10 to 20 μg mℓ^{-1}) was able to degrade the fungicide to the extent of 36% in 6 days. Nakanishi and Oku[269] confirmed greater uptake of toxicant by sensitive fungi such as *Rhizoctonia solani* and showed that quintozene added at 25 ppm to a liquid culture of *F. oxysporum* was not only reduced to PCA, but was also converted to pentachloromethyl benzene. According to Ko and Farley,[270] reduction is favored in soils by flooding and the resulting PCA was reported to be stable in both moist and submerged soil. The formation of PCA extends the activity of PCNB in soils, since PCA also showed some toxicity to *R. solani*. There is some evidence that pentachlorophenol is also a metabolite of PCNB. An additional factor which may account for extended activity of PCNB in soils is that some soil organisms are capable of oxidizing PCA to PCNB.[4] Pentachlorothioanisole (PCTA) has also been detected as a degradation product of PCNB.[271] Among the compounds identified in extracts of PCNB-treated greenhouse soils[272] were PCNB, PCA, PCTA, tetrachloronitrobenzone (TCNB), tetrachloroaniline (TCA), tetrachlorothioanisole (TCTA), hexachlorobenzene (HCB), and pentachlorobenzene (QCB). HCB, QCB, and PCNB are impurities in technical PCNB.[272] Tetrachloroanisole and TCTA are presumably degradation products of PCA and PCTA, respectively.

PCNB was rapidly lost from soils made anaerobic by flooding or exclusive of oxygen.[273] Losses are mainly due to microbial degradation. Loss rates were inversely related to the content of organic matter. Organic amendments increased the rate of PCNB loss under flooded conditions, but not at field capacity. Degradation in flooded soils was retarded by free oxygen or combined O_2 in nitrate form, but not by MnO_2 or Fe_2O_3. Rate of loss varied directly with temperature from 15 to 35°C. Regarding PCP, some degradation was observed with *Trichoderma virgatum*, but no metabolites were identified.[274] Other investigations indicated that the disappearance of PCP from soils was more rapid in watered than unwatered soils and that decomposition proceeded more rapidly in soils high in organic matter than low in organic matter.[275] It is believed that PCP degradation proceeds by both biological and chemical means.[276] Seven bacterial species which degraded PCP have been isolated.[221] Pure culture of *T. virgatum*, growing in malt extract, was shown to methylate PCP to pentachloroanisole.[277] From an aerobic enrichment culture with PCP as sole organic substance, Chu and Kirsch[278] obtained a coryneform bacterium which could break down PCP completely; 73% of the (^{14}C) pentachlorophenol appeared as $^{14}CO_2$ when exposed for 24 hr to washed cell suspensions in buffer. Similar results were obtained by Watanabe, who reported that in soil perfused with 40 ppm PCP the chlorine atoms were liberated completely after 3 weeks. From this soil, *Pseudomonas*, like bacteria, were isolated which were able to grow aerobically on PCP as a sole source of carbon.[221] Kaufman[4] reported isolation of methyl ether of PCP and the dimethyl ether of tetrachlorohydroquinone from the culture media of these bacteria further obtained methyl ethers of three tetrachlorophenols in extracts of PCP-treated soil. The isolation of tetra- and trichlorophenols from PCP-treated soils has also been demonstrated.[276] Three tetrachlorophenols and four trichlorophenols in PCP-treated soils were observed.[4,276] Kuwatsuka[276] detected 2,3,4,5-, 2,3,4,6-, and 2,3,5,6-tetrachlorophenols and 2,3,6-, 2,4,6-, 2,3,5-, and 2,3,4- and/or 2,4,5-trichlorophenol in PCP-treated soils. Major products were 2,3,4,5-tetrachlorophenol and 2,3,6- and 2,4,6-trichlorophenols. The other phenols were present in only trace amounts. No di- or monochlorophenols were reported.

The effects of several soil factors on PCP degradation in soil have been discussed.[276] Temperature, aeration, and organic matter were closely related to the rate of PCP dissipation from soil. Cation-exchange capacity and soil pH were only partially related to the PCP degradation, whereas soil texture, clay content, degree of base saturation, and free iron oxides were not closely related to the rate of PCP degradation. At higher temperatures PCP disappeared more rapidly from soils under anaerobic (flooded) conditions than under aerobic conditions. The half-life of PCP at an initial concentration of 100 ppm was 10 to 40 days at 30°C under flooded conditions, whereas almost 100% was detected in aerobic soil after 2 months. PCP degraded more rapidly in high-organic-matter soils than in soils low in organic matter. Pure culture of *Escherichia coli* rapidly reduced ^{14}C-DCNA (2,6-dichloro-4-nitroanilin) to DCPD (2,6-dichloro-*p*-phenyle-nediamine) as a major metabolite, whereas for *Pseudomonas capacia*, ADCAA (4-amino-3,5-dichloroacetanilide) was the main product.[279] Apart from these two compounds, four minor metabolites were also detected in cultures. Reduction to DCPD by *E. coli* was greatly enhanced under anaerobic conditions. Further breakdown was observed for a rod-shaped bacterium isolated from soil repeatedly treated with DNCA. In nutrient solution, it completely transformed 20 ppm of (^{14}C) DCNA within 5 days, 25 to 50% of the label being recovered as [14]CO.[138]

Chloroneb is slowly converted to a nontoxic metabolite in culture solutions of *Rhizoctonia solani*.[280] The metabolite was subsequently identified as 2,5-dichloro-4-methoxyphenol and apparently arises by cleavage of the O–CH[3] linkage by dealkylation. Of 23 organisms grown in the presence of chloroneb, 13 demethylated the fungicide to produce DCMP.[4] A *Fusarium* species was most effective and about 50% of the added chloroneb was degraded to DCMP in 5 days. there were eight organisms, particularly *Trichoderma viride* and *Mucor ramannianus*, capable of converting DCMP back to chloroneb. *Cephalosporium gramineum, Rhizoctonia solani,, M. ramannianus,* and *Fusarium* could both methylate and demethylate DCMP to produce chloroneb and 2,5-dichlorohydroquinone, respectively. The microbial methylation of DCMP to chloroneb may have accounted in the past for the relative stability and long-term effectiveness of chloroneb in soil.[4]

d. Captan

Captan, Captafol and Folpet react readily with thiols such as cysteine and glutathione at pH levels in excess of 4.0 to 5.0. Phthalimide, thiophosgene, dithiophosgene, tetrahydrophthalimide, carbonyl sulfide, and hydrogen sulfide are degradation products obtained from reactions of captan and Folpet with microbial cell components.[281-285] The trichloromethylthio groups and thiophosgene were believed to be intermediates in the formation of 2-thioxo-4-thiazolidine carboxylic acid, another degradation product. *Saccharomomyces cerevisiae* degraded captan to thiophosgene and di(thiophosgene).[283]

e. Systemic Fungicides

Carboxin was degraded rapidly in soil.[286] Although hydrolysis of carboxin would produce aniline, this mode of degradation has not been observed in either soil or aqueous systems. The major route of degradation in both systems appears to be via the formation of the sulfoxide and subsequently the sulfone. The sulfone was observed only in aqueous systems. The oxidation of carboxin in soil to its sulfoxide is completed 2 weeks after application under greenhouse conditions, but no sulfone was detected. The first product, the sulfoxide, is much less active than the parent material, carboxin. Acid pH ranges of 2 to 4 favored a more rapid oxidation of carboxin in aqueous systems. Cultures of *Rhizopus japonicus* could convert carboxin into carboxin sulfoxide and carboxin sulfone (oxycarboxin). When the cultures were maintained under aerobic conditions, both the sulfoxide and sulfone were produced.[287] When *R. japonicus* was cultured under limited anaerobic conditions, the sulfoxide, but not the sulfone, was produced. However, under the partial anaerobic conditions, a third metabolite was produced. This metabolite had an identical UV maximum as a compound observed under similar conditions and had been recognized as a substituted anilide.[4] Agnihotri et al.[188] also observed degradation of Vitavax by several fungi during their growth. Of the 15 common soil fungi tested, only *Aspergillus niger, A. tamarii,* and *T. viride* were able to break down Vitavax into Plantvax, sulfoxide, and two unidentified compounds. The two ascomycetes tested, viz. *C. nigricolor* and *V. vasinfecta,* failed to degrade Vitavax. *P. aphanidermatum* and *R. bataticola* grew luxuriously but failed to degrade Vitavax. Oxycarboxin, in addition to being a metabolite of carboxin, is itself a systemic fungicide. It degraded rapidly and extensively in hydroponic solutions.[4] The degradation was concomitant with bacterial growth and an increase in pH. A strain of *Pseudomonas* from red sandy loam soil was isolated which was capable of degrading carboxin and oxycarboxin.[288] The bacterium hydrolyzed oxycarboxin to aniline and oxidized carboxin to its sulfoxide and to the sulfone. The sulfone derivative was further hydrolyzed liberating aniline.

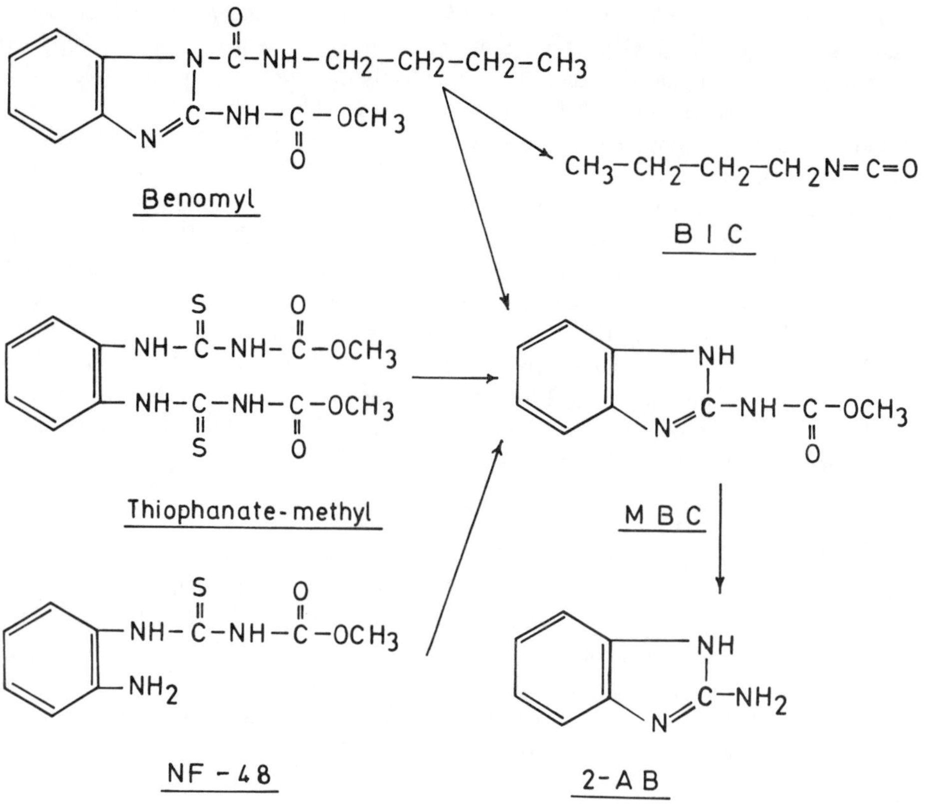

FIGURE 13.   Main precursors of MBC and possible metabolites in soil. (From Van Wambeke, E., *Soil Disinfestation*, Mulder, D., Ed., Elsevier Scientific Publishing Company, Amsterdam, 1979, 351. With permission.)

Further degradation of aniline by the organism resulted in the accumulation of ammonia which was partly oxidized to nitrite that accumulated in the medium. Engelhardt et al.[289] reported the hydrolysis of carboxin by a crude enzyme preparation of *B. sphaericus* which resulted in the liberation of aniline. It seems likely that aniline is further broken down in soil, because soil bacteria of the species *Nocardia* can grow on a medium with carboxin as the sole source of carbon and nitrogen.[291]

In aqueous solution, benomyl is quite rapidly hydrolyzed to produce MBC, which is probably the actual toxicant for fungi.[292] Because benomyl is so unstable, it has been proposed that MBC is the actual toxicant in plant tissues removed from the point of benomyl application. A more polar, but unidentified produce of MBC has also been isolated. In soil, benomyl is completely converted to MBC in a few hours; this metabolite is further degraded to nonfungicidal metabolites, and four species of bacteria and two of fungi were isolated that can affect this degradation.[293] Rapid loss of benomyl in soils was observed, which could not be attributed to leaching or to plant uptake, but to microbial degradation.[294]

Benomyl was degraded to MBC and 2-aminobenzimidazole (2-AB) in bare and turf-field soil.[182] From studies by Baude et al.[295] with radioactive-labeled benomyl in aqueous suspension at pH 7.3, TLC separation followed by radioscanning indicated four metabolites of benomyl, namely, MBC, 2-AB, 2-(3-butylureido)-benzimidazole or BUB, and 3-butyl-*S*-triazina(1,2a)benzimidazole-2,4(1*H*, 3*H*)dione, or STB. MBC precursors and two of their possible metabolites in soil are shown in Figure 13. Rouchaud et al.[296] described the different kinds of reactions in biochemical media for benomyl or MBC and tested same metabolites involved which could be identified, namely, 2-AB, benzimidazole, *O*-aminobenzonitrile, aniline, and *O*-phenylene diamine. Although enzymatic systems are generally involved, other factors may also be responsible for conversion. An *Achromobacter* isolate was reported to degrade benomyl.[221] Mixed cultures of carbendazim-degrading bacteria were obtained on a medium with benomyl as a source of carbon.[297] Working with ring-labeled carbendazim, Fuchs and Bollen[297] observed the formation of 2-aminobenzimidazole and of an unknown compound. From the cultures, $^{14}CO_2$ evolved probably as a result of ring splitting of 2-aminobenzimidazole. The formation of the S-hydroxy derivative of carbendazim by fungi was observed independently by Japanese and Dutch workers.[298,299]

Likewise, thiophanate-methyl also underwent rapid conversion to MBC.[300] Thiophanate-methyl is unstable in soil. Studies with variously ^{14}C- and ^{35}S-labeled compounds revealed that after 1 week only 1% of the parent compound was left.[301] The major metabolite was MBC. Its concentration increased at first, but later declined. From the chromatographic data given, it may be deduced that 2-aminobenzimidazole is also formed. The rate of conversion of thiophanate-methyl was four times faster in soil at pH 7.4 than in soil at pH 5 to 6. Steam treatment reduced this rate showing that microbial activity might be a factor in these degradations.[300] At a higher temperature the degradation of the MBC formed occurred faster.[301] Regarding transformations of thiophanates by pure cultures of microorganisms, little information is available in the literature that several species of fungi can accelerate the conversion of thiophanate-methyl into MBC.[299-302]

f. Miscellaneous

Fumigant chemicals may be degraded by chemical and biological means. The rate of degradation was found to be rather slow for EDB.[61,65,303] It was found that after 31 days only 5% and after 172 days 91% of the active substance had been destroyed in loam soil, whereas only 9% had been eliminated in a sandy loam soil during the same period under alkaline conditions. There is faster hydrolysis of EDB than under acid conditions. The active substances can alkylate the organic matter. There is only slight decomposition of the EDB applied in a liquid form to dry peat.[303] The degradation is favored in wet soil.[63] Little information is available on the biological decomposition of fumigants containing bromine in the soil. Castro and Belser[304] described a biodehalogenation of EDB, DBCP, and 2,3-dibromobutane. A series of excellent papers by Castro et al.[304-308] followed the chemistry of biological and nonbiological degradation of halogenated fumigants and their breakdown products. *Cis*- and *trans*-1,3-dichloropropene (I) hydrolyze in moist soil to the corresponding 3-chloryl alcohols (II) which are biocidal. In the laboratory experiments, cultures of a species of *Pseudomonas* isolated from soil previously enriched with (I) eventually converted the residue steps to Cl^- and CO_2. These steps follow: (II) was converted into formalacetic acid (III) with the release of Cl^-. Rapidly, after the previous slow step, (III) is decarboxylated into CO_2. Munnecke[10] was of the opinion that their work is notable in that the data are confirmed by using material balance calculations, rate-curve analyses, stoichiometry, and stereochemistry.

Several strains of *T. viride* capable of metabolizing allyl alcohol were isolated, and its rapid detoxification in soil was conclusively demonstrated.[4] A number of bacteria, including *P. fluorescens, P. putida,* and *Nocardia corallina,* were also capable of utilizing allyl alcohol for growth.[309] Although no degradation products have been described, the work of Legator and Racusen[310] may be worthwhile mentioning. They reported that the toxicity of allyl alcohol to *R. solani* was attributable to its in vivo conversion to the aldehyde acrolein, which reacts readily with thiols. The rate of Dexon disappearance from soil is initially rapid, but slows down with decreasing concentration, Dexon degradation takes place in the presence of reduced diphosphopyridine nucleotide (DPNH) by *Rhizoctonia.*[311] Addition of organic nutrients such as glucose or straw facilitated a more rapid degradation.[172] They believed that a higher incubation temperature (30°C) and moisture content may have accounted for the more rapid degradation. However, no degradation product was characterized. A culture of *P. fragi* was isolated from Dexon-treated soil.[312] It was unable to use Dexon as a source of carbon or nitrogen for growth, but accumulated *N*-, *N*-dimethyl-*p*-phenylene diamine in a glucose mineral salt medium treated with 100 ppm Dexon. This diamine is as toxic to *P. aphanidermatum* as Dexon itself. A reductase responsible for the formation of the diamine has been isolated from the bacterium.[313]

2. Photodecomposition

The action of sunlight may chemically alter and destroy pesticides in the environment. The rate of breakdown and the potential of the products to contaminate the environment influence the usefulness of the parent compound. The practical significance of photodecomposition, or degradation by the direct influence of light, as a process affecting pesticides in soils is certainly less clear than others.[4] It can only occur at the soil surface; therefore, only those compounds not soil incorporated or those which have moved to the surface during drying will be degraded. Most research deals with reactions occurring in solutions, often organic rather than aqueous. Further, more UV lamps are generally used as an energy source, despite almost no radiation of this frequency actually reaching the earth's surface. Since environmental photochemistry is complicated by the possible interaction of the reacting molecule with many environmental components, it has become necessary to conduct laboratory experiments to obtain comparative data for determining the effect of factors that might modify products and rates of reaction. There are two fundamental laws of photochemistry: (1) the Grotthus-Draper law which states that only the light adsorbed by a molecule is responsible for reaction, and (2) the Stark-Einstein law which states that the absorption of light by a molecule is a one-quantum process, so that the sum of the primary quantum yields must be unity.[4] Elaborate experimental methods are required for the determination of quantum yields. Main emphasis has been given currently to atmospheric photochemistry where consideration can be given to reactions of organic molecules at low concentrations in the vapor phase.[4] The importance of light energy in the fate of toxicants has become increasingly recognized in the past decade, and the reactions may be unobtrusively photopromoted, if not photoinitiated. The wavelength of UV light extends from 400 to 40 nm, but because of the thin layer of ozone in the earth's atmosphere, practically all the sun's emitted radiation below 290 nm is absorbed before it reaches the surface. Gore et al.[314] demonstrated that of 76 pesticides, 29 had no bands above 290 nm in their UV absorptive spectra, and 25 had only a few. This could be critical in terms of terminal residue considerations. However, because of a not inconsiderable flux density of 10^{16} photons per square centimeter per month, a long-term effect at shorter wavelengths is possible, and indeed, dieldrin with no absorption band as high as 290 nm, does, in fact, breakdown in sunlight.

An organic compound must absorb energy either directly or through a photosensitizer (photochemically excited molecule) for a photoreaction to occur, which may be deliberately introduced or which may already be present. The important fact is that the energy available in the UV region is of the order needed for fission of some typical covalent bonds in organic compounds. The environment of the toxicant molecule is also important, for inter- as well as intramolecular reactions may occur. Thus, adsorption on soil particles may change the maximum absorption wavelength and hence alter photodecomposition energy requirements. The effect of photosensitizers present in soil or in the air may also be extremely important in relation to synergistic effects, on which very little work has been done.

Stimulation of wavelength range and intensity of solar radiation in vitro have generally involved single lamp sources. While photolysis products are generally identical for all light sources, a correlation between artificial and natural sunlight is necessary. Crosby[315] discussed the experimental approach in which low-pressure lamps produce 90% of their energy at 254 nm; medium-pressure lamps have this proportion of the available energy fairly evenly spread over the range of 290 to 366 nm. In vitro experiments are invaluable, in that degradation products can be obtained in relative abundance, enabling their detection, isolation, and identification to be carried out more easily and more confidently than in experiments in vivo.[316]

Woodcock[316] is of the opinion that since pesticides in the atmosphere are exposed to more intensive UV radiation over a wide wavelength range, and to a variety of free radicals produced photochemically from other common pollutants such as sulfur dioxide and nitrogen tetroxide, investigations involving gaseous phase-irradiation seem highly desirable and many more researches of this type surely will be instigated. Crosby and Li[317] discussed the problems of conducting experiments in sunlight under natural environmental conditions. Contamination, variable light intensity, and volatilization are only a few of the problems. Since photodecomposition does occur under the less energetic sunlight, sensitizers have been suggested as occurring in soil.[124] These substances resemble catalysts in that they are light activated, then transmit this energy to the pesticides before returning to the ground state. Riboflavin and ferric salts are known sensitizers.

Solvent, concentration, and aeration are important factors affecting rates of photolysis. Rates of photochemical reaction are seldom influenced by temperature. For environmental studies of photolysis, water is the predominant solvent. Organic solvents are frequently limited by their ability to transmit light or by the fact that they themselves may participate in the chemical reactions. In the presence of high concentrations, many complex polymeric substances may be formed. In dilute aqueous solutions, first-order rate laws apply to decomposition rate by photolysis. The course and products of reaction are also influenced by the presence or absence of oxygen. Photooxidation reactions predominate in the presence of oxygen rather than nitrogen.[318] Photochemical studies have been performed on thin layers of pure material deposited on glass plates, silica gel-coated glass plates, or thin layers of soil. Studies conducted on soil surfaces resulted in the slow rates of photolysis as compared to silica gel.[4] However, changes in absorption spectra occur with toxicants deposited on surfaces. Since this type of spectral change is a general phenomenon and is dependent upon the nature of the compound.[319] care must be taken in extrapolating photochemical information from one system to another. Under certain conditions, photodecomposition seems to be a major factor in the degradation of certain fungicides. Logical goals in future pesticide decomposition research would seem to be establishment of the quantitative significance of photodegradation, evaluation of soil parameters influencing photodegradation, and understanding the photochemical mechanisms at the soil surface.

a. Organomercurials

Shiina et al.[320] studied in vitro photodecomposition of organomercury toxicants, using light with wavelength of 290 nm for a period of equivalent to 7 summer days, and reported that those compounds (including phenylmercury acetate and chloride) which do not absorb below 290 nm were little affected, whereas two others with absorption bands of 290 nm were degraded in varying degree. Menzie[206] mentioned that mercury in organomercury fungicides may undergo photoreduction depending on the compound and conditions of photolysis.

b. Organosulfur Compounds

Cruickshank and Jarrow[321] demonstrated degradation of ETU on silica gel by radiation of wave length 285 nm, mainly to ethylene urea, though several other products, including bisimidazolin-2-yl sulfide, were also formed. Ethylene urea has been reported to undergo photoxidation to hydantoin, but the latter compound was not detected among the products. Particularly in the presence of photosensitizers, photodecomposition of ETU was rapid and in aqueous solution a very slow photolysis was markedly accelerated by the presence of compounds such as 1- and 2-acetonaphthone, 1-naphthaldehyde, and methylene blue.[316] Ross and Crosby[322] also found that ETU in aqueous solution was virtually stable to sunlight, but the presence of dissolved oxygen and sensitizers such as acetone or riboflavin was found to cause rapid photolysis to ethylene urea and glycine, possibly via hydantoin; photochemical work with dithiocarbamate fungicides is conspicuously scanty.

Fission of the C-S bond using model compounds in benzene solution was demonstrated as early as 1960, but this work was in connection with photosensitive polymers.[316] Woodcock[316] cited the breakdown of sodium DMDC in aqueous solution at pH 8 under the influence of 254 nm radiation. The products were carbon disulfide and dimethylamine, and of the nine aromatic ketones examined as sensitizers of this photolysis, 2-acetonaphthone was the most effective. Fitton et al.[323] photoirradiated several 4-hydroxybenzyldithiocarbamates under various conditions. They demonstrated that the initial step is the homolytic fission of the C-S bond with the formation of dithiocarbamoyl and benzyl radicals, which then proceed to dimerize, the former giving a thiuram disulfide and the latter either dimerizing to diphenylethane or dissociating with loss of a proton to a *p*-quinone methide. When irridations were carried out in the presence of water, aromatic ketones were also produced probably arising via the *p*-quinone methide.

c. Aromatic Compounds

Photodecomposition of aromatic compounds has been reported and the extent of this reaction as a major route of environmental loss of soil fungicides such technazene (1,2,4,5-tetrachloro-3-nitrobenzene) and quintozene (pentachloronitrobenzene) is important. Crosby and Hamadmad[324] observed that although a solution of quintozene in methanol or hexane or with exposure to sunlight for 7 days became yellow, no decomposition products that could be demonstrated using TLC or gas-liquid chromatography (GLC) irridation of a thin film were similarly negative. However, laboratory studies, in which an examination of quintozene had been exposed in hexane solution to a low-pressure mercury arc lamp radiating at 254 nm, revealed the presence of four decomposition products in the rapidly discolored solution. These were in order of increasing GLC retention time 1,2,4,5-tetrachlorobenzene, QCB, and 2,3,4,6- and 2,3,4,5-tetrachloronitrobenzenes; all four compounds were present after only 5 min irradiation. After 56 hr, they represented 20 to 23% of the initial quintozene, only 7% of which remained, and their relatively constant proportions indicated steady-state photolysis, which could be excluded when a borosilicate filter absorbing below 285 nm was interposed between the lamp and the solution. Although a solution of QCB in hexane or methanol did not discolor after 7 days exposure to sunlight, irradiation to hexane for 3 hr with a mercury lamp resulted in the formation of 1, 2, 4,5-tetrachlorobenzene in 50% yield, together with some 13% of the isomeric 1,2,3,5-tetrachlorobenzene. This seems to confirm the irradiation fate of QCB, the slow disappearance of which corresponded to the appearance of the 1,2,4,5-tetrachloro compound. These results have been considered in terms of both free-radical and ionichydride transfer mechanisms.[324] The effect of the electron-attracting substituent on the position of reductive dechlorination is consistent with the former, involving hydrogen abstraction from the solvent.[325] This is also supported by the fact that very similar proportions of identical products are obtained whether the irradiation was carried out in hexane, cyclohexane, acetone, or ethanol, and the detection of 1,2- and 3-chlorohexanes in equal amounts when hexane was used as solvent is also in agreement with radical formation. Study of the photoreduction of some substituted nitrobenzenes in organic solvents under the influence of a high-pressure mercury lamp using rapid scan electron spin resonance (ESR) suggests that radicals of the type PhN(O)OR (where Ph = phenyl and R = solvent radical) are involved.[316] An ionic hydride transfer mechanism involving zwitterionic intermediates, and analogous to similar photosubstitution reactions involving the cyanide ion,[326] can also be involved to explain the products formed from quintozene, although the absence of products such as pentachloroaniline is contrary to the known affinity of the nitrogroup for hydride ions.

Stability of fenaminosulf (sodium-*p*-dimethylaminobenzenediazosulfonate) in alkaline solution has been demonstrated. Exposure of its orange-colored aqueous solution to sunlight destroys its fungistatic activity with the progressive formation of *p*-dimethyl aminophenol and colored products. Two-stage decomposition has been reported by Hills and Leach.[146] Photoreduction of the diazonium group, with evolution of nitrogen and subsequent oxidation, was followed ultimately by polymerization of the phenol formed. Crosby and Moilanen[327] confirmed the rapid hydrolysis of fenaminosulf in aerated water, and this was another example of photonucleophilic reaction. *p*-Dimethylaminophenol was obtained in 99% yield, the remaining 1% being accounted for as dimethylaniline. Alconero[328] observed Dexon to be fairly persistent in soil, despite its highly susceptibility to photodecomposition.

Extensive studies have been made to PCP photolysis in aqueous systems. It provides a good example of the importance of irradiation conditions to photolytic pathways. Using a low-pressure mercury lamp at approximately room temperature in each instance, Mitchell[329] observed that PCP remained unchanged when irradiated on filter paper, while Crosby and Hamadmad[324] observed that only 2,3,5,6-tetrachlorophenol was formed in hexane solution. On the other hand, tetrachlorophenol, chloronil, and a large portion of humic acid polymer appeared in water at slightly acidic pH. Furthermore, irradiation of PCP by sunlight in water led to a loss of aromaticity and formation of alicyclic compounds. Haitt et al.[330] had earlier reported that the decomposition followed first-order kinetics, but they did not mention the identity of the products formed. Kuwahara and co-workers[331] observed tetrachlororesorcinol, chloranilic acid, and other oxidized products after irradiation in water at slightly alkaline pH. Octachlorodibenzo-*p*-dioxin was also detected under these conditions.[332] It was confirmed that an aqueous solution of sodium pentachlorophenate was decomposed when subjected to solar irradiation, the solution turning purple and the concentration being reduced to 50% in 10 days.[332] They showed the presence of six products which did not include the previously reported 2,3,5,6-tetrachlorophenol. Four compounds were previously unknown, and in addition there was considerable resignification. It is interesting to note that the photodegradation products were in general more fungitoxic than sodium pentachlorophenate, but less phytotoxic.

d. Heterocyclic Compounds

The harmful effect of light on captan was demonstrated[316] and its total destruction by UV light of wavelength 254 nm was reported by Mitchell,[329] though none has mentioned specific degradation products. It was observed that 50% of the fungicide was transformed into inactive material after 3 days of sunshine, or after 6 hr at a distance of 30 cm from a "lampe germicide".[316] He also found the effect of light on the related compounds Folpet and Captafol and demonstrated that after 8 hr irradiation, the extents of degradation were 33, 64, and 88% for Folpet, captan, and Captafol, respectively.

e. Systemic Fungicides

Kilgore and White[334] found that exposure to sunlight enhanced the conversion of benomyl to carbendazim. However, Buchenauer[335] expressed doubts about this effect. It has been demonstrated that the UV absorption spectrum of a sample of carbendazim exposed to sunlight, but protected from rain, showed a distinct change after several months. This clearly demonstrated that carbendazim is not absolutely photostable.[316] Some 80% decomposition of carbendazim in methanol took place in 4 days, using a Hanovia 100-W medium pressure lamp in laboratory irradiation experiments. Products identified were dimethyl oxalate, in addition to mono- and dicarbomethoxyguanidines, and evidence was also there for the presence of an acid salt of quanidine. There was no evidence of any products of dehydro-dimerization such as have been observed in the case of benzimidazole itself by Cole et al.[336] They demonstrated the formation of 2,4- and 2,5-benzimidazoles by assuming the inter-mediacy of a benzimidazol-2-yl radical, either generated directly as a result of UV radiation or by rearrangement of a performed benzimidazol-1-yl radical. Instability of carbendazim under high-intensity radiation has also been reported.[316] In phosphate citrate buffer (pH 5) more than 90% was transformed in 2 hr into another fungitoxic compound with a biomodel dosage response pattern of inhibition, but the metabolite was not stable and completely reverted to carbendazim in 7 days. Stronger radiation and high pH enhanced photoalteration of benomyl and carbendazim.[335] Therefore, while carbendazim loses no fungitoxicity in solution at pH 5 or 7, it is inactivated at pH 9 with the formation of 2-aminobenzimidazole, though its formation perhaps represents photo-assisted hydrolysis rather than true photodecomposition.

Evidence of some change on irradiation of triforine in methanolic solution has been reported under nitrogen using a medium-pressure lamp.[316] This sensitivity to UV (254 nm) and sunlight has recently been investigated. Buchenauer[335] demonstrated that in aqueous solution 50% inactivation took place in 3 and 30 hr, respectively, while 25% decomposition of solid triforine was caused by irradiation on glass for 80 hr. Irradiation in aqueous solution followed by bioautographic examination using *Cladosporium cucumerinum* showed that after expo-sures up to 4 hr, a photosensitive fungitoxic metabolite was present which was no longer detectable after 16 hr. Acceleration of photodecomposition by the presence of photosensi-tizers such as riboflavin and xanthene has been observed.

Photochemical decomposition of the thiophanates was demonstrated by Soeda et al.[337] who used [14]C-(ring)-labeled thiophenate-methyl. Two products were obtained and identified, viz., carbendazim and dimethyl-*O*-phenylene-4,4′-bisallophanate. Buchenauer and associates[338] also observed photochemical transformation of thiophanates, and found in-creased conversion to carbendazim or the ethyl analogue. When aqueous solutions were irradiated either in sunlight or by UV light of 254 nm, no conversion took place as a result of incubation in the dark. The greater rate of photochemical transformation, which was found on glass as compared to leaves, confirmed the observation made by Soeda et al.,[337] and it was assumed to be a higher concentration of the fungicide suspension on the surface of leaves.

f. Miscellaneous

Other toxicants also appear sensitive to light. Exposures of chloranil to UV resulted in a two-step reduction. In the presence of UV, dichlone reacted with pyridine nucleotides and underwent reduction.[316] Solutions of dichlone bleach upon standing in glass bottles in light[339] and the effectiveness of fungicidal seed treatment are greatly reduced upon exposure to the sun. White et al.[340] demonstrated that even in subdued daylight, benzene solutions of dichlone underwent photochemical replacement of the chlorines with phenyl groups via free-radical reactions.

IV. FACTORS AFFECTING FUNGICIDAL EFFICIENCY

Soil is a big sink for accumulation of fungitoxicants which interact with the biophysical functions of cultivated soils. Biologically, soil is a very active unit and very variable in its characteristics. In recent years, a great deal of interest has been generated to unveil some of the interaction that occurs in the soil due to man-made factors. However, information with regard to fungitoxicants and their interactions with soil characteristics and plant growth, vis-à-vis plant disease control, has not been investigated in detail. There are several fungicides which are widely used to control different seed and soilborne plant diseases including damping-off/root rots. Soil drench and seed treatment are the two major methods of application to combat soil and soilborne pathogenic fungi infecting roots and stems of crops.

The strategy of chemical control of plant diseases is based on the principle of immobilizing the pathogen so that it does not interact with the host to cause the disease. The process of pathogenesis involves the interaction between host, pathogen, and the environment. In the disease control studies with fungitoxicants, the role of physicochemical environment has not been studied in detail. Some information, however, is available with regard to modification in the toxicity of fungicides in vitro by chemicals like carbohydrates,[341,342] phenols, thiol and other compounds,[343-346] and also metal ions.[347,348] Similarly, in vitro effects of temperature and pH of the substrate medium on the activity of some fungitoxicants have been studied.[345,349,350] However, little information is available with regard to in vivo interactions of these factors on the disease-controlling ability of fungicides.

Since the soil is a highly complex and biologically active substrate through which the fungicide acts against fungi in soil and plants,[351] fungitoxicants often give variable success in controlling seedling diseases of crops in diverse agroclimatic regions of the world. While looking for reasons for discrepancies in the uniform effectiveness of fungicides for controlling seedling diseases on a crop, differential soil characteristics have been implicated as one the reasons.[10,352-354] The studies have shown that different fungicides act best within specific reactive ranges of temperature, moisture, pH, and other physical characteristics of the soil.[345,351,354-356] The fungitoxicants might also react chemically with the fertilizers and thereby either potentiate or antagonize their effect. Altered reactivity of the fungus towards fertilizer and fungitoxicants when both the components are present together is also likely to occur. Moreover, nutrient status of the soil alters the disease potential of the pathogen. There is a direct correlation between inoculum density and the amount of fungitoxicant needed to immobilize it.[35] Therefore, fungitoxicants would be expected to express variable disease-controlling potentials under such conditions. The above changes in the components of the disease triangle and also the fungitoxicants will logically reflect on the disease-controlling behavior of fungitoxicants. Under continuous influence of a chemical-tolerant fungus, strains will be naturally selected among these possibly existing and among those produced by mutation and new genetic combinations. Variable reversal of soil fungistasis under the influence of fungicides has also been demonstrated.[357] Other factors, like the action of antagonistic microbes which are also known to be influenced by certain fungitoxicants, decomposition of fungitoxicants, and penetration of chemicals in plant residues, may also influence the chemical control in soil. Influence of these and several other factors on disease-controlling potential of chemicals in soil was discussed in the preceding pages.

A. Physical and Chemical Properties of Soil
1. Soil Types
When a number of fungicides were compared for their efficacy in agar and in soils, large differences were found.[29] The efficiency of some materials such as captan, nabam, Vapam, and some mercury compounds was somewhat better when mixed in fine sand loam than in agar, but oxyquinolines, ferbam, maneb, and semesan were partially inactivated by the soil.[12] Strong inactivation of captan by soil was reported by Picci.[358] Addition of sand in clay soil increased the efficiency of many fungicides.[359] Organic matter (1%) such as starch or hay had the opposite effect with captan, ferbram, PCNB, semesan, thiram, zineb, and ziram, but not with actidione and Vancide.[51] Alexander[360] demonstrated that captan, PCNB, and zineb controlled a *Rhizoctonia* disease in sand only, while semesan and Panogen controlled it in both soil and sand. Inhibition of *Rhizoctonia* by seed treatment with the chemicals was greater in sand than in compost.[38] Domsch[39] also found that peat amendment in soil-inactivated chinosol, especially strongly coarse sand, seemed to inactivate the fungistasis of captan and thiram by allowing the growth of the fungi without intimate contact with the

fungitoxicants. Methylmercury iodide, in spite of being in some way bound (resisting leaching), remained highly fungicidal in soil, while methoxyethylmercury chloride (MEMC) and PMA quickly lost their toxicity.[361] Domsch[38] found that fungitoxicity with an organic Hg compound varied due to sorption by different soils.

In peat-sand mixtures, the sorption increased sharply with the increase of the proportion of peat. The efficacy of carbendazim seed treatment in controlling mungbean mortality caused by *Rhizoctonia solani* decreased markedly, with increasing silt content, clay content, and with a parallel increase in CEC.[351] The fact that carbendazim seed treatment was more effective at low dosage in sandy soil and poor in clay loam could be correlated with the different characteristics and adsorptive capacities of the soil, in which the proportions of silt, clay, organic matter, and CEC varied. In clay loam soil, it gave only a partial disease control, perhaps because of high organic matter and CEC, together with higher proportions of clay and silt. The precise role of CEC and organic matter or other components, alone or in combinations in modifying the efficacy of carbendazim, is not yet clearly understood. Kataria and Grover[354] have reported that benomyl, thiophanatemethyl, chloroneb, and quintozene were relatively more effective in sandy loam than loam or clay loam, while there was not much difference in the disease incidence caused by *R. solani* in these soils. The different adsorptive capacity of the sandy loam soil due to its low clay content, organic matter, and cation exchange capacity may be responsible for this as compared to loam and clay loam with high adsorptive capacity.[359] Singh[187] conducted an experiment to discover the efficacy of MBC in nine soil types for disease control. The various soil types were inoculated with *R. solani* and seeds treated with MBC were sown in each soil type. The maximum disease (87.5%) developed in hill soil and minimum (77.7%) in silty clay loam. The disease control was best in silty loam (bhabhar) and least in silty clay loam. On the basis of percent disease control with MBC, the nine soils in decreasing order can be arranged as silty loam (bhabhar), sandy loam, loam (hill), fine sandy loam, loam, silty loam (alluvial), clay loam, and silty clay loam.

2. Temperature

An interaction between control of cotton necrosis and temperature has been observed.[362] Vancide 51, while giving excellent control at 21°C, failed at 28°C. Captan seemed to be more effective against the pathogens prevalent at 28°C than against those at 21°C.[12] In other tests PCNB and zineb applications gave poor control at low temperatures.[363] The effectiveness of seed treatment may depend on temperature,[363-365] but not always.[38] The fungitoxicity of Karathane is reduced by high temperature.[366] Protectants otherwise effective sometimes fail at low temperature.[367] Exposing protectants to steaming (in soil) reduced the potency of thiram and phaltan, but not of ferbam, captan, and zineb.[151] In pot tests, maximum protection of mungbean, maximal against damping-off caused by *R. solani*, was obtained when seeds were treated with carbendazim at 1 g a.i. kg^{-1} (as a wettable powder) and sown in river sand kept at 20°C rather than at 25, 30, or 35°C.[351] In pot trials using mungbean, long melon, eggplant, common pea, and sugar beet as seed treatments, benomyl and thiophanatemethyl gave optimum control at 20°C and pH 7.6 against seedling mortality caused by *R. solani*.[354] On the other hand, Singh[187] reported that MBC worked better against *Rhizoctonia* seedling mortality of soybean at 25°C soil temperature than at 30 or 35°C. Best disease control at lower temperature can be explained on the basis of higher seed uptake at this temperature.

3. pH

pH of the substrate modifies the toxic behavior of fungicides by affecting their action on specific enzymes and influence on the cell wall permeability.[349] Increasing the pH of soil from 4.6 to 7;1 increased the efficiency of nabam and Vancide 51 in sterile soils in tests by Rushdi and Jeffers.[359] Similar pH changes decreased efficiency of actidione. At still higher pH (8.0) nabam may again lose its efficiency of PCNB against *Pythium* and of $CuSO_4$ against *Phytophthora cinnamomi*.[369] Liming is reported to reduce the efficiency of certain fungicides.[370] Copper oxyquinoline is most fungicidal at neutrality and sorbic acid at low pH.[371] Domsch[38] found that the fungitoxicity of a Hg compound was minimum at pH 5.9 and increased towards pH 4.1 and 7.4. Chloranil is most fungicidal at the pH range 3.8 to 4.8; it is almost void of activity at pH7.[372] Kataria and Grover[354] demonstrated that best disease control was obtained at pH 7.6 with benomyl and TPM and at pH 5.4 for chloroneb and PCNB.

4. Water Stress

Fungus spores held at low relative humidity are much more resistant to toxicants than if exposed at high humidities. The reason frequently given is that the metabolic activity of the organism is higher and, hence, more susceptible to the toxic material. This may not always be the case, because it may be related as much to sorption, membrane passage, and solution of the gas at the membrane air interphase. Dried spores of *Alternaria solani* were exposed to MB gas at various relative humidities. At very low humidities it was possible to kill the spores even with concentrations as high[373] as 20% MB by volume. The susceptibility of the spores rose sharply with increasing humidity, but fell off slightly above 90% relative humidity. There appeared to be a close correlation between the amount of bromine ions sorbed by the spores and the relative humidity at which the spores were exposed. If spores were held in an atmosphere of nitrogen, water vapor, and MB, the spores were as sensitive as they were in the air. These results indicate that susceptibility of spores is directly related to uptake, not necessarily to spore metabolism. Miller[374] listed five processes involved in membrane passage: (1) mass flow through pores, (2) diffusion (which requires no specific membrane-permeating material relationship), (3) facilitated diffusion (specific structural relationship between membrane and material), (4) active transport (requires relationship of membrane to material and metabolic energy), and (5) pinocytosis. Fungitoxicants accumulate in the cytoplasmic particles of fungi.[375] Therefore, it would appear that the effect of moisture content or relative humidity is related more to penetration and permeation of the spores and mycelia than to metabolism. High humid conditions affect the reaction in soil, and it had been shown[376] that insecticides are more toxic at high relative humidity, because they are less sorbed by soil than under conditions of low humidity. Kataria and Grover[354] also demonstrated that soil receiving irrigation every 24 hr resulted in maximum disease control with all the fungicides tested. Singh[187] reported that maximum disease development (80.7%) occurred at the 24-hr watering interval and was minimum at the 12-hr interval. The disease control was best (69.2%) at the 24-hr interval and least (30.0%) at the 12-hr interval of watering.

5. Influence of Fertilizers and Organic Amendments

In intensive agriculture, application of fertilizer is a must to realize the gains from the genetic potential of high yielding varieties. Retrieval from energy-intensive fertilizer usage to manure derived from bioorganic agricultural wastes is now advocated by exponents of the energy crisis owing to large-scale energy shortages. Fungicides are likely to be used along with these amendment customs to control plant diseases. The use of these practices alters the microedaphic conditions and the nutrient and microbial status of soil. Some of these factors are known to modify fungitoxicity. The interaction of these amendments on fungitoxicity had been ignored only until recently.[377] Yadav observed that soil application of urea, potassium nitrate, and ammonium nitrate lead to a variable reduction in the disease-controlling potential of fungicides against *R. solani* in cowpea. Nitrate nitrogen was more reactive than the amide or ammonium form in lowering the fungitoxic action. Carbendazim was potentiated by two phosphatic fertilizers, viz., monocalcium phosphate (MCP) and potassium dihydrogen phosphate (PDHP). MCP improved and PDHP reduced the activity of Agallol, where as the reverse was true with quintozene. Potassic fertilizers reversed the action of carbendazim considerably. Soil amendments with biogas slurry (BGS), farmyard manure (FYM), and mahua cake led to marked decrease in the disease-controlling potential of all the fungicides. The benefits from Agallol and quintozene treatment were totally lost in the presence of BGS and FYM. Absolute nullification of the disease-controlling potential of Agallol was observed with mahua cake, whereas efficiency of quintozene and carbondazin was reduced to the tune of 75 and 50%, respectively.

The behavior of fungitoxicants was greatly influenced by various fertilizer materials and the magnitude of alteration was governed by the nature of fungitoxicants and fertilizers.[377] Similarly, Bandyopadhyay and co-workers[357] demonstrated noteworthy reduction in efficiency of disease-controlling potential of fungitoxicants used as seed treatments on cowpea and cotton against seedling rot caused by *R. solani* in the presence of organic manures. MEMC was most vulnerable to their effect followed by quintozene and carbendazim. Mahua cake (*Madhuca indica*) was very potent, the extract of it more so, in reducing the activity of fungitoxicants tested than other amendments. They are of the opinion that the composition of mahua cake has a fairly large content of carbohydrates, in addition to other nutrients which probably serve as a food base for the multiplication and increased virulence of the pathogen, thereby enhancing its potency to cause more disease.[357] This is further substantiated by the fact that the addition to the soil of extracts of mahua cake which contained the water-soluble materials, including the nutrients, was potentially more detrimental to the benefit expected to be derived from the fungicides. This extract also contained a harmful compound(s), which is primarily phytotoxic to the cotton root system. Soil amendments with BGS and FYM have considerable nutritive impact in soil, as these consist of high amounts of organic carbon, humic substances, and many major and micronutrients. BGS was more active in reducing the disease-controlling potential of fungicides than FYM on both crops. This is probably because quicker release of nutrients from BGS at higher rates completely nullified the activity of MEMC and quintozene. Both the manures contain large amounts of humic substances which have a high capability to form complexes with organic chemicals.[357] When BGS was depleted of humic substances by alkaline extraction and added to soil, considerable improvement in its efficiency occurred. This suggested reduction in adsorptive sites and, consequently, greater availability of fungitoxicants for disease control. Adsorption on humus of carbendazim,[353] quintozene,[377] and organomercurials[378] has already been reported. However, soil amended with humic acid did not have much influence on quintozene, whereas the activity of the other two fungitoxicants was significantly affected. Amendment with *Sesbania aculeata*, like other green manure crops, adds less organic manure to the soil than is provided by BGS and FYM,[379] and the fungicidal activity is less affected. Singh[187] also conducted a detailed study on this aspect using carbendazim as a test chemical, and his observation was more or less similar to the results of Bandyopadhyay et al.[357] He reported that high amounts of FYM markedly reduced efficacy of the fungicide. He had experimentally proved how green manuring reduces disease control with MBC. Green manuring reduces fungicidal uptake by seeds and inactivation of fungicide by manures may be attributed to humic acid (HA).

Regarding nitrogenous fertilizers, carbendazim treatment had minimum disease when no nitrogen was added to the soil.[187] However, carbendazim failed to retain an adequate level of disease control in the presence of nitrogenous fertilizers. Augmentation of urea reduced the efficacy of carbendazim. Urea amendment did not have much effect on disease in the quintozene-treated set as compared to unamended controls.[187] Potassium nitrate at 200 μg g^{-1} soil decreased the disease in the presence of quintozene as compared to 100 μg g^{-1} soil. As in carbendazim treatment, urea and potassium nitrate increased the disease in the presence of MEMC so much so that the disease-controlling potential of MEMC was totally nullified in the presence of urea at 200 μg g^{-1} soil. Nevertheless, ammonium nitrate at a lower rate slightly potentiated the activity of MEMC.

Singh[187] also demonstrated the effect of different fertilizers, e.g., urea, diammonium phosphate (DAP), superphosphate (P$_2$O$_5$), and murate of potash (K$_2$O), on the disease-controlling potential of MBC seed treatment. Disease control with MBC was good in soil amended with urea, while it was poor in P$_2$O$_5$ or K$_2$O amended soil. He could not explain the exact mechanism by which these fertilizers increase or decrease efficacy of the fungicide.

It may be concluded that variable fertilization and manurial schedules may participate in altering in vivo fungitoxicity. If such a situation exists it is of value to take a fresh look at the relevance of blanket recommendation of fungicides; in particular, crop and nutrient amendment also need due consideration while choosing chemicals for disease control.

B. Influence of Host Species

Host species also influence the fungicidal interaction with the pathogen. Kataria and Grover[354] noticed that TPM controlled *R. solani* efficiently on mungbean and long melon, benomyl on mungbean, eggplant, and sugar beet, and PCNB was best on sugar beet. On brinjal, chloroneb, PCNB, and benomyl were least effective in controlling the disease. Participation of host (foliage) in modifying the toxicity of fungitoxicants has been documented by a few workers.[342,380]

C. Penetration into Plant Residues

An important source of pathogenic inoculum is probably kept protected in plant residues in soil.[12] It has been demonstrated that most *Pythium* colonies obtained from nursery soil started from spores surrounded by clay or humic particles. Therefore, most successful control of damping-off may only be possible with fungicides or strong fungistats capable of penetrating such microenvironments. Corden and Young[24] reported the penetration of various fungicides into cut and buried pieces of infected tomato stems. *Fusarium oxysporum* f. *lycopersici* survived well in the pieces of soils mixed with such fungicides as captan, dichlone, and phaltan at 400 ppm; nabam reduced the survival at 400 ppm and 1-chloro-2-nitropropane, allyl alcohol, Vapam, Mylone, MIT, and 2-chloro-3-(tolylsulfonyl) propionitrile at 100 ppm. Most effective were methylmercury oxyquinolate, which reduced the survival at 25 ppm, and 1,2-dicyano-1,2-dichloroethylene, which eliminated the fungus at 25 ppm. Penetration into dried pieces was better than into fresh ones. According to Vaartaja,[12] this kind of penetration is probably correlated with water solubility and volatility of the compounds, properties which unfortunately also enhance the disappearance of the compound from the soil.

D. Tolerance of Sclerotia

Domsch[381] conducted a detailed experiment and noticed that higher rates of Vapam are needed to kill sclerotia than mycelia. The concentrations in a test were the following (ppm):

	Mycelium	Sclerotia
Botrytis glanthina	100	500
Rhizoctonia solani	100	500
Sclerotinia sclerotiorum	300	200

Partyka[382] demonstrated that Vapam and Mylone drench may kill the sclerotia of *S. sclerotiorum*, while PCNB only inhibits apothecia formation. Tuzet killed the sclerotia of *S. rolfsii* at 200 ppm, while semesan was able to kill only the mycelia, and captan and PCNB had fungicidal effects.[383] Abeygunawardena and Wood[384] demonstrated that captan has controlled a disease caused by *S. rolfsii* (when mixed into the soil) at 2000 ppm while being ineffective against the sclerotia. In a soil column test[385] drenches of captan and thiram killed only the mycelium of *S. cepivorum*, while MIT, Mylone, and Vapam were also fungicidal to the sclerotia. It was reported that thick sclerotia of *R. solani* sorb ethyl mercury chloride and mercuric chloride. Sorption is mainly in the surface tissue only. Water and chelation may gradually remove the sorbed mercury. Due to this problem in killing dormant sclerotia of *Rhizoctonia*, these should be activated (with water sprinkling) before applying the fungicides.[386-388] Microsclerotia of *Verticilliums alboatrum* needs 200 ppm nabam, 50 ppm actidione, 500 ppm Vapam to be killed, whereas *Phytophthora cinnamomi* was killed at much lower rates of drench under the same conditions.[29]

E. Importance of Fungistasis

The activity of many common soil "fungicides" is fungistatic, not truly fungicidal, and they could be called as fungistats.[12] This appears true at least for the following fungitoxicants at the rates commonly used: ferbam,[389] captan[39,40,390,391] Chinosol,[39] PCNB,[133,382,392] nabam,[393] and thiram.[23,40,389] It was concluded that such soil fungistats as captan, thiram, and zineb would be phytotoxic if applied at the fungicidal rates in soil, but they are utilized successfully at the fungistatic rates. It is usual to cause only 50% inhibition of the pathogen and nevertheless obtain good disease control.[12,39,40,390] Two possible phenomena have been suggested to explain this:[12] (1) systemic activity and (2) further inhibition of the pathogens by antagonistic soil organisms not affected or even encouraged by the "fungicide". The second phenomenon seems to be more important. However, this biological control could not be manipulated as yet. Considerable attention has been devoted, recently, to these indirect effects of fungistats, but the knowledge is still insufficient to make firm conclusions.[394-396]

Virtanen et al.[397] isolated chlorogenic acid phenols and benzoxazolinone compounds from plants, and demonstrated that they increase the resistance of potato, wheat, rye, etc. to certain fungi. Benzoxazolinones are fungistatic[398] to many species; however, these, as well as many other antibiotics from higher plants, failed in screening tests with pine seeds and *R. solani*.[12] Probably each natural fungistat within a plant cannot markedly inhibit the pathogen when operating alone. The same may be true for most potentially antagonistic effects by saprophytes. However, when these two main effects and their components happen to be properly complemented with a suitable artificial effect, then sufficient control results. Tammen et al.[151] applied fungicides in steamed soil to use them as fungistats against reinfestation by *P. spinosum*. Good control was obtained with thiram (at 200 ppm for 26 days) and Dexon (50 to 142 ppm for over 36 days). Granular Dexon (500 ppm) gave complete control for 92 days, and captan gave prolonged but inconsistent partial control. However, in some experiments (e.g., captan, phantan at 120 ppm, thiram at 200 ppm) the disease control was better in nonsteamed soil. Elsaid and Sinclair[399] obtained similar results with copper 1,2-napthoquinone-2-oxime. Vaartaja[12] was of the opinion that an explanation of such results is the additional biological control by soil saprophytes.

F. Action of Antagonistic Microbes

Influence of fungitoxicants on antagonistic microorganisms does affect their disease-controlling potential. There are several important groups of antagonists like *Trichoderma*, *Streptomyces*, *Gliocladium*, and *Penicillium*, which are known to be tolerant to fungitoxicants. The role of fungitoxicants on indirect/integrated control of soilborne plant pathogens will be discussed in another chapter. Here, discussions will be confined particularly to the tolerance of antagonists against fungicides. Richardson[31] reported the protection of *Pythium* damping-off of spinach and cucumber with thiram in soil even after degradation of the compound to a low concentration level not affecting pathogens. It was explained by the natural microbiological control exerted by the saprophytic flora which was unaffected or enhanced by thiram. This flora included bacteria and species of *Penicillium* and *Trichoderma*. Domsch,[33,39,390,400] too, obtained the control of *Rhizoctonia* and *Pythium* damping-off with captan rates below the fungicidal rates. It was also due to partial biological control. Occasional disease control in soil with nonfungitoxic compounds, as reported with the nematicide, Nemagon, in the absence of pathogenic nematodes.[401] Beneficial selective action has been demonstrated for captan, ferbam, maneb, nabam, thiram, Vapam, and zineb.[394-396,402-408] Several antagonistic fungi are sensitive to nabam, with the exception of *Penicillium nigricans*.[394] Especially beneficial selectivity was shown by Vapam and allyl alcohol. *Trichoderma* is a common antagonist to many damping-off fungi and it tolerates relatively well several fungitoxicants including chloropicrin, EDB,[409] MB,[410,411] formaldehyde,[412-414] trinitrotoluene,[414] allyl alcohol,[408,416] carbon disulfide,[409,412] Ceresan, nabam,[409,410] Vapam, and thiram.[31,404] One of the most strongly antagonistic groups of soil fungi, *Streptomyces* spp., may be specially tolerant to PCNB,[417,418] Vapam,[390,410,418] nabam,[410] and allyl alcohol.[394,395] *Streptomyces* spp. has tolerated captan and Dexon better than did *R. praticola*, and thiram and pyrogallol better than did *Pythium*.[420] Effects of fungitoxicants on actinomycetes will be discussed elsewhere.

Gliocladium spp. and *Penicillium* spp. are other important groups of antagonists which tolerated fungitoxic materials better than several pathogens.[404,420,421] However, they were susceptible to PCNB, zineb, and Karathane. Among the species of *Penicillium*, *P. nigricans* may be particularly useful, as it produces powerful antifungal compounds. This highly antagonistic[422] fungus tolerated all the tested fungitoxicants, namely, captan, a combination of thiram and arsenic compounds, nabam, Vapam, and allyl alcohol, better than did any other fungus species. This fungus also tolerates organic Hg compounds[402,423] and CS_2[412] relatively well. *Penicillium* and *Trichoderma* seem to be selected after soil treatments with Vapam[394,402,409] and *Trichoderma* after fumigation with MB.[12] After foliage application of urea[424] *Penicillium* spp. increased in the rhizosphere of rice. Antagonistic actinomycetes have been observed to increase at some times and decrease at others in the corn rhizosphere after applications of urea.[425] Bacterial population in the rhizosphere may increase after foliage treatment with chloronil, dichlone, and tetramethylammonium bromide.[426] Bacteria may tolerate protectants and enhance after application of captan,[358,390,391,400,407,415] ferbam,[403] formalin,[427] MB,[410] nabam,[390,400] TCNB,[428] thiram,[31,404,406,407,415,429] Vapam,[410] and especially of allyl alcohol.[394,395,400] At least some of these bacteria were demonstrated as potential antagonists of *Pythium* and *Rhizoctonia*.[406,429,430] Domsch[390,394,395] has demonstrated increases of *Verticillium dahliae* and *Thielaviopsis basicola* after applications of captan and thiram in compost soil. Smith[427] observed increase of *Pythium* after application of formalin, D-D, or dichloropropane, which are primarily nematicides. *R. solani* and *P. ultimum* tolerate ethylmercury phosphate better than do common soil saprophytes.[431] It was reported that among strains of *Aspergillus niger,* the cause of rot of groundnuts, some are tolerant to Hg and increased in number after applications of mercury compounds.[432]

Extensive research is necessary on all the microbiological effects that fungicides may have in soil in order to obtain the maximum benefit from these. The selective action of certain chemicals and the resulting indirect biological control may greatly facilitate disease control and prevent the pathogens from developing tolerance. According to Vaartaja,[12] seed treatments and band applications may utilize biological control in the following way. The seedlings avoid the harm from the locally concentrated toxicant, because with the food store of the seed, they rapidly extend their roots to nontreated soil. Pathogens are suppressed locally. As the concentration of the toxin gradually decreases from the treated area onward, it may be just right for selective action and biological control by one or several organisms at one or several zones around the treated area. This may explain the reduced control when seed treatments were supplemented with soil treatments.[1] Inconsistent results with fungitoxicants in soil and the often found decreases in control in sterilized soil[151,399] may be due to the dynamic and unstable nature of the biological control involved. The same may be true where applications enhanced the disease.

G. Interaction of Dose and Inoculum Potential

The effectiveness of a fungitoxicant is directly related to its concentration and the duration of exposure, provided that the concentration of infectious materials is the same and that it is uniformly susceptible to the fungitoxicant.[1] In soil this is not generally the case, making the inoculum potential of the soil the third great variable in effectiveness of a fungitoxicant. Toxicant effectiveness may be compared on the basis of the ratio of the concentration in parts per million applied to soil, and the number of days before half of the original activity is gone. Thus, a 10:1 ratio would be obtained with a fungicide having a half-life of 50 days with an initial concentration of 500 ppm. Richardson and Munnecke[35] summarized the difficulties in determining effective fungicide dosage in relation to inoculum concentration in soil and determined a quantitative relationship between the fungicide dosage required to control a soilborne disease and the concentration of mycelial inoculum in soil. They observed the incidence of preemergence damping-off in pea seedlings grown in nonsterile soil, artificially infested with a series of a concentration of inoculum of *P. irregularae* or *Rhizoctonia solani* and treated with a dosage series of thiram and MMD. Parallel linear dosage-control curves were obtained which indicated that the logarithm of the fungicide dosage at the 50% control levels increases proportionately with the logarithm of the inoculum density. In other words, it can be said that the amount of fungicide required to control a soilborne disease is proportional to the degree of infestation of the soil by the pathogen. The linear log inoculum/log dosage curve provides a basis for comparing either the effectiveness of various fungicides against a single test organism or the relative sensitivity of different pathogens to the same fungicide. These findings underline the importance of standardizing the inoculum density in any soil fungicide evaluation test. Some interesting observations pertinent to this were demonstrated by Bald and Jefferson.[433] With an increased dose of Vapam on corms treated with New Improved Ceresan (NIC), incidence of bacterial scab of gladiolus fell, rose, and fell again. They explained their results as being due to a combination of residual activity of NIC combined with higher dosages of Vapam, causing failure or erratic development of a protective microflora in the rhizosphere of growing plants. These results and explanation were similar to those reported by Gibson[432] earlier. They fitted their data to a poison distribution series. This showed that to obtain infection of 50% of the susceptible plants it would take 60 infectious particles, but to raise it to 99% would take 400 particles. In natural soil fumigation, soil heterogenicity would tend to extend the range of doses that are less than 100% lethal. In a highly infested soil only an efficient fungicide is worthwhile.[1] More work is needed on this aspect for the judicious application of fungitoxicants.

H. Adaptation of Pathogens

Under continuous influence of a chemical, tolerant fungus strains will naturally be selected among those possibly existing and among those produced by mutation and new genetic combination. This aspect will be discussed at length in a separate chapter.

I. Combinations of Protectants

It has usually been recommended to use two or more toxicants in combination.[12] This, of course, is not a new idea; Erwin and co-workers[434,435] obtained good disease control with a mixture of captan + Phorate. The systemically acting Phorate was effective against

Rhizoctonia, but if used alone it increased the disease caused by *P. debaryanum*. Phorate appeared to inactivate Ceresan 200. In experimentally infested soil, mixtures such as Dexon + PCNB have increased the control against mixed inoculum (*Pythium + Rhizoctonia*); each fungicide alone failed against one of the inoculum components.[436] Better damping-off control with single than with various combined treatments in alkaline loam was obtained by Vaartaja and co-workers.[12] It appeared that including the second toxicant, especially PCNB, nullified the biological control that would have resulted from the selectivity of the first. Harrison[437] observed that soil treatments with ferbam or Vancide 51 reduced the control obtained from certain seed treatments alone. Jacks[438] reported reduced emergences after applications of mixed fungitoxicants that individually increased emergence. When Labonte[439] used various compounds for seed and soil treatments, there appeared to be increase of post emergence damping-off with certain combinations of captan, thiram, and PCNB.

In the control of damping-off of cotton, it was a regular practice to apply two or even three different fungicides. Combining two or preferably three of them, usually PCNB + captan + zineb or thiram, seems to be more effective than single ones when applied in various soil and weather conditions against a variety of soil pathogens.[119,368,440-442] However, no combination has been outstanding consistently.[119,368,443,444] A mixture of $CuSo_4$ + captan + PCNB has controlled a persistent *Fusarium* disease of tomato.[445] The best control of bean cankers was obtained with a mixture of captan + PCNB, although the disease was caused by *R. solani* alone.[446] Mixtures of omadines + PCNB have sometimes, but not always,[443] been successful against *Pythium* and *Rhizoctonia*.[447,448] Silvar is reported to enhance synergistically the effectiveness of Hg, Cu, and that of Streptomycin.[449] DTC and Chinosol show synergism at certain combinations.[450-452] The activity of maneb is decreased by Terramycin.[453] Mixing sulfur or kaonite with captan has caused phytotoxicity.[449] Thiocarbamates may destroy captan.[454] Surfactants may cause synergistic fungitoxicity, e.g., alkylbenzol sulfate with thiram;[455] certain surfactants are strongly fungistatic.[456]

J. Effects of Recontamination

Natural soil contains a multitude of organisms in continuously changing equilibria which form a biological buffering system that tends to prevent any one segment of the population from gaining ascendancy. When this buffer is modified by fungitoxicants, the buffer capacity is greatly reduced, and toxicant treatment may be completely obscured by the subsequent reintroduction of a parasite.[1] This relationship was elegantly shown by Ferguson.[457] He treated a naturally infested gladiolus field with MB gas and measured the disease response of plants grown from planting stock obtained from various sources. When pathogen-free stock was used the effectiveness of MB increased with increased dose. When the grower's stock was used, although it had no obvious disease symptoms at planting time, the best treatment was obtained from the soil which had not been treated with MB. This phenomenon of obscuring the effectiveness of soil fungitoxicants by reintroduced pathogens is now well known, but it was unappreciated in the past. Munnecke[1] was of the opinion that results from research should be scrutinized with this factor in mind.

K. Increase of Disease

Occasionally soilborne diseases are increased by application of fungicides. This is quite common after single applications, especially when soil sterilization is approached or achieved and nothing is done to prevent reinfestations.[12] This has been demonstrated with MB,[458], Hg[431] compound, captan,[394] and thiram.[394,439] In studies of *Thielaviopsis*, after incubation in fumigated soil, the inoculum was not recovered when 100 ppm or less of Vapam had been applied. Contaminating antagonists inhibited *Thielaviopsis* in the soil samples. When 200 ppm had been applied, *Thielaviopsis* was recovered without contamination. An increase of *Sclerotium* blight has been found after an application of D-D and Nemagon.[459] According to Vaartaja,[12] they may have reduced the biological buffering of the soil, but they have not eliminated *S. rolfsii.*

Grossmann and Steckhan[460] observed that *Pythium* disease of cauliflower was appreciably enhanced by the common insecticide, chlordane. Aldrin had a similar but smaller effect. Disease increase took place with *Pythium*, while *Rhizoctonia* and *Ophiobolus* diseases decreased. Similarly, Phorate insecticide has enhanced *Pythium* disease while it controlled *Rhizoctonia* disease.[435] Increased damping-off of pine has taken place after chlordane application.[461] After application of PCNB[462] increased disease of cotton seedlings by *Fusarium* and *Colletotrichum* has also been reported.[12] Gibson et al.[463] found an increase in disease after applications of PCNB as a result of an observed decrease in *Penicillium* sp. antagonistic to *Pythium*. In another case[464] in which repeated heavy soil treatments with thiram at first gave good control of *Pythium*, *Rhizoctonia*, and *Phytophthora cactorum,* there was considerable late damping-off which could not be solely attributed to phytotoxicity. Complex microbiological interactions were suggested. In another trial[12] increased disease took place with PCNB and D-113. Good control given by Dexon was reversed when this was applied together with PCNB. Ranney and Bird[368] demonstrated an increase of cotton diseases from a mixture of captan + PCNB + thiram, which usually gives a good control. Volger[465] found more damping-off when thiram seed treatment was combined with various soil treatments. Thiram sometimes increased postemergence damping-off of red pine.[439] A combination of captan + thiram enhanced damping-off by *Thielaviopsis basicola* in a greenhouse, while that by *Rhizoctonia*, *Pythium*, and *Phytophthora* was decreased.[430]

Several compounds enhanced damping-off of cotton in tests of Elsaid and Sinclair,[399] especially copper 1,2-naphthoquinone-2-oxime in sterilized soil (but not in unsterilized soil). An increase in turf disease has been demonstrated after applications of As and Hg compounds.[401] Several antibiotics have enhanced *Rhizoctonia* disease in potato stems.[12] Streptomycin[466] has been found to enhance potato decay by *Fusarium* and *Phoma*, while decreasing bacteria. Many more increases of disease after applications of protectants perhaps were never published or were routinely attributed to phytotoxicity. Water-soluble Hg and As compounds may induce Zn deficiencies.[467]

L. The Increased Growth Response Phenomenon

Frequently, better plant growth, known as increased growth response (IGR), is obtained in fungicide-treated soils, even in the absence of pathogens.[1] It is frequently explained as being due to the inhibition of unrecognized root parasites, to release of nutrients (mostly nitrogen), or to the chemical itself.[468] Domsch[469] showed whether or not IGR was due to the suppression of semipathogenic fungi by studying the effects of 25 saprophytic soil fungi on growth of five test plants. Some fungi were stimulatory, some were inhibitory, but the total was algebraically zero for plant response to the whole group. His opinion was that "semiparasitic" potencies of fungi are unlikely to cause marked inhibition in plants unless a saprophyte becomes dominant. It might be possible that fungi similar to the ones he studied could survive chemical treatments and be responsible for IGR. If a species classified above as an "inhibitor" survived, decreased plant growth would ensue, but the reduction probably would be ascribed to an inefficient control or the known pathogens. If a "stimulator" survived, however, the presence would be noted as an IGR.[1]

Nitrification processes in soil were stressed in the late 19th century with the assumption that total plant growth was a reliable indication of increased microbial activity. It has been known that soil possessed the capability of converting ammonia N to nitrate N and that fungitoxicants stopped the process. Fungicide-treated soil sometimes may have an unfavorable ratio of ammonium to nitrate nitrogen and subsequent plantings may be abnormal. Not all plants are susceptible to this toxicity, and it is possible that some plants may thrive in the presence of the different nitrogen balance. This factor may explain some IGR, also.

Ureas, formaldehyde, calcium cyanamide, ammonium, nitric acid, urea, etc. are used as both fertilizers and fungitoxicants.[12] In soils poor in N, such materials of course stimulate growth of plants. Similarly, Mylone, captan, PCNB, and N-containing fungicides may act as fertilizers in very poor soils if applied in large quantities.[12] Zineb, maneb, copper compounds,[470] etc. may stimulate growth of plants at least if these are deficient in microelements.[421] Thulin et al.[472] demonstrated a fourfold growth for *Larix* sp. after chloropicrin application. Palmer and Hacskaylo[473] found similar increases in growth of southern pines. Stimulatory effects of methyl bromide lasted for 2 years, and that of Vapam and Bedrench only 1 year. Foster[474] reported increase in growth and survivals in plantations for southern pines grown in beds treated with MB.

Sometimes IGR is due to control of associated parasites, such as nematodes, and these relationships have been discussed in conjunction with fungal pathogens.[1] The quantity of soil nutrients and their utilization by plants may be altered. Smith[475] demonstrated that treatments of Hawaiian soils with MB or D-D resulted in improved utilization, but no uptake of iron by pineapple. Phosphorous content increased also, but uptake was reduced. Therefore, the growth-stimulating effect of soil sterilization can be explained in various ways by reducing weeds,[476] nullifying toxins in soil,[476] release of macro- or micronutrients into a form more available to plants,[477-480] destruction of nematodes or fungi that would otherwise cause considerable, but not easily detectable, destruction of fine roots.[481-483] The same may be true in many nurseries.[414,484-488] In most of these cases the main pathogens were probably nematodes. Sometimes, the complex microbiological effect of fungi (*Thielaviopsis* sp. and *Pythium* spp.[422,489]) may also be involved. Stimulation of pine seedlings with allyl alcohol in prairie soils is sometimes similar in kind and magnitude to that with mycorrhizeal inoculum.[490] Fumigation increased total nutrient uptake by pine seedlings from a sandy soil.[491] Wilde et al.[492] demonstrated a peculiar enhancement in growth of pine seedlings. This took place after the seedlings had been transplanted to prairie soil from a sandy soil that had been fumigated with allyl alcohol. This probably involved complex microbial changes. Mader[493] found that thiram applications enhanced the uptake of N, Na, and K by three seedlings. Some DTCs have been found to act as toxins.[494,495]

M. Residual Effects

Thus, PCNB at 60 to 100 lb/acre causes some phytotoxicity to potato seedlings the year after application.[496] The damage, however, was offset by the continued control of *Rhizoctonia*. Materials containing C, H, O, and N will ultimately decompose enough to lose all their effects. The residue would be H_2O. CO_2, and simple N-containing, mostly useful compounds. Materials releasing chloride are also harmless except when used in high quantities in certain soils rich in salts. However, gradual accumulation of As, Br, S, I, and heavy metal may cause long-lasting toxicity in soil. Various plants are very different in their sensitivity to these elements.[483] Citrus, for instance, was readily damaged by Br residue when certain citrus soils were treated with any agent, chemical or physical, which effectively "sterilizes" the soil; a very haphazard and difficult-to-reproduce stunting of citrus plants resulted, lasting a few weeks to a year.[1] It was associated with an inhibition of phosphorous absorption and reduction of copper and zinc uptake, although extractable soil phosphorus was high. Plants were stunted and phosphorus was depleted in leaves, but other ions were taken up and sodium and boron occasionally reached toxic levels in the tissues. Beans seem susceptible to As residue.[497] Damage has been observed only in certain soils after prolonged heavy application. Soil treated with Ceresan (5 to 11 lb/acre) for 5 years was without toxicity to wheat or soybeans[498] and even nodulation was not affected. Likewise, no harmful, microbiological effects have been found from 90 lb/acre of zineb, maneb, or nabam,[499] though nitrification was somewhat reduced. Abundant use of sulfur decreases pH;[497] yield of apples, therefore, was smaller in sulfur than in fermate-treated trees.[500] According to Vaartaja,[12] the residue problem should be negligible with such compounds as captan, Dexon, Mylone, nabam, thiram, and Vapam if an appropriate time interval is allowed for decomposition.

There are several other factors like decomposition, leaching and penetration, effect on mycorrhizae, and nitrogen cycle which are known to affect fungicidal efficiency. However, these factors have already been discussed elsewhere.

V. TECHNIQUES USED TO ENHANCE EFFICACY
OF FUNGICIDES IN SOIL

Control of plant disease is arrived at through (1) escape of infection, (2) suppression of inoculum potential, (3) improving host resistance, or (4) improving recovery from infection.[501] Whether by design or accident, most techniques devised to increase soil fungicide action have been aimed at reducing inoculum density of the pathogen in soil. A few of the novel ways to use fungicides or to handle soil to control plant diseases are simultaneous fumigating and planting of seeds,[502,503] use of anhydrous ammonia as a fungicide as well as fertilizer,[504] and use of a fungicide in freshly steamed soil to prevent recontamination losses.[505] According to Munnecke,[10] an alteration or adaptation of old methods may make possible a more efficient use of a soil fungicide.

A. Increasing the Dose of the Fungicide
Although there are well-known limits to the concentration that may be used, one of the easiest ways to obtain greater penetration of soil and increased kill by a soil chemical is to enhance the concentration applied. Munnecke et al.[506] demonstrated that increasing the dose of MB from 1 to 4 lb/100 ft^2 of surface beneath an impervious (mylar) cover gave a pronounced enhancement of penetration of a sandy loam soil. The effect was more pronounced as soil depth increased. At the 1-ft depth the concentration of MB using 4 lb was approximately 2.5 times higher than the treatment using 1 lb. The maximum concentration attained at 6 ft, however, was 12 times higher.

B. Placement of the Fungicide in Soil
Goring,[7] in his review article, discussed at length the geometry of diffusion patterns of fumigants applied to the soil, pointing out the differences obtained when a fumigant is applied on a point, line, or plane source in the soil. The pattern of the placement of the fumigant greatly affects the behavior of the fumigant in soil. Most fumigations are designed to treat the upper 18 in. of soil where most pathogens exit.[10] In the case of *Armillaria mellea*, all root pieces infected with this pathogen are capable of becoming infected and remaining in the soil as a potential source of inoculum. Deep vertical distribution of a fumigant is of utmost importance with such pathogens. Therefore, a fundamental difference in strategy is involved when treating soil to be planted to strawberry to control root-rotting fungi or *Verticillium* compared to treating land infested with *A. mellea* before planting to a woody perennial. Fumigants like MB-chloropicrin mixtures are often applied as liquid in continuous-flow applications through tubes attached to chisels drawn by tractors. If the stratum to be treated is mostly the upper 2 ft of soil, the chisels are set 8 to 10 in. apart and 6 to 8 in. deep. If the upper levels of the soil are not so important, but it is necessary to fumigate the soil as deep as possible, the chisels are set 3 to 5 ft apart and as deep as the terrain and equipment permits.[10] While applying fumigants by hand application, holes may be bored or punched into the soil 3 to 5 ft deep and fungicides applied to the bottom and quickly covered with soil. With MB, a safe way to do this is to chill 1-lb cans in a freezer or on dry ice. When cans are punctured and dropped in the hole, no gas escapes before the hole is filled with the soil and tamped.[10] The hand-applied charges might be useful in treating ''problem'' soils that contain deep plow soles, hardpans, or clay strata that prevent the downward

diffusion of fumigants. In such cases holes may be bored through the obstruction layers and charges of liquid MB or carbon disulfide placed beneath the layers. It may be necessary to use shallow injections to fumigate the areas above the obstructing layers after the hole is filled, since the gases will diffuse horizontally and vertically beneath the cap. The problem of getting toxic concentrations of the gases in the clay layer itself is yet to be solved.[10]

The judicious placement of nonvolatile fungicides can increase their activity, also. Broadcasting of granular materials like PCNB, captan, or Dexon over the surface and incorporating them to the top 4 to 6 in. by rototillers is common in floriculture practice. Within-furrow applications, positioning PCNB in bands immediately above the seed increases effectiveness. The principle involved with most nonvolatile fungicides is that the fungicide must be in close proximity to the pathogen to be effective, since diffusion through water in soil is so slow.[10]

C. Confinement of Fungicides in Soils following Treatment

The greatest impetus to successful field use of highly volatile soil fumigants such as MB was the development of cheap polyethylene sheeting and machines capable of applying fumigants and covering a large field in one operation.[507] Unfortunately, the polyethylene is permeable to MB and other gases as compared to other types of films. Waack and co-workers[508] demonstrated that the permeability of films (P = cc gas (STP)/sec/cm^2 for 1 mm thickness) to MB gas as follows: polyethylene at 20°C, 12.5×10^8 P; saran at 30°C, 0.13×10^8 P; and mylar at 30°C, 0.0022×10^8 P. Since other films nowadays are prohibitively expensive or unsuited for use in field, a substitute should be found for polyethylene. It has been shown that when the relatively impermeable mylar or saran covers are used in lieu of polyethylene, the dose of MB may be greatly minimized in the field, retaining the same concentrations of gas deep in the soil. If impervious traps were available commercially, deep penetration of soil from relatively shallow applications (12 to 18 in.) would be possible.[10]

The role of thickness of polyethylene covers has been studied regarding increasing the effectiveness of soil fumigants. The effect of confining MB-chloropicrin mixtures in soil using polyethylene covers 1 or 4 mil thick has been compared.[509] Soil infested with *Phytophthora* in plots was buried, then treated for 4 days with the fumigants; the cover was removed and the samples retrieved 12 days later. Doses of 0.25, 0.5, 0.75, and 1 lb/100 ft^2 were applied to the surface beneath the covers. They observed 100% kill of *Phytophthora* at all depths to 4 ft with the 1-lb dose under the 4-mil trap. However, only approximately 60% of the samples were killed at the same depth under the 1-mil trap. Another work of Munnecke and co-workers[10] on this aspect was the effect of polyethylene covers 1, 4, or 6 mm thick on penetration of MB (66%) and chloropicrin (33%) mixtures in a commercial application of strawberry land. The concentration of MB attained at 6 in. depth after 60 hr for the 300-lb/acre application was as follows: 1 mil, 200 ppm; 4 mil, 900 ppm; and 6 mil, 2000 ppm. The concentration observed with the 150-lb/acre treatment under a 6-mil cover was 1300 ppm. Thus, increasing the thickness from 4 to 6 mm resulted in more than doubling the concentration in the soil. Sandy loam soil dried to less than 15% moisture (by weight) was uniformly treated with 4 lb MB applied as a point source 5 ft deep. The gas diffused laterally and vertically very rapidly so that even though some escaped to the air at the surface the concentration × time (CT) products were lethal to *A. mellea* in a volume over 12 ft in diameter and 9 ft deep. In another experiment at the same location the dry soil was irrigated so that the upper 2 ft were wet and 4 lb MB was applied as a point source 4 ft deep. The wet soil layer at the top greatly slowed the diffusion of the gas to the surface and much higher concentrations of gas were obtained laterally and downward than in the unwetted soil. The CT values at the upper 6 to 12 in. were lethal for *A. mellea*.[10] These experiments were made under ideal conditions, but he mentioned that if soil is properly managed, it is possible to obtain excellent deep fumigation without using a cover.

D. Alteration of Soil before or after Treatment with Fungicides

The deep plowing technique to control parasites in the upper portion of the soil profile has been combined with fumigant treatments.[510] This relatively simple and cheap method may be used to greatly increase the effectiveness of compounds which act as gaseous diffusants. Watson[511] reported that combining crop residues, flooding, and anaerobic fermentation to attempt to rid soil of *Pyrenochaeta terrestris* indicates the practices which may be used in combination with soil treatments. With many biological control situations it has been a radical upset in the existing ecological equilibrium (which is necessary before an antagonist may be successfully introduced or stimulated to increase in the environment) to maintain a suitably high population to either control or eliminate the unwanted organisms.[512] Integration of both chemicals and biological agents for maximum effectiveness has been well recognized in recent years. This aspect has already been discussed elsewhere in detail. Munnecke[1] was of the opinion that a fertile approach would be to try to develop soil organisms which are resistant to a given fungicide, and also antagonistic to soil pathogens. In such a case it may be possible to keep the population of antagonists at high levels in soil by frequent applications of a fungicide.

Wilhelm[503] demonstrated how ingenuity and knowledge of the physical behavior of fungicides in soil may be used in new and effective ways. He mentioned that it is possible to plant melon seeds and apply chloropicrin simultaneously for the control of *Pythium* damping-off. It was possible because melon seeds are relatively resistant to chloropicrin, (and chloropicrin acts on the fungi when the seeds are dormant) and the gas would be dissipated by the time the seeds were in the susceptible stage. The application of these facts allowed the use of this very efficient fumigant in this novel way, which shows that old materials can be used in new ways by ingenious investigators.[1] The manipulation of field soil moisture may be one of the most efficient means of enhancing fungicide efficacy. Moisture greatly influences the susceptibility to fungicides of pathogens in soil. Munnecke et al.[513] determined the dosage response affected by soil moisture by using a carefully controlled continuous flow of MB through columns of soil infested with *P. ultimum* or *R. solani*. The effect of soil moisture on control of damping-off was pronounced in soil infested with *P. ultimum*, but not so pronounced in soil infested with *R. solani*. Most effective control of *P. ultimum* was obtained in moderately moist soil, next in very wet soil, and least in very dry soil. With *R. solani*, best control of damping-off was obtained in moderately moist soil. In contrast to results with *P. ultimum*, control of damping-off due to *R. solani* in very wet soil and in very dry soil was similar and only slightly poorer than that obtained in moderately moist soil. Another report of Monro et al.[514] on ethylene oxide, propylene oxide, and MB fumigations against *Synchytrium endobioticum* indicated, as is usually the case, less kill in dry than in wet soil. However, they attributed the decreased kill in dry soil to an increased absorbance of the toxicants. They were of the opinion that there is additional evidence in other reports to indicate that resistance of organisms in dry soil is inherent in the organism, rather than in sorptive factors of the soil itself. Soil moisture also affects chemical reaction, and soil distribution in liquid as well as in gaseous phases.[10] Moisture content is probably the chief limiting factor in deep penetration of soils by fumigants. Few deeper fine-textured soils never dry sufficiently to allow adequate diffusion of gases such as MB or carbon disulfide to be fungitoxic to *A. mellea*.[10] This problem can be corrected partially by growing a cover crop of Sudan grass or safflower. The upper 3 to 4 ft may be dried by withholding irrigation water during the hot, dry season in California, but plants are needed to withdraw water from the deeper levels. Since cover crops most efficiently withdraw water late in the growing season, it is essential to delay fumigating as long as possible, but before the fall rains occur.

E. Formulation of Fungicides

Toxicants have been altered to enhance effectiveness. Surfactants have been claimed to increase fungicidal activity in soil.[515,516] Soil infected with benomyl plus surfactant F or Tween 2 reduced Dutch elm disease 79 or 97% on trees treated prior to inoculation.[515] The role of the surfactants was not determinable, since treatments without surfactants were not demonstrated. A synergistic effect is claimed occasionally for mixtures of fungicides, as in the case of the use of mixtures of 66% MB and 33% chloropicrin for control of *Verticillium*.[10] A potentially useful development with MB has been formulation of the gas in thixotropic gel preparation. This preparation solidifies upon release in the soil and the gas volatilizes less rapidly than if liquid MB were released. It has been observed that it may be possible to use gel preparations without covering the soil with traps. Soil fungicides may be used in a similar fashion to the controlled release of fertilizers in pelleted form. Some of the physicochemical problems involved with pelleted pesticides have been discussed.[517] Mills and Schreiber[518] have worked extensively on the use of latex-coated pellets for the control of root rot of wheat. The preparations did not control root rot, nor did they accurately release the fungicides. They stated that if time-release pellets are to be used effectually, a mechanism of release independent of moisture needs to be devised.

F. Using Microbes to Enhance Fungicide Activity

Control of pathogens by the use of soil chemicals should be aimed with a minimum upset of the natural environment. It is almost impossible to eradicate a pathogen in the field except in soil held in containers. If an eradicative dose of a fungicide is used, it leads to numerous problems. According to Munnecke,[10] the reduction of inoculum density of the pathogen to economically controllable levels appears to be the key to use of soil fungicides. Wilhelm[507] has already reviewed chemical treatments and inoculum potential of soil and discussed the subject at length. Wilhelm[507] cited the studies carried out on southern *Sclerotium* rot of sugar beet and club rot of crucifers in relation to dosage of a fungicide and inoculum density. Horsfall and Dimond[519] have drawn an interesting analogy between the fungicide dosage-response curve and inoculum potential curves, where pathogen propagule number is linked to dosage of a fungicide. Ludwig[520] pointed out that a pathogen is more easily inhibited when its population density is low, and that less chemical is required to control a disease under such conditions. Baker[521] amplified these views in relation to the mathematics involved in the relationship of inoculum density and control of soil fungicides.

Richardson and Munnecke[35] clearly demonstrated a quantitative relationship between the dosage required to control a soilborne disease and the concentration of mycelial inoculum in soil. The incidence of preemergence damping-off was observed in pea seedlings grown in nonsterile soil artificially infested with a series of concentrations of inoculum of *P. irregulare* or *R. solani* and treated with a dosage series of thiram or MMDD. Parallel linear dosage control curves were obtained which indicated that the logarithm of the fungicide dosage at the 50% control level increases proportionately with the logarithm of the inoculum density.

It is a common procedure for plant pathologists to test a dosage series of a fungicide against a fixed population of the test organism. In the examination of solid fungicides it is essential, and perhaps more important, to test against a range of inoculum potentials. The development of the mathematics of systems in which the dosages of the two components, pest and pesticides, are varied might lead to rewarding conclusions. More research, however, is needed on these aspects. The judicious crop rotation leading to reduction in inoculum density as well as to excellent disease control is well known. There has been lack of research on whether a fungicide should precede or follow nonsusceptible plants in a crop rotation. It is possible that fungicide applied after a susceptible host has been cropped might be more efficacious than the same fungicide applied before the susceptible host is planted. Lewis and Papavizas[522] and Papavizas and Lewis[523] reported that decomposition products of cabbage, kale, mustard, turnips, and Brussels sprouts reduced root rot of peas caused by *Aphanomyces euteiches* and that vapors from decomposition contained a numbers of potentially fungicidal compounds such as mercaptans, sulfides, and isothiocyanates. They observed that substances like Vapam and dazomet, which release isothiocyanate in soil, controlled the disease when applied to the soil as a drench at concentration of 50 to 200 ppm. The treatments were enhanced by enclosing the soil with a plastic cover. The study of products of host breakdown deserves more emphasis, with the aim of stimulating formation of fungitoxicants in the soil.[10] It is possible that disease control could be greatly increased by covering the soil immediately after plowing under the crop refuse to confine any fungitoxic vapors that possibly might be produced.

REFERENCES

1. **Munnecke, D. E.,** Fungicides in the soil environment, in *Fungicides, an Advance Treatise,* Vol. 1, Torgeson, D.S., Ed., Academic Press, New York, 1967, 510.
2. **Fuchs, W. A.,** Entseuchungs massnahmen, in *Handbuch der Pflanzenkrankheiten,* Vol. 6, Appel, O., Ed., Parey, Berlin, 1952, 144.
3. **McNew, G. L.,** The future for fungicides, *Agric. Chem.,* 8, 44, 1953.
4. **Kaufman, D. D.,** Soil fungicide interactions, in *Antifungal Compounds,* Siegel, M. R. and Sisler, H. D., Eds., Marcel Dekker, New York, 1977, 1.
5. **Carson, R.,** *Silent Spring,* Houghton Mifflin, Boston, 1962.
6. **Burchfield, H. P.,** Performance of fungicides on plants, in soil physical, chemical and biological considerations, *Plant Pathology,* Vol. 3, Horsfall, J. G. and Diamond, A. E., Eds., Academic Press, New York, 1960, 477.
7. **Goring, C. A. I.,** Theory and principles of soil fumigation, *Adv. Pest Control Res.,* 5, 47, 1962.
8. **Goring, C. A. I.,** Physical aspects of soil in relation to the action of soil fungicides, *Annu. Rev. Phytopathol.,* 5, 285, 1967.
9. **Hoffmann, G. M. and Malkomes, H. P.,** The fate of fumigants, in *Soil Disinfestation,* Mulder, D., Ed., Elsevier, New York, 1979, 291.
10. **Munnecke, D. E.,** Factors affecting the efficacy of fungicide in soil, *Annu. Rev. Phytopathol.,* 10, 375, 1972.
11. **Woodcock, D.,** Microbiological detoxification and other transformations, in *Fungicides, an Advanced Treatis,* Torgeson, D. C., Ed., Academic Press, New York, 1967, 613.
12. **Vaartaja, Q.,** Chemical treatment of seed beds to control nursery diseases, *Bot. Rev.,* 30, 1, 1964.
13. **Ashley, M. G. and Leigh, B. L.,** The action of methamsodium in soil. I. Development of an analytical method for the determination of methyl isothiocyanate residues in soil, *J. Sci. Food Agric.,* 14, 148, 1963.
14. **Burchfield, H. P.,** Comaparative stabilities of Dyrene, 1-fluoro-2, 4-dinitrobenzene, Dichlone and Captan in a silt loam soil, *Contrib. Boyce Thompson Inst.,* 20, 205, 1959.
15. **Hughes, J. T.,** Preliminary observations on the conversion of sodium N-methyldithiocarbamate (methamsodium) to methyl isothiocyanate in soil, in *Annu. Rep. Glasshouse Crops Res. Inst.,* p. 108, 1960.
16. **Jurinak, J. J. and Inouye, T. S.,** Analysis of ethylene dibromide in soils by gas chromatography, *Soil Sci. Soc. Am. Proc.,* 27, 602, 1963.
17. **Kotter, K., Willenbrink, J., and Junkmann, K.,** Der Abbau von S³⁵-markiertem Methylsenfol in verschiedenen Boden, *Z. Pflanzenkr. Pflanzenschutz,* 68, 407, 1961.
18. **Moje, W., Munnecke, D. E., and Richardson, L. T.,** Carbonyl sulphide, a volatile fungitoxicant from Nabam in soil, *Nature (London),* 202, 831, 1964.
19. **Munnecke, D. E., Domsch, K. H., and Eckert, J. W.,** Fungicidal activity of air passed through columns of soil treated with fungicides, *Phytopathology,* 52, 1298, 1962.
20. **Turner, N. J.,** Decomposition of Soidum *N*-Methyldithiocarbamate of Vapam in Soil, Ph.D. thesis, Oregon State University Corvallis, 1962.
21. **Turner, N. J. and Corden, M. E.,** Decomposition of sodium N-methyl dithiocarbamate in soil, *Phytopathology,* 53, 1388, 1963.
22. **Willenbrink, J., Schulze, E., and Junkmann, K.,** Über die Abgabe von S³⁵-Markiertem Methylsenfol aus dem Boden an die Luft und seine Aufnahme in die Tomatenpflanze, *Z. Pflanzenkr. Pflanzenschutz,* 68, 92, 1961.

23. **Chinn, S. H. F. and Ledingham, R. J.,** A laboratory method for testing the fungicidal effect of chemicals on fungal spores in soil, *Phytopathology,* 52, 1041, 1962.
24. **Corden, M. E. and Young, R. A.,** Evaluation of eradicant soil fungicide in the laboratory, *Phytopathology,* 52, 503, 1962.
25. **Domsch, K. H. and Schicke, P.,** Erfolgskontrolle fungizider Bodenbehandlungen, Nachrbl Dent, *Pflanzenschutzdienst (Stuttgart),* 12, 121, 1960.
26. **Klotz, L. J., Dewolfe, T. A., and Baines, R. C.,** Laboratory method for testing effectiveness of soil disinfectants, *Plant Dis. Rep.,* 43, 1174, 1959.
27. **Munnecke, D. E. and Ferguson, J.,** Methylbromide for nursery soil fumigation, *Phytopathology,* 43, 375, 1953.
28. **Newhall, A. G.,** An improved method of screening potential soil fungicides against *Fusarium oxysporum* f. *cubense, Plant Dis. Rep.,* 42, 677, 1958.
29. **Zentmyer, G. A.,** A laboratory method for testing soil fungicides with *Phytophthora cinnamomi* as test organisms, *Phytopathology,* 45(Abstr.), 298, 1955.
30. **Munnecke, D. E.,** A biological assay of non-volatile diffusible fungicides in soil, *Phytopathology,* 48, 61, 1958.
31. **Richardson, L. T.,** The persistence of thiram in soil and its relationship to the microbiological balance and damping-off control, *Can. J. Bot.,* 32, 335, 1951.
32. **Jacks, H. and Smith, H. C.,** Soil disinfection. XII. Effect of fumigants of growth of soil fungi in culture, *N. Z. J. Sci. Technol.,* 6, 69, 1952.
33. **Domsch, K. H.,** Die Prufung von Bodenfungiciden. II. Pilz Boden Wirt-Fungicide-Kombinationen, *Plant Soil,* 10, 132, 1958.
34. **Richardson, L. T.,** Effect of insecticide-fungicide combinations on emergence of peas and growth of damping-off fungi, *Plant Dis. Rep.,* 44, 104, 1960.
35. **Richardson, L. T. and Munnecke, D. E.,** Effective fungicide dosages in relation to inoculum concentration in soil, *Can. J. Bot.,* 42, 301, 1964.
36. **Reinhart, J. H.,** A method of evaluating fungicides in the soil under controlled conditions, *Plant Dis. Rep.,* 44, 648, 1960.
37. **Mauer, C. L., Baker, R., Phillips, D. J., and Danielson, L.,** Evaluation of applied soil fumigants with soil microbiological sampling tube, *Phytopathology,* 52, 957, 1962.
38. **Domsch, K. H.,** Zur Wirkung von Beizmitteln gegen Bodenpilze, *Phytopathol. Z.,* 25, 311, 1956.
39. **Domsch, K. H.,** Die Prufung von Bodenfungiciden. I. Pilz-Substrat-Fungicid-Kombinationen, *Plant Soil,* 10, 114, 1958.
40. **Domsch, K. H.,** Die wirkung von Bodenfungiciden. I. Wirstoff spektrum, *Z. Pflanzenkr. Pflanzenschutz,* 65, 385, 1958.
41. **Goring, C. A. I.,** Physical soil factors and soil fumigation action, in *Root Diseases and Soil-Borne Pathogens,* Toussoun, T. A., Bega, R. V., and Belson, P. E., Eds., University of California Press, Berkeley, 1970, 229.
42. **Goring, C. A. I.,** Fumigants, fungicides and nematicides, in *Organic Chemicals in the Soil Environment,* Vol. 2, Goring, C. A. I. and Hamaker, J. W., Eds., Marcel Dekker, New York, 1972, 569.
43. **Hartley, G. S.,** Herbicide behaviour in the soil. I. Physical factors, and action through the soil, in *The Physiology and Biochemistry of Herbicides,* Audus, L. J., Ed., Academic Press, New York, 1964, 111.
44. **Knight, B. A. G., Coutts, J., and Tomlinson, T. E.,** The sorption of ionised pesticides by soil, in *Sorption and Transport Processes in Soils,* S. C. I. Monogr. No. 37, Society of Chemical Industry, London, 1970, 54.
45. **Osgerby, J. M.,** Sorption of un-ionised pesticides by soils, in *Sorption and Transport Processes in Soils,* S. C. I. Monogr. No. 37, Society of Chemical Industry, London, 1970, 63.
46. **Hartley, G. S.,** Physicochemical aspects of the availability of herbicides in soil, in *Herbicides and the Soil,* Woodford, E. K. and Sagar, G. R., Eds., Blackwell Scientific, Oxford, 1960, 63.
47. **Jurinak, J. J. and Volman, D. H.,** Application of the Brunauer, Emmett, and Teller equation of ethylene dibromide adsorption by soil, *Soil Sci.,* 83, 87, 1957.
48. **Phillips, F. T.,** the aqueous transport of water soluble nematicides through soils. I. The sorption of phenol and ethylene dibromide solutions and the chromatographic leaching of thenols in soils, *J. Sci. Food Agric.,* 15, 444, 1968.
49. **Terent'ev, S. N.,** Sorption of methyl bromide by various materials, *Mater. Sess. Zakavkaz, Sov. Koord, Nauchno-Issled, Rab. Zashch. Rast.* p. 721, 1967.
50. **Grim, R. E.,** *Clay Mineralogy,* 1st ed., McGraw-Hill, New York, 1953, 384.
51. **Jurinak, J. J.,** Adsorption of 1,2-dibromo-3-chloropropane vapor by soils, *J. Agric. Food Chem.,* 5, 598, 1957.
52. **Call, F.,** The mechanism of sorption of ethylene dibromide on moist soils, *J. Sci. Food Agric.,* 8, 630, 1957.

53. **Stark, F. L., Jr.,** Investigations of chloropicrin as a soil fumigant, *Cornell Univ. Agric. Exp. Stn. Mem.,* 278, 1, 1948.
54. **Call, F.,** Soil fumigation. IV. Sorption of ethylene dibromide on soils at field capacity, *J. Sci. Food Agric.,* 8, 137, 1957.
55. **Allen, M. W. and Raski, D. J.,** The effect of soil type on the dispersion of soil fumigants, *Phytopathology,* 40, 1043, 1950.
56. **Leistra, M.,** Distribution of 1,3-dichloropropene over the phases in soil, *J. Agric. Food Chem.,* 18, 1124, 1970.
57. **Leistra, M.,** Diffusion and adsorption of the nematicide 1,3-dichloropropene in soil, *Agric. Res. Rep.,* 769, 105, 1972.
58. **Siegel, J. J., Erickson, A. E., and Turk, L. M.,** Diffusion characteristics of 1-3 dichloropropene and 1-2 dibromoethane in soils, *Soil Sci.,* 72, 333, 1951.
59. **Youngson, C. R., Goring, C. A. I., and Noveroske, R. L.,** Laboratory and greenhouse studies on the application of fumazone in water and soil for control of nematodes, *Down Earth,* 23, 27, 1967.
60. **Read, W. H., Hughes, J. T., and Smith, R. J.,** Some recent investigations with chemical soil sterilants, *Proc. 1st Br. Insect. Fungic. Conf.,* p. 361, 1961.
61. **Wade, P.,** Soil fumigation. I. The sorption of EDB by soils, *J. Sci. Food Agric.,* 5, 184, 1954.
62. **Wade, P.,** Soil fumigation. III. The sorption of ethylene dibromide by soils at low moisture contents, *J. Sci. Food Agric.,* 6, 1, 1955.
63. **Brown, A. L., Jurinak, J. J., and Martin, P. E.,** Plant growth as affected by ethylene dibromide fumigation of soils at two moisture levels, *Proc. Soil Sci. Soc. Am.,* 23, 311, 1959.
64. **Hanneson, H. A.,** Movement of carbondisulphide vapour in soils as affected by soil types, moisture content, and compaction, *Hilgardia,* 16, 503, 1945.
65. **Hanson, W. J. and Nex, R. W.,** Diffusion of ethylene dibromide in soils, *Soil Sci.,* 76, 209, 1953.
66. **Fuhr, I., Bransford, A. V., and Silver, S. D.,** Sorption of fumigant vapours by soils, *Science,* 107, 274, 1948.
67. **Frissel, M. J.,** The adsorption of some organic compounds, especially herbicides, on clay minerals, *Versl. Landbouwkd. Onderz.,* 67, 1, 1961.
68. **Craft, A. S.,** Movement of herbicides in soils and plants, *Proc. West. Weed Control Conf.,* 18, 43, 1962.
69. **Harris, C. I. and Warren, G. F.,** Adsorption and desorption of herbicides by soil, *Weeds,* 12, 120, 1964.
70. **Munnecke, D. E. and Martin, J. P.,** Release of methylisothiocyanate from soils treated with mylone (3,5-dimethyltetrahydro-1, 3, 5-2H thiadiazone-2-thione), *Phytopathology,* 54, 941, 1964.
71. **Kreutzer, W. A.,** Soil treatment, in *Plant Pathology,* Vol. 3, Harsfall, J. G. and Diamond, A. E., Eds., Academic Press, New York, 1960, 431.
72. **Call, F.,** Soil fumigation. V. Diffusion of ethylene dibromide through soils, *J. Sci. Food Agric.,* 8, 148, 1957.
73. **Youngson, C. R., Baker, R. G., and Goring, C. A. I.,** Diffusion and pest control by methyl bromide and chloropicrin applied to covered soil, *J. Agric. Food Chem.,* 10, 21, 1982.
74. **Hemwall, J. B.,** A mathematical theory of soil fumigation, *Soil Sci.,* 88, 184, 1959.
75. **Hemwall, J. B.,** Theoretical consideration of soil fumigation, *Phytopathology,* 52, 1108, 1962.
76. **Kendrick, J. B., Jr., and Zentymer, G. A.,** Recent advances in control of soil fungi, *Adv. Pest Control Res.,* 1, 219, 1957.
77. **Hemwall, J. B.,** Theoretical considerations of several factors influencing the effectivity of soil fumigants under field conditions, *Soil Sci.,* 90, 157, 1960.
78. **Peachey, J. E. and Chapman, M. R.,** Chemical control of plant nematode, *Commonw. Bur. Helminthol. Technol. Commun.,* 36, 119, 1966.
79. **Hill, G. D., McGahen, J. W., Baker, H. M., Finnerty, D. W., and Bingeman, C. W.,** The fate of substituted urea herbicides in agricultural soils, *Agron. J.,* 47, 93, 1955.
80. **Sheets, T. J.,** Review of disappearance of substituted urea herbicides from soil, *J. Agric. Food Chem.,* 12, 30, 1964.
81. **Kearney, P. C., Sheets, T. J., and Smith, J. W.,** Volatility of seven S-triazines, *Weeds,* 12, 83, 1964.
82. **Kimura, K. and Miller, V. L.,** The degradation of organomercury fungicides in soil, *J. Agric. Food Chem.,* 12, 253, 1964.
83. **Kolbezen, M. J., Munnecke, D. E., Wilbur, W. D., Stolzy, L. H., Abu El-Haj, F. J., and Szuszkiewicz, T. E.,** Factors that effect deep penetration of field soils by methyl bromide, *Hilgardia,* 42, 465, 1974.
84. **McKenry, M. V. and Thomason, I. J.,** Dosage values obtained following pre-plant fumigation for perennials. II. Using special methods of applying methyl bromide and 1,3-dichloropropene nematicides, *Pestic. Sci.,* 7, 535, 1976.
85. **Bliss, D. E.,** Soil disinfestation in citrus orchards against *Armillaria* root rot, *Phytopathology,* 38(Abstr.), 913, 1948.

86. **Turner, N. J.,** A simple laboratory technique for determining the diffusion rates and patterns of lathal concentrations of nematicides in soil, *Contrib. Boyce Thompson Inst.,* 23, 19, 1965.

87. **Baines, R. C., Foote, F. J., and Martin, J. P.,** Fumigate soil before replanting to control citrus nematode, *Calif. Citrogr.,* p. 41, 1956.

88. **Vanachter, A. and Van Assche, C.,** The influence of soil temperature and moisture control on the effect of soil fumigant, *Neth. J. Plant Pathol.,* 76, 240, 1970.

89. **Van Assche, C., Van den Broeck, H., and Vanachter, A.,** Einfluss der relativen Bodenfeuchtigkeit auf die Evolution chemischer Bodenentseuchungsmittel, *Mitt. BBA Berlin,* 132, 50, 1969.

90. **Holges, L. R. and Lear, B.,** Effect of time of irrigation on the distribution, 1,2-dibromo-3-chloropropane in soil after shallow infection, *Pestic. Sci.,* 4, 795, 1973.

91. **Gerstl, Z., Mingelgrin, V., and Yaron, B.,** Behavior of Vapam and methyl isothiocyanate in soils, *J. Soil Sci. Soc. Am.,* 41, 545, 1977.

92. **Hogan, R. M.,** Movement of carbon disulfide vapor in soils, *Hilgardia,* 14, 83, 1941.

93. **Jensen, H. J., Caveness, F. E., and Mulvey, R. H.,** A modification of Thorne's technique for examining soil diffusion patterns of nematicides, *Plant Dis. Rep.,* 38, 680, 1954.

94. **Throne, G.,** Diffusion patterns of soil fumigants, *Proc. Helminthol. Soc. Wash. D.C.,* 18, 18, 1951.

95. **French, C. A.,** Handbook of the Destructive Insects of Victoria, Vol. 1 (Part 2), Brain, R. S., Ed., Government Printer, Melbourne, Aust., 1893, 222.

96. **Thomas, H. E. and Lawyer, L. O.,** The use of carbon disulfide in the control of *Armillaria* root rot, *Phytopathology,* 29, 827, 1939.

97. **Schmidt, C. T.,** Dispersion of fumigants through soil, *J. Econ. Entomol.,* 40, 829, 1947.

98. **McClellan, W. D., Christie, J. R., and Horn, N. L.,** Efficacy of soil fumigants as affected by soil temperature and moisture, *Phytopathology,* 39, 272, 1949.

99. **Newhall, A. G.,** Volatile soil fumigants for plant disease control, *Soil Sci.,* 61, 67, 1946.

100. **O'Bannon, J. H. and Tomerlin, A. T.,** Studies on soil penetration of fumigants for the control of the burrowing nematodes (*Radopholus similis*), *Proc. Soil Crop Sci. Soc. Fla.,* 28, 299, 1968.

101. **Drosihn, U. G., Stephan, B. R., and Hoffmann, G. M.,** Untersuchungen uber die Bodenentseuchung mit Methylbromid, *Z. Pflanzenkr. Pflanzenschutz,* 75, 272, 1968.

102. **Hoffmann, G. M.,** Untersuchungen uber die Fusarium wilke der Aster (*Callistephus chinensis* Nees), *Gartenbauwissenschaft,* 28, 319, 1963.

103. **Hoffmann, G. M. and Zinkernagel, V.,** Zur Abtotung von Dauerorganen pathogener Bodenpilze durch Bodenentseuchung mit, *Terabol Gartenwelt,* 72, 291, 1972.

104. **Kundu, H. L.,** Effects of fumigation with methylbromide on certain coleopterous larvae. I. Various responses, *Proc. Zool. Soc. Calcutta,* 16, 77, 1963.

105. **Knavel, D. E., Walkins, H., and Herron, J. W.,** The influence of soil temperature, soil moisture and soil compaction on the diffusion of methylbromide, *Proc. Am. Soc. Hortic. Sci.,* 87, 573, 1965.

106. **Latham, A. J.,** The soil-column method and other techniques for evaluating soil fungicides, *Diss. Abstr.,* 22, 3342, 1962.

107. **Schicke, P.,** Die Fungizide und Keimhemmende Wirkung eines unverdunnten Praparates auf Basis Na-N-Methyl-dithiocarbamate im Boden, Nachrichtenbl, *Dtsch. Pflanzenschutz Dienstes Braunschw.,* 14, 142, 1962.

108. **Van den Brande, J., D'Herde, J., and Kips, R. H.,** Verspreiding van dichloropropaan dichloorpropeen in verschillende grondsoorten, *Meded. Landbouwhogesch, Gent,* 22, 377, 1957.

109. **Van Assche, C., Vanachter, A., and Van den Broeck, H.,** Chemische Bodenentseuchung durch Methylbromid, *Z. Pflanzenkr. Pflanzenschutz,* 75, 394, 1968.

110. **Wilhelm, S. and Ferguson, J.,** Soil fumigation against *Verticillium albo-atrum, Phytopathology,* 43, 593, 1953.

111. **Drosihn, U. G.,** Untersuchungen uber das physikalische und chemische verhalten von Methylbromid als Bodenentseuchungsmittel, Dissertation, Technical High School, Hannover, 1967, 58.

112. **Monro, H. A. U., Buckland, C. T., and King, J. E.,** Preliminary observation, on the use of the thermal conductivity method for the measurement of methyl bromide concentrations in ship fumigation, *Annu. Rep. Entomol. Soc. Ont.,* p. 71, 1953.

113. **Hoffmann, G. M., Stephan, B. R., and Dorsihn, U. G.,** Methylbromide als Bodenentseuchungsmittel. Fungizide wirkung und Anwend ung bei Nelken, *Gartenwelt,* 68, 71, 1968.

114. **Vanachter, A. and Van Assche, C.,** Involved van de Bodemvochtigheid en het gehalte Organische stof op de werking van Methylbromide en van het Mengsel Methylbromide met Chloropicrin, *Tuinbouwberichten,* 32, 54, 1968.

115. **Munnecke, D. E., Kolbezen, M. J., and Stolzy, C. H.,** Factors affecting field fumigation of citrus soils for control of *Armillaria mellea, Proc. 1st Int. Citrus Symp.,* 3, 1273, 1968.

116. **Higgins, J. C. and Pollard, A. G.,** Studies in soil fumigation. II. Distribution of carbon disulphide in soil fumigated under various conditions, *Ann. Appl. Biol.,* 24, 895, 1937.

117. **Ichikawa, S. T., Gilpatrick, J. D., and McBeth, C. W.,** soil diffusion pattern of 1, 2-dibromo-3-chloropropane, *Phytopathology,* 45, 576, 1955.
118. **Bird, L. S.,** Consistency in performance in covering soil fungicides for control of the cotton seedling disease complex in Texas, *Phytopathology,* 55, 497, 1965.
119. **Maier, C. R.,** In-the-furrow application of soil fungicides for control of cotton seedling diseases, *Plant Dis. Rep.,* 45, 276, 1961.
120. **Richardson, L. T. and Munnecke, D. E.,** A bioassay for volatile toxicants from fungicides in soil, *Phytopathology,* 54, 836, 1964.
121. **Linstrom, O.,** Fungicide and dye distribution in liquid seed treatment, *J. Agric. Food Chem.,* 8, 217, 1960.
122. **Adams, R. S., Jr.,** Soxhlet extraction of simazine from soil, *Soil Sci. Soc. Am. Proc.,* 38, 689, 1966.
123. **Thomas, G. W.,** Kinetics of chloride desorption from soils, *J. Agric. Food Chem.,* 11, 201, 1963.
124. **Helling, C. S., Kearney, P. C., and Alexander, M.,** Behavior of pesticides in soils, in *Advances in Agronomy,* Vol. 23, Brady, N. C., Ed., Academic Press, New York, 1971, 147.
125. **Freed, V. H. and Haque, R.,** Adsorption, movement and distribution of pesticides in soil, in *Pesticide Formulations,* Valkenburg, W. V., Ed., Marcel Dekker, New York, 1973, 441.
126. **Munnecke, D. E.,** Movement of non-volatile diffusible fungicides through columns of soil, *Phytopathology,* 51, 593, 1961.
127. **Helling, C. S.,** Movement of S-triazines herbicides in soils, *Residue Rev.,* 32, 175, 1970.
128. **Helling, C. S.,** Pesticide mobility in soils. II. Application of soil thin layer chromatography, *Soil Sci. Soc. Am. Proc.,* 35, 737, 1971.
129. **Helling, C. S. and Turner, B. C.,** Pesticide mobility: determination by soil thin layer chromatography, *Science,* 162, 562, 1968.
130. **Stipes, R. J. and Oderwald, D. R.,** Soil thin layer chromatography of fungicides, *Phytopathology,* 60, 1018, 1970.
131. **Helling, C. S., Dennison, D. G., and Kaufman, D. D.,** Fungicide movement in soil, *Phytopathology,* 64, 1091, 1974.
132. **Owens, R. G.,** Organo sulfur compounds, in *Fungicides, an advance Treatise,* Vol. 2, Torgeson, D. C., Ed., Academic Press, New York, 1969, 147.
133. **Kendrick, J. B., Jr. and Middleton, J. T.,** The efficacy of certain chemicals as fungicides for a variety of fruit, root and vascular pathogens, *Plant Dis. Rep.,* 38, 350, 1954.
134. **Pitblado, R. E. and Edgington, L. V.,** Movement of benomyl in field soils as influenced by acid surfactants, *Phytopathology,* 62, 513, 1972.
135. **Hagimoto, H.,** The herbicidal activity of 2-amino-3-chloro-1,4-naphthoquinone. I. Herbicidal activity and water depth, *Weeds Res.,* 9, 296, 1969.
136. **Raabe, R. D. and Hurlimann, J. H.,** Penetration of dexon into greenhouse soil mixes, *Phytopathology,* 55, 1072, 1965.
137. **Rhodes, R. C., Belasco, I. J., and Pease, H. L.,** Determination of mobility and adsorption of agro chemicals on soils, *J. Agric. Food Chem.,* 18, 524, 1970.
138. **Groves, K. and Chough, K. S.,** Fate of fungicide, 2,6-dichloro-4-nitroaniline (DCNA) in plants and soils, *J. Agric. Food Chem.,* 18, 1127, 1970.
139. **Lapidus, L. and Amundson, N. R.,** Mathematics of adsorption in beds. VI. The effect of longitudinal diffusion in ion exchange and chromatographic columns, *J. Phys. Chem.,* 56, 984, 1952.
140. **Frehse, H. and Anderson, J. P. E.,** Pesticide residues in soil-problems between concept and concern, in *Pesticide Chemistry, Human Welfare and the Environment,* Vol. 4, Miyamoto, J. and Kearney, P. C., Eds., Pergamon Press, New York, 1983, 23.
141. **Frehse, H.,** Persistence of insecticides and herbicides, in *Proc. BCPC Symp.,* Monogr. No. 17, British Crop Protection Council, London, 1976.
142. **Greenhalgh, R., Baron, R. L., Desmoras, J., Engst, R., Esser, H. O., and Klein, W.,** *Pure Appl. Chem.,* 52, 2563, 1980.
143. **Domsch, K. H.,** Soil fungicides, *Annu. Rev. Phytopathol.,* 2, 293, 1964.
144. **Munnecke, D E.,** The persistence of non-volatile diffusible fungicide in soil, *Phytopathology,* 48, 525, 1958.
145. **Daines, R. H.,** Some principles underlying the fungicidal action of mercury in soils, *Phytopathology,* 26, 90, 1936.
146. **Hills, F. J. and Leach, L. D.,** Photochemical decomposition and biological activity of *p*-dimethylamino-benzenediazo sodium sulfonate (Dexon), *Phytopathology,* 52, 51, 1962.
147. **Urbschat, E.,** Gegen Boden-Und Blatt-Pilze hochwirksame, Azoverbindungen, *Angew,. Chem.,* 72, 981, 1960.
148. **Hartzfeld, F. G.,** Terrachlor, a new soil fungicide, *Agric. Chem.,* 21, 31, 1957.
149. **Hildebrand, A. A.,** Soil treatment with arasan for the control of black root of sugarbeet seedlings, *Proc. Can. Phytopathol. Soc.,* 16, 16, 1949.

150. **Scheffer, R. P. and Haney, W. J.,** Causes and control of root rot in Michigan greenhouses, *Plant Dis. Rep.,* 40, 520, 1956.
151. **Tammen, J., Muse, D. P., and Haas, J. H.,** Control of *Pythium* root diseases with soil fungicides, *Plant Dis. Rep.,* 45, 858, 1961.
152. **Vermeire, A. and Welvaert, W.,** Werkingsduur van Fungiciden in de boden verh, *Pijksstat. Plantenz. Gent,* 13, 1, 1962.
153. **Domsch, K. H.,** Die wirkung von Bodenfungiziden. II. Wirkungsdauer, *Z. Pflanzenkr. Pflanzenschutz,* 65, 651, 1958.
154. **Edwards, C. A.,** Insecticide residues in soils, *Residue Rev.,* 13, 83, 1966.
155. **Foy, C. L., and Bingham, S. W.,** Some research approaches towards minimizing herbicidal residues in the environment, *Residue Rev.,* 29, 105, 1969.
156. **Hamaker, J. W.,** Mathematical prediction of cumulative levels of pesticides in soil, *Adv. Chem. Ser.,* 60, 122, 1966.
157. **Alexander, M.,** *Soil Biology, Reviews of Research,* UNESCO Publication, New York, 1969, 209.
158. **Brown, A. W. A.,** Fungicides and the soil microflora, in *Ecology of Pesticides,* John Wiley & Sons, New York, 1978.
159. **Agnihotrudu, V. and Mithyantha, M. S.,** *Pesticide Residues: A Review of Indian Work,* Rallis India, Bangalore, 1978, 1973.
160. **Griffith, R. L. and Mathews, S.,** The persistence in soil of the fungicidal seed dressings Captan and Thiram, *Ann. Appl. Biol.,* 64, 113, 1969.
161. **Munnecke, D. E. and Mickail, K. Y.,** Thiram persistence in soil and control of damping-off caused by *Pythium ultimum, Phytopathology,* 57, 969, 1967.
162. **Mehrotra, R. S., Tanwar, R. D. S., and Kakralia, O. P.,** Persistence and microbial degradation of fungicides and their microbiological and biochemical effects in soil, in *Recent Advances in Plant Pathology,* Hussain, A., Singh, K., Singh, B. P., and Agnihotri, V. P., Eds., Print House India, Lucknow, 1983, 109.
163. **Vyas, S. C.,** Studies on persistence of fungicides in important crops, Final Report ICAR Project, Indian Council of Agricultural Research, New Delhi, 1980, 41.
164. **Indulkar, A. S. and Grewal, J. S.,** Studies on chemical control of *Sclerotium rolfsii, Indian Phytopathol.,* 23, 455, 1970.
165. **Chinn, S. H. F.,** Effect of eight fungicides on microbial activities in soil as measured by a bioassay method, *Can. J. Microbiol.,* 19, 771, 1973.
166. **Munnecke, D. E. and Moore, B. J.,** Fungicidal activity in soil in relation to time, concentration, and *Penicillium* population, *Phytopathology,* 57, 823, 1967.
167. **Saha, J. G., Lee, Y. M., Tinline, R. D., Chinn, S. H. F., and Austenson, H. M.,** Mercury residues in cereal grains from seed soil treated with organomercury compounds, *Can. J. Plant Sci.,* 50, 597, 1970.
168. **Sinha, A. P., Agnihotri, V. P., and Singh, K.,** Persistence of aretan in soil and its effect on soil bioecosystems and their related biochemical activity, *Proc. Indian Natl. Sci. Acad. Part B,* 45, 261, 1979.
169. **Agnihotri, V. P.,** Persistence of captan and its effects on microflora, respiration, and nitrification of a forest nursery soil, *Can. J. Microbiol.,* 17, 377, 1971.
170. **Pimentel, D.,** Ecological Effects of Pesticides on Non-target Organisms, Office of Science and Technology, Executive Office of the President, U.S. Government Printing Office, Washington, D.C., 1971, 220.
171. **Alconero, R. and Hagedorn, D. J.,** The persistence of Dexon in soil and its effects on soil microflora, *Phytopathology,* 58, 34, 1968.
172. **Karanth, N. G. K. and Vasantharajan, N. N.,** Persistence and effect of Dexon on soil respiration, *Soil Biol. Biochem.,* 5, 679, 1973.
173. **Michail, S. H., Elarosi, H., Abd-El-Rehim, M. A., and Shohda, W. T.,** Persistence of certain systemic and non-systemic fungicides in the soil, with special reference to the method of application, *Phytopathol. Mediterr.,* 14, 138, 1975.
174. **Caseley, J. C.,** The loss of three chloronitrobenzene fungicides from the soil, *Bull. Environ. Contam. Toxicol.,* 3, 180, 1968.
175. **Wang, C. H. and Broadbent, F. E.,** Kinetics of losses of PCNB and DCNA in three California soils, *Soil Sci. Soc. Am. Proc.,* 36, 742, 1972.
176. **Kher, A. K., Vyas, K.M., and Sakasena, S. B.,** Persistence of various fungicides in soil, *Indian Phytopathol.,* 37, 574, 1984.
177. **Hine, R. B., Johnson, D. L., and Wenger, C. J.,** The persistency of two benzimidazole fungicides and their fungistatic activity against *Phymatotrichum omnivorum, Phytopathology,* 59, 798, 1969.
178. **Erwin, D. C., Garber, R. H., Carter, L., and Dewolge, T. A.,** Studies on thiabendazole and benlate as systemic fungicides against *Verticillium* wilt of cotton in the field, *Proc. 1969 Beltwida Cotton Prod. Res. Conf. 29th Cotton Dis. Counc.,* (Abstr.), 29, 1969.

179. **Smith, P. M. and Worthing, C. R.,** The effect of soil sterilization on efficiency of soil applied benomyl and carbendazim for the control of some tomato diseases, *Proc. 7th Br. Insect. Fungic. Conf.,* p. 202, 1973.

180. **Raynal, G. and Ferrari, F.,** Persistence of soil incorporated with benomyl and its effect on soil fungi, *Phytiatr. Phytopharm.,* 22, 259, 1973.

181. **Netzer, D. and Dishon, I.,** Persistence of benomyl and thiophonate compounds in soil and various plants following soil application, *Phytoparasitica,* 1, 33, 1973.

182. **Baude, F. J., Pease, H. L., and Holt, R. F.,** Fate of benomyl on field soil and turf, *J. Agric. Food Chem.,* 22, 413, 1974.

183. **Solel, Z., Sandler, D., and Dinoor, A.,** Mobility and persistence of carbendazim and thiabendazole applied to soil drip irrigation, *Phytopathology,* 69, 1273, 1979.

184. **Sinha, A. P., Agnihotri, V. P., and Singh, K.,** Persistence of carbendazim in soil and its effect on rhizosphere fungi of sugarbeet seedlings, *Indian Phytopathol.,* 33, 21, 1980.

185. **Aharonson, N. K.,** Adsorption, mobility and persistence of thiabendazole carbamate in soils, *J. Agric. Food Chem.,* 23, 720, 1976.

186. **VanWainbeke, E. and Van Assche, C.,** The importance of some factors involved with the persistence of carbendazim (MBC) in soil, *Meded. Fac. Landbouwwet. Rijksuniv. Gent,* 41, 1405, 1976.

187. **Singh, P.,** Effect of Some Soil Factors on Uptake, Persistence, Adsorption and Disease Controlling Potential of Carbendazim, Ph.D. thesis, G. B. Pant University of Agriculture and Technology, Pantnagar, India, 1984, 102.

188. **Agnihotri, V. P., Singh, K., and Budhraja, T. R.,** Persistence and degradation of Vitavax in soil and sugarcane setts and its effect on soil fungi, *Proc. Natl. Sci. Acad. U.S.A.,* 39, 561, 1974.

189. **Austin, D. J., Allford, P., and Harris, D. C.,** Metalaxyl in soil, in *Report, East Malling Research Station, Maidstone, England,* 1981, 94.

190. **Edney, K. L. and Chambers, D. A.,** The use of metalaxyl to control *Phytophthora syringae* rot of apple fruits, *Plant Pathol.,* 30, 167, 1981.

191. **Benson, D. M.,** Efficacy and *in vitro* activity of two systemic acylalnines and ethazole for control of *Phytophthora cinnamomi* root rot of azalea, *Phytopathology,* 69, 174, 1979.

192. **Benson, D. M.,** Chemical control of rhododendron dieback caused by *Phytophthora heveae, Plant Dis.,* 64, 684, 1980.

193. **Tripathi, R. K. and Singh, U. S.,** Metalaxyl: physicochemical and biological properties, in *Recent Advances in Plant Pathology,* Hussain, A., Singh, K., Singh, B. P., and Agnihotri, V. P., Eds., Print House India, Lucknow, 1983, 201.

194. **Paulus, A., Snyder, M., Gafney, J., Nelson, J., and Otto, H.,** New systemic fungicide controls downy mildew of broccoli, *Calif. Agric.,* 32, 12, 1978.

195. **Domsch, K. H.,** Erfahrungen mit neueren fungiciden Wirkstoffen im Boden, *Tagungsber. Dtsch. Akad. Landwirtschaftswiss. (Berlin),* 41, 61, 1961.

196. **Domsch, K. H.,** Prufgang für *Thielaviopsis* — und *Fusarium* aktive wirkstoffe, *Z. Pflanzenkr. Pflanzenschutz,* 69, 1, 1962.

197. **Volger, C.,** Verfahren der Bodenentseuchung und ihre Bedeutung fur die Anzucht von Forstpflanzen, *Schriftenr. Forstl. Fak. Univ. Goettingen,* 26, 1, 1962.

198. **Wilson, J. D.,** Initial and subsequent control of radish yellows by various treatments during eight successive plantings, *Phytopathology,* 47, 538, 1957.

199. **Young, R. A. and Tolmsoff, W. J.,** Current season and residual effects of Vapam soil treatments for control of *Verticillium* wilt of potatoes, *Plant Dis. Rep.,* 42, 437, 1958.

200. **Dawson, J. H.,** Longevity of dodder control by soil-applied herbicides in the greenhouse, *Weed Sci.,* 17, 295, 1969.

201. **Anderson, J. P. E.,** Soil moisture and the rates of biodegradation of diallate and triallate, *Soil Biol. Biochem.,* 13, 155, 1981.

202. **Woodcock, D.,** Microbial degradation of synthetic compounds, *Annu. Rev. Phytopathol.,* 2, 321, 1964.

203. **Kaufman, D. D., Plimmer, J. R., Kearney, P. C., Blake, J., and Guardia, F. S.,** Chemical versus microbial decomposition of amitrole in soil, *Weed Sci.,* 16, 266, 1968.

204. **Plimmer, J. R., Kearney, P. C., Kaufman, D. D., and Guardia, F. S.,** Amitrole decomposition by free radical generating systems and by soils, *J. Agric. Food Chem.,* 15, 996, 1967.

205. **Gray, R. A.,** Rate of Vapam decomposition in different soils and other media, *Phytopathology,* 52, 734, 1962.

206. **Menzie, C. M.,** Metabolism of pesticides, *U.S. Fish Wildl. Serv. Spec. Sci. Rep. (Washington, D.C.),* p. 127, 1969.

207. **Burschel, P. and Freed, V. H.,** The decomposition of herbicides in soils, *Weeds,* 7, 157, 1960.

208. **Moje, W.,** The chemistry and nematicidal activity of organic halides, *Adv. Pest Control Res.,* 3, 181, 1960.

209. **Kaufman, D. D., and Kearney, P. C.,** Microbial degradation of S-triazine herbicides, *Residue Rev.,* 32, 235, 1970.
210. **Kaufman, D. D.,** *Pesticides in Soil and Water,* Guenzi, W. D., Ed., Soil Science Society of America, Madison, Wis., 1974, 133.
211. **Audus, L. J.,** Microbiological breakdown of herbicides in soil, in *Herbicides and the Soil,* Woodford, E. K. and Sagar, G. R., Eds., Blackwell Scientific, Oxford, 1960, 1.
212. **Alexander, M.,** Persistence and biological reactions of pesticides in soils, *Soil Sci. Soc. Am. Proc.,* 29, 1, 1965.
213. **Alexander, M. and Lustigman, B. K.,** Effect of chemical structure on microbial degradation of substituted benzens, *J. Agric. Food Chem.,* 14, 410, 1966.
214. **Horvath, R. S.,** Cometabolism of the herbicide 2,3,6-trichlorobenzoate, *J. Agric. Food Chem.,* 19, 291, 1971.
215. **Horvath, R. S. and Alexander, M.,** Cometabolism: a technique for the accumulation of biochemical products, *Can. J. Microbiol.,* 16, 1131, 1970.
216. **Sheets, T. J. and Harris, C. I.,** Herbicide residues in soils and their phytotoxicities to crops grown in rotations, *Residue Rev.,* 11, 119, 1965.
217. **Schuldt, P. H., Burchfield, H. P., and Bluestone, H.,** Stability and movement studies on the new experimental nematicide 3,4-dichlorotetra-hydrothiophene-1,1-dioxide in soil, *Phytopathology,* 47, 534, 1957.
218. **Hamaker, J. W., Youngson, C. R., and Goring, C. A. I.,** Prediction of the persistence and activity of TORDON herbicide in soils under field conditions, *Down to Earth,* 23, 2, 1967.
219. **Kearney, P. C. and Kaufman, D. D.,** *Degradation of Herbicides,* Marcel Dekker, New York, 1969.
220. **Alexander, M. and Aleen, M. I. H.,** Effect of chemical structure on microbial decomposition of aromatic herbicides, *J. Agric. Food Chem.,* 9, 44, 1961.
221. **Kaars Sijpesteijn, A., Dekhuijzen, H. M., and Vonk, J. W.,** Biological conversion of fungicides in plant and microorganisms, in *Antifungal Compounds,* Siegel, M. R. and Sisler, H. O., Eds., Marcel Dekker, New York, 1977, 91.
222. **Russel, P.,** Inactivation of phenyl mercuric acetate in groundwood pulp by a mercury resistant strain of *Penicillium roqueforti* Thom., *Nature (London),* 176, 1123, 1955.
223. **Spanis, W. C., Munnecke, D. E., and Solberg, R. A.,** Biological breakdown of two organomercurial fungicides, *Phytopathology,* 52, 455, 1962.
224. **Spanis, W. C.,** Bacterial Detoxication of Organic Mercurial Fungicides, Ph.D. thesis, University of California, Los Angeles, 1963.
225. **Russell, P.,** The inactivation of organomercurial fungicides by sulphur compounds in groundwood made from salt-water stored pulpwood, *Nor. Skogind.,* 4, 3, 1960.
226. **Booer, J. R.,** The behavior of mercury compounds in soil, *Ann. Appl. Biol.,* 31, 340, 1944.
227. **Jernelov, A.,** Conversion of mercury compounds, in *Chemical Fallout: Current Research on Persistent Pesticides,* Miller, M. W. and Berg, G. G., Eds., Charles C Thomas, Springfield, Ill., 1968, 68.
228. **Jensen, S. and Jernelov, A.,** Biological methylation of mercury in aquatic organisms, *Nature (London),* 223, 753, 1969.
229. **Wood, J. M., Kennedy, F. S., and Rosen, G. C.,** Synthesis of methyl mercury compounds by extracts of a methanogenic bacterium *Nature (London),* 220, 173, 1968.
230. **Furakawa, F., Suzuki, T., and Tonomura, K.,** Decomposition of organic mercurial compounds by mercury resistant bacteria, *Agric. Biol. Chem.,* 33, 128, 1969.
231. **Furukawa, K. and Tonomura, K.,** Enzyme system involved in the decomposition of phenyl mercuric acetate by mercury resistant *Pseudomonas, Agric. Biol. Chem.,* 35, 604, 1968.
232. **Spranger, W. J., Spigarelli, J. L., Rose, J. M., and Miller, H. P.,** Methylmercury: bacterial degradation in lake sediments, *Science,* 180, 192, 1973.
233. **Spranger, W. J., Spigarelli, J. L., Rose, J. M., Flippin, R. S., and Miller, H. H.,** Degradation of methylmercury by bacteria isolated from environmental samples, *Appl. Microbiol.,* 25, 488, 1973.
234. **Nelson, J. D., Blair, W., Brinckman, F. E., Colwell, R. R., and Iverson, W. P.,** Biodegradation of phenylmercuric acetate by mercury-resistant bacteria, *Appl. Microbiol.,* 26, 321, 1973.
235. **Matsumura, F., Gotoh, Y., and Boush, G. M.,** Phenyl mercuric acetate metabolic conversion by microorganisms, *Science,* 173, 49, 1971.
236. **Tonomura, K., Maeda, K., Futai, F., Nakagami, T., and Yamada, M.,** Stimulative vaporization of PMA by mercury resistant bacteria, *Nature (London),* 217, 644, 1968.
237. **Dekhuijzen, H. M.,** Transformation in plants of sodium dimethyldithiocarbamate into other fungitoxic compounds, *Nature (London),* 191, 198, 1961.
238. **Dekhuijzen, H. M.,** The systemic action of dimethyldithiocarbamates in cucumber scab caused by *Cladosporium cucumerinum* and the conversion of these compounds by plants, *Neth. J. Plant Pathol.,* 70, 1, 1964.

239. **Kaars Sijpesteijn, A. and Kaslander, J.,** Metabolism of fungicides by plants and microorganisms, *Outlook Agric.,* 4, 119, 1964.

240. **Ludwig, R. A., Thorn, G. D., and Miller, D. M.,** Studies on the mechanism of fungicidial action of disodium ethylene bisdithiocarbamate (nabam), *Can. J. Bot.,* 32, 48, 1954.

241. **Ludgwig, R. A. and Thorn, G. D.,** Chemistry and mode of action of dithiocarbamate fungicides, *Adv. Pest Control Res.,* 3, 219, 1960.

242. **Lopatecki, L. E. and Newton, W.,** The decomposition of dithiocarbamate fungicides with special reference to the volatile products, *Can. J. Bot.,* 30, 131, 1952.

243. **Morehart, A. L. and Crossan, D. F.,** Studies on the ethylenebis dithiocarbamate fungicides, *Del. Univ. Agric. Exp. Stn. Bull. (New York),* 357, 243, 1965.

244. **Owens, R. G.,** Metabolism of fungicides and related compounds, *Ann. N. Y. Acad. Sci.,* 160, 114, 1969.

245. **Rich, S. and Horsfall, J. G.,** Gaseous toxiciants from organic sulphur compounds, *Am. J. Bot.,* 37, 643, 1950.

246. **Vonk, J. W. and Kaars Sijpesteijn, A.,** Tentative identification of 2-imidazoline as a transformation product of ethylenebisdithiocarbamate fungicides, *Pestic. Biochem. Physiol.,* 1, 163, 1971.

247. **Kaars Sijpesteijn, A., Kaslander, J., and Vander Kerk, G. J. M.,** On the conversion of sodium dimethyldithiocarbamate into its α-aminobutyric acid derivative by micro-organisms, *Biochem. Biophys. Acta,* 62, 587, 1963.

248. **Kaars Sijpesteijn, A. and Vander Kerk, G. J. M.,** Fate of fungicides in plants, *Annu. Rev. Phytopathol.,* 3, 127, 1965.

249. **Sisler, H. D. and Cox, C. E.,** Effects of tetramethyl thiuram disulfide on metabolism of *Fusarium roseum, Am. J. Bot.,* 41, 338, 1954.

250. **Sisler, H. D. and Cox, C. E.,** Release of carbon disulfide from tetramethylthiuram disulfide by fungi, *Phytopathology,* 41, 565, 1951.

251. **Yip, G., Onley, J. H., and Haward, S. F.,** Residues of maneb and ethylene thiourea on field-sprayed lettuce and kale, *J. Assoc. Offic. Anal. Chem.,* 54, 1373, 1971.

252. **Raghu, K., Murthy, N. B. K., Kumarsamy, R., Rao, R. S., and Sane, P. V.,** *Proc. Joint FAO/IAEA Division of Atomic Energy in Food and Agriculture, Vienna,* 1975, 137.

253. **Owens, R. G. and Rubinstein, J. H.,** Chemistry of the fungicidal action of thiram and ferbam, *Contrib. Boyce Thompson Inst.,* 22, 241, 1964.

254. **Ayanaba, A., Verstraete, W., and Alexander, M.,** Formation of dimethynitrosamine carcinogen and mutagen, in soil treated with nitrogen compounds, *Soil Sci. Soc. Am. Proc.,* 37, 565, 1973.

255. **Kaars Sijpesteijn, A. and Vonk, J. W.,** Microbial conversions of dithiocarbamate fungicides, *Meded. Fac. Landbouwwet. Gent,* 35, 799, 1970.

256. **Vonk, J. W. and Kaars Sijpesteijn, A.,** Studies on the fate of ethylene-bisdithiocarbamate fungicides and their decomposition products, *Ann. Appl. Biol.,* 65, 489, 1970.

257. **Engst, R. and Schnaak, W.,** Untersuchungen zum metabolismus der fungiziden Athylen-bis-dithiocarbamaten Maneb and Zineb. III. *Z. Lebensm. Unters. Forsch.,* 143, 99, 1970.

258. **Kaufman, D. D. and Fletcher, C. L.,** *2nd Int. Congr. Plant Pathol. Minneapolis 1973 Abstr.,* p. 1018, 1973.

259. **Vander Kerk, G. J. M.,** The present state of fungicide research, *Meded. Landbouwhogesch. Opzoekng Stat Gent,* 21, 305, 1956.

260. **Llyd, G. A.,** The elimination of methylisothiocyanate from soil after treatment with metham sodium, *J. Sci. Food Agric.,* 13, 309, 1962.

261. **Turner, N. J., Corden, M. E., and Young, R. A.,** Decomposition of Vapam in soil, *Phytopathology,* 52, 756, 1962.

262. **Von Kotter, K., Willenbrink, J., and Junkmann, K.,** Der Abbau von ^{35}S-markiertem Methylsenol in verschie denen Boden, *Z. Pflanzenkr. Pflanzenschutz,* 68, 407, 1961.

263. **Gray, R. A. and Streim, H. G.,** Identification of a nonvolatile phytotoxic impurity in vapam and preventing its formation, *Phytopathology,* 52(Abstr.), 734, 1962.

264. **Drescher, N. and Otto, S.,** Uber den Abbau von Dazomet in Boden, *Residue Rev.,* 23, 49, 1968.

265. **Torgeson, D. C., Yoder, D. M., and Johnson, J. B.,** Biological activity of nylone breakdown products, *Phytopathology,* 47, 536, 1957.

266. **Kotter, C.,** Untersuchungen zum Abbau von ^{35}S-markiertem Monomethylthioharnstoff in Boden, *Z. Pflanzenkr. Pflanzenschutz,* 72, 684, 1965.

267. **Munnecke, D. E., Martin, J. P., and Moore, B.,** Effect of ammonium humate and clay preparations on release of methylisothiocyanate from soil treated with fungicides, *Phytopathology,* 57, 572, 1967.

268. **Chacko, C. I., Lockwood, J. L., and Zabik, M.,** Chlorinated hydrocarbon pesticides: degradation by microbes, *Science,* 154, 893, 1966.

269. **Nakanishi, T. and Oku, H.,** Metabolism and accumulation of pentachloronitrobenzene by phytopathogenic fungi in relation to selective toxicity, *Phytopathology,* 59, 1761, 1969.

270. **Ko, W. H. and Farley, J. D.,** Conversion of PCNB to pentachloroaniline in soil and the effect of these compounds on soil microorganisms, *Phytopathology,* 59, 64, 1969.

271. **Kaufman, D. D.,** Pesticides metabolism, in *Pesticides in the Soil: Ecology Degradation, and Movement, Int. Symp. on Pesticides in Soil,* Michigan State University, East Lansing, 1970, 144.

272. **deVos, R. H., ten Noever Brauw, M. C., and Olthof, P. D. A.,** Residues of pentachloronitrobenzene and related compounds in greenhouse soils, *Bull. Environ. Contam. Toxicol.,* 11, 567, 1974.

273. **Wang, C. H. and Broadbent, F. E.,** Effect of soil treatments on losses of two chloronitrobenzene fungicides, *J. Environ. Qual.,* 2, 511, 1973.

274. **Cserjesi, A. J.,** . The adoption of fungi to pentachlorophenol and its biodegradation, *Can. J. Microbiol.,* 13, 1243, 1967.

275. **Young, H. C. and Carroll, J. C.,** The decomposition of pentachlorophenol when applied as a residual preemergence herbicide, *Agron. J.,* 43, 504, 1951.

276. **Kuwatsuka, S.,** Degradation of several herbicides in soils under different conditions, *Environmental Toxicology of Pesticides,* Matsumuro, F., Boush, G. M., and Misato, T., Eds., Academic Press, New York, 1972, 385.

277. **Cserjesi, A. J. and Johnson, E. L.,** Methylation of pentachlorophenol by *Trichoderma virgatum, Can. J. Microbiol.,* 18, 45, 1972.

278. **Chu, J. P. and Kirsch, E. J.,** Metabolism of pentachlorophenol by an axenic bacterial culture, *Appl. Microbiol.,* 23, 1033, 1965.

279. **Van Alfen, N. K. and Kosuge, T.,** Microbial metabolism of the fungicide 2,6-dichloro-4-nitroaniline, *J. Agric. Food Chem.,* 22, 221, 1974.

280. **Hock, W. K. and Sisler, H.D.,** Metabolism of chloroneb by *Rhizoctonia solani* and other fungi, *J. Agric. Food Chem.,* 17, 123, 1969.

281. **Lukens, R. J. and Sisler, H. D.,** Chemical reactions involved in the fungitoxicity of captan, *Phytopathology,* 48, 235, 1958.

282. **Lukens, R. J. and Sisler, H.,** 2-Thiazolidinethione-4-carboxylic acid from the reaction of captan with cysteine, *Science,* 127, 650, 1958.

283. **Lukens, R. J.,** Thiophosgene split from captan by yeast, *Phytopathology,* 53, 881, 1963.

284. **Siegel, M. R. and Sisler, H. D.,** Fate of the phthalimide and trichloromethylthio ($ScCl_3$) moieties of folpet in toxic action of cells of *Saccharomyces pastorianus, Phytopahtology,* 58, 1123, 1968.

285. **Seigel, M. R. and Sisler, H. D.,** Reactions of folpet with purified enzymes, nucleic acids and subcellular components of *Saccharomyces pastorianus, Phytopathology,* 58, 1129, 1968.

286. **Chin, W. T., Stone, G. M., and Smith, A. E.,** Degradation of carboxin (Vitavax) in water and soil, *J. Agric. Food Chem.,* 18, 731, 1970.

287. **Wallnofer, P. R., Koniger, M., Safe, S., and Hutzinger, O.,** The metabolism of the systemic fungicide carboxin (Vitavax) by *Rhizopus japonicus, Int. J. Environ. Anal. Chem.,* 2, 37, 1972.

288. **Balasubramanya, R. H. and Patil, R. B.,** Degradation of carboxin and oxycarboxin by a species of *Pseudomonas* isolated from soil, *Madras Agric. J.,* 63, 505, 1976.

289. **Engelhardt, G., Wallnofer, P. R., and Plapp, R.,** Degradation of linuron and some other herbicides and fungicides by a linuron inducible enzyme obtained from *Bacillus sphaericus, Appl. Microbiol.,* 22, 284, 1971.

290. **Engelhardt, G., Wallnofer, P. R., and Plapp, R.,** Purification and properties of an aryl acylamidase of *Bacillus sphaericus,* catalyzing the hydrolysis of various phenylamide herbicide and fungicides, *Appl. Microbiol.,* 26, 709, 1973.

291. **Bochoffer, R., Oltmanns, O., and Lingens, F.,** Isolation and characterization of a *Nocardia* like soil bacterium growing on carboxamlide fungicides, *Arch. Microbiol.,* 90, 141, 1973.

292. **Clemons, G. P. and Sisler, H. D.,** Formation of afungitoxic derivatives from benlate, *Phytopathology,* 59, 705, 1969.

293. **Helweg, A.,** Persistence of benomyl in different soil types and microbial break-down of the fungicide in soil and agar culture, *Tijdsskr. Planteavl.,* 77, 232, 1973.

294. **Worthing, C. R.,** *Rep. Glasshouse Crops Res. Inst. for 1974,* p. 69, 1976.

295. **Baude, F. J., Gardiner, J. A., and Han, C. Y.,** Characterization of residues on plants following foliar spray applications of benomyl, *J. Agric. Food Chem.,* 21, 1084, 1973.

296. **Rouchaud, J., Decallonne, J., and Meyer, J.,** La metabolisation des fongicides systemiques dans les plantes, in *Semanie d'Etude Agriculture et Hygiene des Plantes,* Martin, P., Ed., Faculte des Sciences Agronomiques de L'Etat, Gembloux, 1975, 475.

297. **Fuchs, A. and Bollen, G. J.,** Benomyl, after seven years, in *Systemfungizide,* Zyr, H. and Polter, C., Eds., Akademic-Verlag, Berlin, 1975, 121.

298. **Davidse, L. C.,** Metobolic conversion of methyl benzimidazol-2-yl carbamate (MBC) in *Aspergillus nidulans, Pestic. Biochem. Physiol.,* 6, 538, 1974.

299. **Yasuda, Y., Hashimoto, S., Soeda, Y., and Noguchi, T.,** Metabolism of thiophanate-methyl by pathogenic fungi and antifungal activity of its metabolites, *Ann. Phytopathol. Soc. Jpn.,* 39, 49, 1973.

300. **Fleeker, J. R., Lacy, H. M., Schultz, I. R., and Honkom, E. C.,** Persistence and metabolism of thiophanate methyl in soil, *J. Agric. Food Chem.,* 22, 592, 1974.

301. **Noguchi, T.,** *Environmental Toxicology of Pesticides,* Matsumura, F., Oush, G. B., and Misato, T., Eds., Academic Press, New York, 1972, 607.

302. **Vonk, J. W. and Kaars Sijpesteijn, A.,** Methyl benzimidazole-2-yl carbamate, the fungitoxic principle of thiophanate-methyl, *Pestic. Sci.,* 2, 160, 1971.

303. **Wade, P.,** Soil fumigation. II. The stability of EDB in soil, *J. Sci. Food Agric.,* 5, 288, 1954.

304. **Castro, C. E. and Belser, N. O.,** Biodehalogenation, reductive dehalogenation of the biocides ethylene dibromide, 1,2-dibromo-3-chloropropane and 2,3-dibromobutane in soil, *Environ. Sci. Technol.,* 2, 779, 1968.

305. **Belser, N. Q. and Castro, C. E.,** Biodehalogenation — the metabolism of the nematicides-cis- and trans-3-chlorallyl alcohol by a bacterium isolated from soil, *J. Agric. Food Chem.,* 19, 23, 1971.

306. **Castro, C. E. and Belser, N. O.,** Hydrolysis of cis- and trans-1,3-dichloropropene in wet soil, *J. Agric. Food Chem.,* 14, 69, 1966.

307. **Castro, C. E. and Bartnicki, E. W.,** Biodehalogenation, epoxidation of halohydrins, expoxide opening and transhalogenation by a *Flavobacterium* sp., *Biochemistry,* 7, 3213, 1968.

308. **Castro, C. E. and Bartnicki, E. W.,** Biological cleavage of carbon-halogen bonds metabolism of 3-bromopropanol by *Pseudomonas* sp., *Biochem. Biophys. Acta,* 100, 384, 1965.

309. **Jensen, H. L.,** Allylalcohol as a nutrient for microorganisms, *Nature (London),* 183, 903, 1959.

310. **Legator, M. and Racusen, D.,** Mechanism of allylalcohol toxicity, *J. Bacteriol.,* 77, 120, 1959.

311. **Tolmsoff, W. J.,** Biochemical basis for biological specificity of Dexon (p-dimethylamino-benzenediazo sodium sulfonate) as a fungistat, *Phytopathology,* 52, 755, 1962.

312. **Karanth, N. G. K., Bhat, S. G., Vaidyanathan, C. S., and Vasantharajan, V. N.,** Conversion of Dexon (p-dimethylaminobenzenediazo sodium sulfonate) to N, N-dimethyl phenylenediamine by *Pseudomonas fragi* BK9, *Appl. Microbiol.,* 27, 43, 1974.

313. **Karanth, N. G. K., Bhat, S. G., Vaidyanathan, C. S., and Vasantharajan, V. N.,** p-Dimethylami-nobenzenediazo sodium sulfonate (Fenaminosulf)-reductase from *Pseudomonas fragi* BK9, *Pectic. Biochem. Physiol.,* 6, 20, 1976.

314. **Gore, R. C., Hannah, R. W., Pattacini, S. C., and Porro, T. J.,** Infrared and ultraviolet spectra of seventy six pesticides, *J. Assoc. Offic. Anal. Chem.,* 54, 1040, 1971.

315. **Crosby, D. G.,** Experimental approaches to pesticide decomposition, *Residue Rev.,* 25, 1, 1969.

316. **Woodcock, D.,** Nonbiological conversions of fungicides, in *Antifungal Compounds,* Vol. 2, Siegel, M. R. and Sisler, H. D., Eds., Marcel Dekker, New York, 1977, 209.

317. **Crosby, D. G and Li, M. Y.,** Herbicide photodecomposition, in *Degradation of Herbicides,* Kearney, P. C. and Kaufman, D. D., Eds., Marcel Dekker, New York, 1969, 321.

318. **Plimmer, J. R., Klingebiel, V. I., and Hummer, B. E.,** Photooxidation of DDT and DDE, *Science,* 167, 67, 1970.

319. **Leermakers, P. A., Thomas, H. T., Weis, L. D., and Jones, F. C.,** Spectra and photochemistry of molecules adsorbed on silica gel. IV, *J. Am. Chem. Soc.,* 88, 5075, 1966.

320. **Shiina, H., Nishiyama, R., Ichihashi, M., and Fujikawa, K.,** *Nippon Nogei Kagaku Kaishi,* 39, 481, 1964.

321. **Cruickshank, P. A. and Jarrow, H. C.,** Ethylene thiourea degradation, *J. Agric. Food Chem.,* 21, 333, 1973.

322. **Ross, R. D. and Crosby, D. G.,** Photolysis of ethylenethiourea, *J. Agric. Food Chem.,* 21, 335, 1973.

323. **Fitton, A. O., Hills, J., Qutob, M., and Thompson, A.,** Studies in the dithiocarbamate series. IV. Photolysis of some 4-hydroxybenzyl dithiocarbamate, *J. Chem. Soc., Perkin Trans. III,* 2658, 1972.

324. **Crosby, D. G. and Hamadmad, N.,** The photoreduction of pentachlorobenzenes, *J. Agric. Food Chem.,* 19, 1171, 1971.

325. **Wolf, W. and Karasch, N.,** Photolysis of iodoaromatic compounds in benzene, *J. Org. Chem.,* 30, 2493, 1965.

326. **Letsinger, R. L. and McCain, J. H.,** Photoinduced substitution. III. Replacement of aromatic hydrogen by cyanide, *J. Am. Chem. Soc.,* 88, 2884, 1966.

327. **Crosby, D. G. and Moilanen, K. W.,** Photonucleophilic reactions of pesticides, in *Environmental Toxicology of Pesticides,* Matsumura, F., Bousch, G. M., and Misato, T., Eds., Academic Press, New York, 1972, 473.

328. **Alconero, R.,** Persistence of Dexon and its effect on soil microflora, *Phytopathology,* 56, 869, 1966.

329. **Mitchell, L. C.,** The effect of ultraviolet light (2537A) on 141 pesticide chemicals by paper chromatography, *J. Assoc. Offic. Agric. Chem.,* 44, 643, 1961.

330. **Haitt, C. W., Haskins, W. T., and Olivier, L.,** The action of sunlight on sodium pentachlorophanate, *Am. J. Trop. Med.,* 9, 527, 1960.

331. **Kuwahara, M., Kato, N., and Munakata, K.,** The photochemical reaction of pentachlorophenol. I. The structure of the yellow compounds, *Agric. Biol. Chem. (Tokyo),* 30, 232, 1966.

332. **Crosby, D. G., Wong, A. S., Plimmer, J. R., and Woolson, E. A.,** Photodecomposition of chlorinated to dibenzo-p — dioxins, *Science,* 173, 748, 1971.
333. **Munakata, K. and Kuwahara, M.,** Photochemical degradation products of pentachlorophenol., *Residue Rev.,* 25, 13, 1969.
334. **Kilgore, W. W. and White, E. R.,** Decomposition of the systemic fungicide 1991 (Benlate), *Bull. Environ. Contam. Toxicol.,* 5, 67, 1970.
335. **Buchenauer, H.,** Non-biological conversions of fungicides, in *Antifungal Compounds,* Siegel, M. R. and Sisler, H. D., Eds., Marcel Dekker, New York, 1975, 209.
336. **Cole, E. R., Crank, G., and Sheikh, A. S.,** Photochemistry of heterocyclic compounds. I. Photodehydrodimerization of benzimidazole, *Tetrahedron Lett.,* No. 32, 2987, 1973.
337. **Soeda, Y., Kosaka, S., and Noguchi, T.,** Identification of alkyl 2-benzimidazole carbamate as a major metabolite of thiophanates fungicides in/on the bean plant, *Agric. Biol. Chem.,* 36, 817, 1972.
338. **Buchenauer, H., Edgington, L. V., and Grossmann, F.,** Photochemical transformation of thiophanatemethyl and thiophanate to alkyl benzimidazol-2-yl carbamates, *Pestic. Sci.,* 4, 343, 1973.
339. **Needle, D. J. and Pollitt, R. J.,** Photolysis of N-2, 4-dinitrophenylamino-acids: structural requirements for the formation of 6-nitrobenzimidazole 1-oxides, *J. Chem. Soc.,* C, 2127, 1969.
340. **White, E. R., Kilgore, W. W., and Mallett, G.,** Phygon: fate of 2,3-dichloro-1, 4-naphthoquinone in crop extracts, *J. Agric. Food Chem.,* 17, 585, 1969.
341. **Dunn, C. L., Brown, K. F., and Montagne, J. Th. W.,** Antagonism between fungicides and water soluble exudate from the leaves of plants, *Phytopathol. Z.,* 64, 112, 1969.
342. **Dunn, C. L., Beynon, K. I., Brown, K. F., and Montagne, J. Th. W.,** The effect of glucose in leaf exudates upon the biological activity of some fungicides, in *Ecology of Leaf Surface Microoganisms,* Preece, T. F. and Dickinson, C. H., Eds., Academic Press, London, 1971, 490.
343. **Ashworth, L. J., Jr. and Amin, J. V.,** A mechanism for mercury tolerance in fungi, *Phytopathology,* 54, 1459, 1964.
344. **Cochrane, V. W.,** *Physiology of Fungi,* John Wiley & Sons, New York, 1958, 524.
345. **Grover, R. K. and Chopra, B. L.,** Factors affecting the toxicity of carboxin and oxycarboxin against *Rhizoctonia solani, Acta Phytopathol. Acad. Sci. Hung.,* 7, 330, 1972.
346. **Mailman, R. E., Hodgson, E., and Huisingh, D.,** Effect of thiols in reversing the inhibition by methyl-1-(butyl carbamoyl)-2-benzimidazole carbamate on *Saccharomyces cerevisiae, Pestic. Biochem. Physiol.,* 1, 401, 1971.
347. **Horsfall, J. G.,** *Principles of Fungicidal Action,* Chronica Botanica, Waltham, Mass., 1956, 274.
348. **Manji, B. T., Bose, E., Ogawa, J. M., and Urin, K.,** Effect of copper on the activity of benomyl, *Phytopathology,* 61, 1323, 1971.
349. **Courson, B. W. and Sisler, H. D.,** Effect of the antibiotic cycloheximide, on the metabolism and growth of *Saccharomyces pastorianus, Am. J. Bot.,* 47, 541, 1960.
350. **Grover, R. K. and Moore, J. D.,** Toximetric studies of fungicides against the brown rot organisms, *Sclerotinia fructicola* and *S. laxa, Phytopathology,* 52, 876, 1962.
351. **Hans, J. K., Tyagi, P. D., Kataria, H. R., and Grover, R. K.,** The influence of soil and other physical factors on the anti-fungal activity of carbendazim against *Rhizoctonia solani, Pestic. Sci.,* 12, 425, 1981.
352. **Aharonson, N. and Kafkafi, U.,** Adsorption, mobility and persistence of thiabendazole and methyl-2-benzimidazole carbamate in soils, *J. Agric. Food Chem.,* 23, 720, 1975.
353. **Aharonson, N. and Solel, Z.,** Accumulation pattern and fungicidal effect of carbendazim and thiabendazole in peanut plants after soil injection, *Pestic. Sci.,* 7, 441, 1926.
354. **Kataria, H. R. and Grover, R. K.,** Some factors affecting the control of *Rhizoctonia solani* by systemic and non-systemic fungicides, *Ann. Appl. Biol.,* 82, 267, 1976.
355. **Malhan, I., Tyagi, P. D., and Grover, R. K.,** Physical factors affecting toxicity of benomyl to *Rhizoctonia solani in vitro, Indian Phytopathol.,* 28, 491, 1975.
356. **Virk, K. S. and Grover, R. K.,** Effect of thiophanate methyl on growth and metabolic activity of *Sphaceloma ampelina, Indian Phytopathol.,* 32, 529, 1979.
357. **Bandyopadhyay, R., Yadav, J. P. S., Kataria, H. R., and Grover, R. K.,** Fungicidal control of *Rhizoctonia solani* in soil amended with organic manures, *Ann. Appl. Biol.,* 101, 251, 1982.
358. **Picci, G.,** Azione dell' SR 406 (N-trichlorometiltiotetraidroftalimide) sui microorganismi del terreno (Italian), *Agric. Ital.,* 56, 376, 1956.
359. **Rushdi, M. and Jeffers, W. F.,** Effect of some factors on efficiency of fungicides in controlling *Rhizoctonia solani, Phytopathology,* 46, 88, 1956.
360. **Alexander, P.,** Etiology and control of poinsettia root and stem rot, *Phytopathology,* 47(Abstr.), 1, 1957.
361. **Takeuchi, H. and Ide, H.,** Studies on the soil fungicides. I. Fungicidal action of the organic mercury compounds in soil, *Ann. Phytopathol. Soc. Jpn.,* 23, 197, 1957.
362. **Ranney, C. D. and Bird, I. S.,** Greenhouse evaluation of in-the-furrow fungicides at two temperatures as a control measure for cotton seedlings necrosis, *Plant Dis. Rep.,* 40, 1032, 1956.

363. **Brinkerhoff, L. A., Brodie, B. B., and Kortsen, R. A.,** Cotton seedling tests with chemical used as protectants against *Rhizoctonia solani* in the green house, *Plant Dis. Rep.,* 38, 476, 1954.

364. **Hanson, E. W.,** Effect of temperature on damping-off and responses to seed treatment in forage legumes, *Phytopathology,* 46(Abstr.), 13, 1956.

365. **Tisdale, W. B., Brooks, A. N., and Townsend, G. R.,** Dust treatments of vegetable seed, *Fla. Agric. Exp. Stn. Bull.,* 413, 32, 1945.

366. **Tamura, H.,** Evaluation of fungicidal effect of organic fungicides. III. Fungitoxicity of dinitro (1-methyl heptyl) phenyl crotonate, and the effect of temperature and pH on it (Japanese), *Tokyo Natl. Inst. Agric. Sci. (B Serv. C) (Phytopathol. Ent.),* 7, 105, 1957.

367. **Kennedy, B. W. and Brinkerhoff, L. A.,** Comparison of four soil fungicides in the greenhouse for the control of seedling diseases of cotton, *Plant Dis. Rep.,* 43, 90, 1959.

368. **Ranney, C. D. and Bird, L. S.,** Influence of fungicides, calcium salts, growth regulators and antibiotics on cotton seedling disease when mixed with the covering soil, *Plant Dis. Rep.,* 42, 785, 1958.

369. **Reavill, M. J.,** Effect of certain chloronitrobenzenes on germination, growth and sporulation of some fungi, *Ann. Appl. Biol.,* 41, 448, 1954.

370. **Colhown, J.,** Biological techniques for the evaluation of fungicides. III. The evaluation of a technique for the evaluation of soil fungicides for the control of club root disease of Brassicae, *Ann. Appl. Biol.,* 41, 290, 1954.

371. **Blocks, S. S.,** Getting the most from fungicide tests, *Agric. Food Chem.,* 7, 18, 1959.

372. **Kelman, A.,** The effect of hydrogen-ion concentration on toxicity of spergon, *Phytopathology,* 37(Abstr.), 12, 1947.

373. **Munnecke, D. E., Ludwig, R. A., and Sampson, R. E.,** The fungicidal activity of methylbromide, *Can. J. Bot.,* 37, 51, 1959.

374. **Miller, L. P.,** Natural membranes, in *Diffusion and Membrane Technology,* Tuwiner, S. B., Ed., Reinhold Publishing, New York, 1962. 345.

375. **Owens, R. G. and Miller, L. P.,** Intercellular distribution of metal ions and organic fungicides in fungus spores, *Contrib. Boyce Thompson Inst.,* 19, 177, 1957.

376. **Barlow, F. and Hadaway, A. B.,** Effect of changes in humidity on the toxicity and distribution of insecticides sorbed by some dried soils, *Nature (London),* 178, 1299, 1956.

377. **Yadav, J. P. S.,** Modifications by Fertilizers and Manures in the Efficiency of Fungicides in Controlling *Rhizoctonia solani,* M.Sc. thesis, Haryana Agriculture University, Hissar, India, 1980, 72.

378. **Frissel, M. J. and Poelstra, P.,** The fate of mercury containing fungicides, in *Soil Disinfestation,* Mulder, D., Ed., Elsevier, Amsterdam, 1979, 337.

379. **Teuschler, H. and Adler, R.,** *The Soil and Its Fertility,* Reinhold Publishing, New York, 1960, 446.

380. **Kovacs, A. and Cucchi, N. J. A.,** Influence of excreted substance from leaves on decomposition of zineb a dithiocarbamate fungicide, *Nature (London),* 204, 1090, 1965.

381. **Domsch, K. H.,** Beitrag zur vapam — wirkung gegen pathogene Bodenpilz, *Nachrichtenbl. Dtsch. Pflanzenschutz,* 10, 152, 1958.

382. **Partyka, R. E.,** The effect of some environmental factors and of certain chemicals on *Sclerotinia sclerotiorum* in the laboratory and in potato fields, *Diss. Abstr.,* 18, 1590, 1958.

383. **Hashioka, Y., Ikegami, and Siraki, S.,** fungicidal control of *Sclerotinia* rot of Chinese milk vetch, with special reference to the seed dressing, *Mem. Fac. Agric. Gifu Univ.,* 5, 210, 1955.

384. **Abeygunawardena, D. V. W. and Wood, R. K. S.,** Effect of certain fungicides on *Sclerotium rolfsii* in the soil, *Phytopathology,* 47, 607, 1957.

385. **Latham, A. J. and Linn, M. B.,** Comparative toxicity of certain fungicides to sclerotia and mycelia of *Sclerotium cepivorum* as determined in soil columns, *Phytopathology,* 52(Abstr.), 17, 1962.

386. **Howard, F. L. and Champlani, B.,** Vulnerability of *Rhizoctonia solani* in relation to turf brown patch control, *Phytopathology,* 44(Abstr.), 493, 1954.

387. **Jacks, H.,** Important factors in seed disinfection, *N. Z. Sci. Rev.,* 14, 119, 1956.

388. **Shurtleff, M. C.,** Control of turf brown patch (*Rhizoctonia solani*), *R. I. Agric. Exp. Stn. Bull.,* 328, 25, 1955.

389. **Ellis, D. E. and Cox, R. S.,** The etiology and control of lettuce damping-off, *N. C. Agric. Exp. Stn. Tech. Bull.,* 93, 33, 1951.

390. **Domsch, K. H.,** Untersuchungen zur wirkung einiger Bodenfungizide, *Mitt. Biol. Bundesanst. Land. Forstw.,* 97, 100, 1959.

391. **Morgon, O. D.,** The use of chemical soil drenches to control *Phytophthora parasitica* var. *nicotianae* and their phytotoxic effects on tobacco and toxicity to other soil flora, *Plant Dis. Rep.,* 43, 755, 1959.

392. **Strecker, B.,** Untersuchungen über die Einwirkung von organischen Fungiziden auf Bodenpilze, *Z. Pflanzenkr. Pflanzenschutz,* 64, 9, 1957.

393. **Ross, J. P.,** Studies on the chemotherapy and physiology of the *Verticillium* disease of peppermint and Chrysanthemum, *Diss. Abstr.,* 17, 11, 1957.

394. **Domsch, K. H.,** Die Wirkung van Bodenfungiciden. IV. Verander ungen im Spektrum der Bodenpilze, *Z. Pflanzenkr. Pflanzenschutz,* 67, 129, 1960.
395. **Domsch, K. H,.** Die wirkung van Bodenfungiciden. V. Empfindlichkeit von Bodenorganismen *in vitro, Z. Pflanzenkr. Pflanzenschutz,* 67, 211, 1960.
396. **Domsch, K. H.,** Das Pilzspektrum einer Bodenprobe. III. Nachweis des Einzelpilze, *Arch. Microbiol.,* 35, 310, 1960.
397. **Virtanen, A. I., Hietala, P. K., and Wahlroos, O.,** Antimicrobial substances in cereals and fodder plants, *Arch. Biochem. Biophys.,* 69, 486, 1957.
398. **Strutz, J., Herbig, H., Poppe, K., and Zinner, H.,** Fungistatische wirkung van Benzazolen, *Naturwissenschaften,* 43, 281, 1956.
399. **Elsaid, H. M. and Sinclair, J. B.,** A new greenhouse technique for evaluating fungicides for control of cotton sore-skin, *Plant Dis. Rep.,* 46, 852, 1962.
400. **Domsch, K. H.,** Die Wirking von Bodenfungiciden. III. Quantitative veranderungen der Bodenflora, *Z. Pflanzenkr. Pflanzenschutz,* 66, 17, 1959.
401. **Madison, J. H.,** The effect of pesticides on turfgrass diseases incidence, *Plant Dis. Rep.,* 45, 892, 1961.
402. **Corden, M. E. and Young, R. A.,** Changes in soil microflora following treatment with fungicides, *Phytopathology,* 51(Abstr.), 64, 1961.
403. **Jaques, R. P., Robinson, J. B., and Chase, F. E.,** Effects of thiourea, ethyl urethane and some dithiocarbamate fungicides on nitrification in fox sandy loam, *Can. J. Sci.,* 39, 235, 1959.
404. **Manten, A., Klopping, H. L., and Vander Kark, G. J. M.,** Investigations on organic fungicides. I. The antimicrobial spectrum of the antifungal substance tetramethyl thiuram disulfide, *J. Microbiol. Serol.,* 16, 45, 1950.
405. **Overman, A. J. and Burgis, D. S.,** Allylalcohol as a soil fungicide, *Phytopathology,* 46, 532, 1956.
406. **Vaartaja, O.,** Microflora on the surface of seedlings as affected by thiram, *Can. Dept. Agric. For. Biol. Div. Bi-Mon. Prog. Rep.,* 10, 3, 1954.
407. **Vaartaja, O.,** Principles and present status of chemical control of seedling disease, *For. Chron.,* 32, 45, 1956.
408. **Yatazawa, M., Persidsky, D. J., and Wilde, S. A.,** Effect of allylalcohol on micropopulation of prairie soils and growth of tree seedlings, *Proc. Soil Sci. Soc. Am.,* 24, 313, 1960.
409. **Martin, J. P., Baines, R. C., and Ervin, J. O.,** Influence of soil fumigation for citrus replants on the fungus population of the soil, *Proc. Soil Sci. Soc. Am.,* 21, 163, 1957.
410. **Hodges, G. S.,** Studies on the black root rot of pine seedlings, *Diss. Abstr.,* 20, 2467, 1960.
411. **Wensley, R. N.,** Microbiological studies of the action of some selected soil fumigants, *Can. J. Bot.,* 31, 277, 1953.
412. **Evans, E.,** Survival and recolonization by fungi in soil treated with formalin or carbon disulphide, *Trans. Br. Mycol. Sco.,* 38, 335, 1955.
413. **Mollison, J. T.,** Effect of partial sterilization and acidification of soil on the fungal population, *Trans. Br. Mycol. Soc.,* 36, 215, 1953.
414. **Warcup, J. H.,** Effect of partial sterilization by steam or formalin on the fungus flora of an old forest nursery soil, *Trans. Br. Mycol. Soc.,* 34, 519, 1951.
415. **Vaartaja, O.,** Screening fungicides for controlling damping off of tree seedlings, *Phytopathology,* 46, 387, 1956.
416. **Overman, A. J. and Burgis, D. S.,** Chemicals which act as combination herbicides, nematicides, and soil fungicides. II. Effect of soil microorganisms, *Fla. State Hortic. Soc. Proc.,* 70, 139, 1957.
417. **Papavizas, G. C., Davey, C. B., and Wood, R. S.,** Comparative effectiveness of some organic amendments and fungicides in reducing activity and survival of *Rhizoctonia solani* in soil, *Can. J. Bot.,* 8, 915, 1962.
418. **Devries, M. L.,** Effect of biocides on biological and chemical factors on soil fertility, *Diss. Abstr.,* 23, 1145, 1962.
419. **Chandra, P. and Bollen, W.,** Effects of nabam and mylone, on nitrification, soil respiration and microbial numbers in four Oregon soils, *Soil Sci.,* 92, 387, 1961.
420. **Vaartaja, O.,** Selectivity of fungicidal materials in agar cultures, *Phytopathology,* 50, 870, 1960.
421. **Cram, W. H. and Vaartaja, O.,** Toxocity of eight pesticides to spruce and caragana seed, *For. Chron.,* 31, 247, 1955.
422. **Martin, J. P., Klotz, L. J., Dewolfe, T. A., and Ervin, J. O.,** Influence of some common soil fungi on growth of citrus seedlings, *Soil Sci.,* 81, 259, 1956.
423. **Welvaert, W. and Veldeman, R.,** Influence of soil disinfectants on the fungus flora (Flemish), *Ghent Landbouwhogesch. Meded.,* 22, 499, 1957.
424. **Ramachandra-Reddy, T. K.,** Foliar spray of urea and rhizosphere microflora of rice (*Oryzae sativa L.*), *Phytopathol. Z.,* 36, 286, 1959.
425. **Horst, R. K. and Herr, L. J.,** Effects of foliar urea treatment on numbers of actinomycetes antagonistic to *Fusarium reseum* f. *cerealis* in the rhizosphere of corn seedlings, *Phytopathology,* 52, 423, 1962.

426. **Halleck, F. E. and Cochrane, V. W.,** The effect of fungistatic agents on the bacterial flora of the rhizosphere, *Phytopathology,* 40(Abstr.), 715, 1950.

427. **Smith, H. C.,** Biology of *Pythium* species in soil, *Cambridge Abstr. Diss.,* 1952-1953, 27, 1954.

428. **Dransfield, M.,** The effects of tetrachloronitrobenzene on soil microflora, *Trans. Br. Mycol. Soc.,* 40, 165, 1957.

429. **Cram, W. H. and Vaartaja, O.,** Rate and timing of fungicidal soil treatments, *Phytopathology,* 47, 169, 1957.

430. **Vaartaja, O., Cram, W. H., and Morgan, G. E.,** Damping-off etiology especially in forest nurseries, *Phytopathology,* 51, 35, 1961.

431. **Gibson, I. A. S.,** An anomalous effect of soil treatment with ethylmercury phosphate on the incidence of damping-off in pine seedlings, *Phytopathology,* 46, 180, 1956.

432. **Gibson, J. A. S.,** Crown rot, on seedling disease of groundnuts caused by *Aspergillus niger.* II. An anomalous effect of organomercurial seed dressing, *Trans. Br. Mycol. Sco.,* 36, 324, 1953.

433. **Bald, J. G. and Jefferson, R. N.,** Interpretation of results from a soil fumigation trial, *Plant Dis. Rep.,* 40, 840, 1956.

434. **Erwin, D. C., Reynolds, H. T., and Garber, M. J.,** Effect of seed treatment of cotton with thimet, a systemic insecticide, on seedling diseases in the field, *Plant Dis. Rep.,* 43, 558, 1959.

435. **Erwin, D. C. and Reynolds, H. T.,** Effect of seed treatments of cotton, with Thimet, a systemic insecticide on *Rhizoctonia* and *Pythium* seedling diseases, *Plant Dis. Rep.,* 42, 174, 1958.

436. **Leach, L. D., Garber, R. H., and Tolmsoff, W. J.,** Selective protection afforded by certain seed and soil fungicides, *Phytopathology,* 50(Abstr.), 643, 1960.

437. **Harrison, A. L.,** Seed treatments and soil drenches for the control of damping-off of tomatoes, *Phytopathology,* 43(Abstr.), 291, 1953.

438. **Jacks, H.,** Effect of seed dressings, applied separately or as mixtures on emergence of certain vegetable seeds, *N. Z. J. Sci. Technol.,* A 34, 206, 1952.

439. **Labonte, G. A.,** Damping-off studies in coniferous seedlings, *Maine For. Serv. Bull.,* 18, 14, 1959.

440. **Bird, L. S., Ranney, C. D., and Watkins, G. M.,** Evaluations of fungicides mixed with the covering soil at planting as a control measure for the cotton seedling disease complex, *Plant Dis. Rep.,* 41, 165, 1957.

441. **Sinclair, J. B.,** Laboratory and greenhouse screening of various fungicides for control of *Rhizoctonia* damping off of cotton seedlings, *Plant Dis. Rep.,* 41, 1045, 1957.

442. **Ranney, C. D. and Hills, A. M.,** A study of the distribution of in-the-furrow applied fungicides, *South. Agric. Workers Proc.,* 55, 218, 1958.

443. **Bird, L. S. and Ranney, C. D.,** Fungicides mixed with the covering soil at planting for cotton seedling disease control, *Tex. Agric. Exp. Stn. Prog. Rep.,* 2003, 3, 1958.

444. **Domsck, K. H.,** Chemical control of cotton seedling disease, *Miss. Agric. Exp. Stn. Inf. Sheet,* 748, 2, 1962.

445. **Borders, H. I.,** Effectiveness of treatments for the control of soil-borne *Fusarium* infections on Manalucie tomatoes, *Plant Dis. Rep.,* 45, 972, 1961.

446. **Kendrick, J. B., Jr., Paulus, A. O., and Davidson, J.,** Control of *Rhizoctonia* stem canker of lima bena, *Phytopathology,* 47, 19, 1957.

447. **Barnes, G. L. and Zerkel, R. S.,** Effectiveness of mixtures of Pyridinethiol derivatives and PCNB (Terractor) for control of a complex of soil fungi, *Plant Dis. Rep.,* 45, 426, 1961.

448. **Hodges, C. S.,** The occurrence of black root rot in some southern pine nurseries, *South. Agric. Workers Proc.,* 54, 221, 1957.

449. **Daines, R. H.,** Foliage fungicides and combination sprayers, *Agric. Chem.,* 12, 32, 1957.

450. **Scardavi, A. and Ciferri, R.,** The antifungal activity of mixtures of non-copper fungicides with S in laboratory experiments (in Italian, English summary), *Not. Mal. Piante,* 57, 3, 1961.

451. **Scardavi, A.,** On the fungicidal efficiency of some mixtures of organic fungicides in laboratory tests (in Italian, English summary), *Not. Mal. Piante,* 56, 33, 1961.

452. **Scardavi, A.,** The synergisitic fungicidal effectiveness of maneb combined with sulfur, barium polysulfide or thiram (in Italian, English summary), *Not. Mal. Piante,* 40, 3, 1962.

453. **Palm, E. T. and Young, R. A.,** The compatibility of certain organic fungicides and antibiotics in treatment mixtures as indicated by stability and phytotoxicity, *Plant Dis. Rep.,* 41, 151, 1957.

454. **Lukens, R. L.,** Chemical and biological studies on a reaction between captan and the dialkyldithiocarbamates, *Phytopathology,* 49, 339, 1959.

455. **Kalbe, L. and Vogel, H.,** Untersuchungen uber die wirkungss teigerung von Fongiziden durch obserflachenaktive Substanzen, *Zentralbl. Bakteriol. 2,* 113, 766, 1960.

456. **Warren, H. L. and Potter, H. S.,** New detergent disinfectants for controlling plant pathogens, *Down Earth,* 17, 2, 1962.

457. **Ferguson, J.,** Reducing Plant Disease with Fungicidal Soil Treatment, Pathogen Free Stock and Controlled Microbial Colonization, Ph.D. thesis, University of California, Los Angeles, 1958.

458. **Cockerile, J.,** Experiments in the control of damping-off at the nursery, Orono, Ontario, *For. Chron.*, 33, 201, 1957.

459. **Rankin, H. W. and Good, J. M.,** Effect of soil fumigation on the prevalence of southern blight on tomatoes, *Plant Dis. Rep.*, 43, 44, 1959.

460. **Grossmann, F. and Steckhan, D.,** Nebenwirkungen einiger Insektizide auf pathogene Bodenpilze, *Z. Pflanzenkr. Pflanzenschutz*, 67, 7, 1960.

461. **Cockerill, J.,** The effect of chlordane and thiram on damping-off and seedling growth, *For. Chron.*, 37, 211, 1961.

462. **Fulton, N. D., Waddle, B. A., and Thomas, R. O.,** Influence of planting dates on fungi isolated from diseased cotton seedlings, *Plant Dis. Rep.*, 40, 556, 1956.

463. **Gibson, I. A. S., Ledger, M., and Boehm, E.,** An anomalous effect of pentachloronitrobenzene on the incidence of damping-off caused by *Pythium* sp., *Phytopathology*, 51, 531, 1961.

464. **Vaartaja, O. and Wilner, J.,** Field tests with fungicides to control damping-off of scab pine, *Can. J. Agric. Sci.*, 36, 14, 1956.

465. **Volger, G. T.,** Versuche über Abwehr und Bekampfung von Kevimlingmykoson an koniferen, *Forstarchiv*, 30, 29, 1959.

466. **Bonde, R. and Malcolmson, J. F.,** Studies in the control of bacterial and fungus decay of potato seed pieces, *Plant Dis. Rep.*, 40, 708, 1956.

467. **Bock, K. R., Robinson, J. B. D., and Chamberlain, G. T.,** Zinc deficiency induced by mercury in coffee arabica, *Nature (London)*, 182, 1607, 1958.

468. **Martin, J. P.,** Influence of pesticide residues on soil microbiological and chemical properties, *Residue Rev.*, 4, 96, 1963.

469. **Domsch, K. H.,** Der Einfluss saprophytischer Bodenpilze auf die jugendentwicklung hoherer Pflanzen, *Z. Pflanzenkr. Pflanzenschutz*, 70, 470, 1963.

470. **De Zeeuw, D. J.,** Fungicide treatment of table beet and spinach seeds for the prevention of damping-off, *Mich. Agric. Exp. Stn. Q. Bull.*, 37, 105, 1954.

471. **Todd, F. A.,** Experiments on tobacco blue mold control, *N.C. Agric. Exp. Stn. Tech. Bull.*, 111, 16, 1955.

472. **Thulin, I. J., Will, G. M., and Bassett, C.,** A pilot trial of soil sterilization in a forest nursery, *N.Z. J. For.*, 7, 88, 1958.

473. **Palmer, J. G. and Hacskaylo, E.,** Additional findings as to the effects of several biocides on growth of seedling pines and incidence of mycorrhizae in field plots, *Plant Dis. Rep.*, 42, 536, 1958.

474. **Foster, A. A.,** Control of black root rot of pine seedlings by soil fumigation in the nursery, *For. Res. Counc. Rep.*, 8, 5, 1961.

475. **Smith, D. H.,** Effect of fumigants on the soil status and plant uptake of certain elements, *Soil Sci. Soc. Am. Proc.*, 27, 538, 1963.

476. **Wensley, R. N.,** The peach replant problem in Ontario. IV. Fungi associated with replant failure and their importance in fumigated and non-fumigated soil, *Can. J. Bot.*, 34, 967, 1956.

477. **Crowther, E., Warren, R. G., and Benzian, B.,** Nutrition problems in forest nurseries, *For. Comm. Rep. For. Res. London*, p. 84, 1954.

478. **Faulkner, R.,** A note on current and recent forestry nursery research in Scotland, *Arbor*, 215, 33, 1956.

479. **Linden, G. and Schicke, P.,** Untersuchungen über die fungicide und herbizide Wirkung von Vapam im Boden unter Berucksichtigung von Eindringtiefe, Adsorption and Karenzzeit, *Ghent Landbouwhogesch. Meded.*, 22, 399, 1957.

480. **Voigt, G. T.,** The effect of fungicides, herbicides, and insecticides on the accumulation of phosphorous by *Pinus radiata* as determined by the use of P^{32}, *Agron. J.*, 46, 511, 1954.

481. **Swart-Fuchtbauer, H.,** Ektoparasitische Nematoden als mogliche Ursache der Bodenmudigkeit im Baumschulen, *Naturwissenschaften*, 41, 148, 1954.

482. **Domsch, K. H., Guba, E. F., and Filgut, C. J.,** The control of damping off of vegetables by formaldehyde and other chemicals, *Mass. Agric. Exp. Stn. Bull.*, 394, 20, 1942.

483. **Martin, J. P., Helmkamp, G. K., and Ervin, J. O.,** Effect of bromide from soil fumigant and from $CaBr_2$ on growth and chemical composition of citrus plants, *Proc. Soil Sci. Soc. Am.*, 20, 209, 1956.

484. **Hollis, J. P., Fielding, M. J., and Wehunt, E. J.,** Action of fumigants on nematodes as related to nematicide specification, *Phytopathology*, 47(Abstr.), 9, 1957.

485. **Hansbrough, T. and Hollis, J. P.,** The effect of soil fumigation for the control of parasitic nematodes on the growth and yield of loblolly pine seedlings, *Plant Dis. Rep.*, 41, 1021, 1957.

486. **Leibundgut, H.,** Bodensterilisation im Forstagartenbetried, *Schweiz. Z. Forstwes.*, 101, 437, 1950.

487. **Molin, N. and Tear, J. J.,** Om jorddesinfektion i skogplanteskolor (Swedish), *Nor. Skogplantesk. Arskri*, 1956, 147, 1957.

488. **Weischer, B.,** Nematoden an Baumschulgewachsen Nachrichtenbl, *Dtsch. Pflanz. Schutzd.*, 8, 34, 1956.

489. **Molin, N., Pesson, M., and Persson, S.,** Root parasites on forest tree seedlings, some exploratory tests of the resistance of germinant seedlings and the virulence of some potential parasites, *Medd. Statens Skogsforskningsinst. (Swed.),* 49, 17, 1960.

490. **Persidsky, D. J. and Wilde, S. A.,** Effect of eradicants on the microbiological properties of nursery soils, *Proc. Wis. Acad. Sci. Art Lett.,* 44, 65, 1955.

491. **Hansbrough, T. and Hollis, J. P.,** The influence of soil fumigation, fertilization, and other cultural factors on nutrient content of pine nursery seedlings and soil, *Phytopathology,* 49(Abstr.), 540, 1959.

492. **Wilde, S. A., Voigt, G. K., and Persidsky, D. J.,** Transmittent effect of allylalcohol on growth of Monterey pine seedlings, *For. Sci.,* 2, 58, 1956.

493. **Mader, D. L.,** Use of forest humus to reduce the toxicity to tree seedlings from insecticides, fungicides, and herbicides in the soil, *Agron. Abstr.,* p. 51, 1956.

494. **Rich, S. and Horsfall, J. G.,** Fungitoxicity of carbamic acid esters, *Conn. Agric. Exp. Stn. Bull.,* 639, 95, 1961.

495. **Gordon, S. A. and Moss, R. A.,** The activity of S-(carboxymethyl) dimethyldithiocarbamate as an auxin, *Physiol. Plant,* 11, 208, 1958.

496. **Van Emden, J. H. and Labruyere, R. E.,** Results of some experiments on the control of common scab *(Rhizoctonia solani)* of potatoes by chemical treatment of the soil, *Eur. Potato J.,* 1, 14, 1958.

497. **Chisholm, D., MacPhee, A. W., and MacEachern, C. R.,** Effect of repeated applications of pesticides to soil, *Can. J. Agric. Sci.,* 35, 433, 1955.

498. **Sakurai, K., Tomiyama, K., Tsutsumi, M., and Takemori, T.,** Effect of yearly soil treatments with agricultural chemicals on the growth behavior of wheat and soybeans (Japanese), *Hokkaido Natl. Agric. Exp. Stn. Res. Bull.,* 72, 16, 1957.

499. **Eno, C. F.,** Field accumulation of insecticide residues in soils, effect of soil applications of carbamate fungicides on the soil microflora, *Fla. Agric. Exp. Stn. Annu. Rep.,* 142, 1957.

500. **Palmiter, D. H.,** Hudson Valley tests of apple fungicides and their long term effects on yield and quality, *Mass. Fruit Grow. Assoc. Rep. 6th Annu. Meet.,* p. 49, 1954.

501. **McNew, G. L.,** Progress in the battle against plant disease, *Sci. Aspects Pest Control,* 1402, 74, 1966.

502. **Johnson, R. C.,** Simultaneous fumigating and planting of sugar beets, *J. Am. Soc. Sugar Beet Technol.,* 15, 379, 1969.

503. **Wilhelm, S.,** Control of *Pythium* damping off of melons by simultaneous seedling and soil fumigation with chloropicrin, *Phytopathology,* 53(Abstr.), 1144, 1963.

504. **Smiley, R. W., Cook, R. J., and Papendick, R. I.,** Anhydrous ammonia as a soil fungicide against *Fusarium* and fungicidal activity in the ammonia retention zone, *Phytopathology,* 60, 1227, 1970.

505. **Engelhard, A. W., Miller, H. N., and DeNeve, R. T.,** Etiology and chemotherapy of *Pythium* root root of chrysanthemums, *Plant Dis. Rep.,* 55, 851, 1971.

506. **Munnecke, D. E., Kolbenzen, M. J., and Stolzy, L. H.,** Factors affecting field fumigation of citrus soils for control of *Armillaria mellea,* in *Proc. 1st Int. Citrus Symp.,* Vol. 3, Chapman, H. D., Ed., University of California, Riverside, 1969, 1839.

507. **Wilhelm, S.,** Chemical treatments and inoculum potential of soil, *Annu. Rev. Phytopathol.,* 4, 53, 1966.

508. **Waack, R., Alex, N. H., Frisch, H. L., Stannett, V., and Szwarc, M.,** Permeability of polymer films to gasses and vapours, *Ind. Eng. Chem.,* 47, 2524, 1955.

509. **Grimm, G. R. and Alexander, A. F.,** Fumigation of *Phytophthora* in sandy soil by surface application of methyl bromide and methyl bromide chloropicrin, *Plant Dis. Rep.,* 55, 929, 1971.

510. **Dallimore, C. E.,** Methods of increasing the efficacy of soil fumigation in experimental fields, *Plant Dis. Rep.,* 39, 511, 1955.

511. **Watson, R. D.,** Eradication of soil fungi by a combination of crop residues, flooding and anaerobic fermentation, *Phytopathology,* 54, 1437, 1954.

512. **Baker, K. F. and Snyder, W. C., Eds.,** *Ecology of Soil-Borne Plant Pathogens,* University of California Press, Berkeley, 1965.

513. **Munnecke, D. E., Moore, B. J., and Haj, F. A.,** Soil moisture effects on control of *Pythium ultimum* or *Rhizoctonia solani* with methyl bromide, *Phytopathology,* 61, 194, 1971.

514. **Monro, H. A. U., Olsen, O. A., and Buckland, C. T.,** Methyl bromide and ethylene oxide fumigation of *Synchytrium endobioticum, Can. J. Plant Sci.,* 50, 649, 1970.

515. **Biehn, W. L. and Diamond, A. E.,** Prophylactic action of benomyl against Dutch elm disease, *Plant Dis. Rep.,* 55, 179, 1971.

516. **Booth, J. A. and Rawlins, T. E.,** A comparison of various surfactants as adjuvants for the fungicidal action of benomyl on *Verticillium, Plant Dis. Rep.,* 54, 741, 1970.

517. **Furmidge, C. G. L., Hill, A. C., and Osgerby, J. M.,** Physicochemical aspects of the availability of pesticides in soil. II. Controlled release of pesticides from granular formulations, *J. Sci. Food Agric.,* 19, 91, 1968.

518. **Mills, J. T. and Schreiber, K.,** Use of latex-coated pellets for control of common root rot of wheat, *Can. J. Plant Sci.,* 51, 347, 1971.

519. **Horsfall, J. G. and Dimond, A. E. A.,** Perspective inoculum potential, *J. Indian Bot. Soc.,* 42, 46, 1963.
520. **Ludwig, R. A.,** The role of chemicals in the biological control of soil-borne plant pathogens, in *Ecology of Soil-Borne Plant Pathogens,* Baker, K. F. and Snyder, W. C., Eds., University of California Press, Berkeley, 1970, 471.
521. **Baker, R.,** Analyses involving inoculum density of soil-borne plant pathogens in epidemiology, *Phytopathology,* 61, 1280, 1971.
522. **Lewis, J. A. and Papavizas, G. C.,** Effect of sulfur-containing volatile compounds and vapours from cabbage decomposition on *Aphanomyces euteiches, Phytopathology,* 61, 208, 1971.
523. **Papavizas, G. C. and Lewis, J. A.,** Effect of amendments and fungicides on *Aphanomyces* root rot of peas, *Phytopathology,* 61, 215, 1971.

Chapter 10

INTERACTIONS BETWEEN FUNGICIDES AND SOIL MICROORGANISMS

I. INTRODUCTION

In the natural environment, no organism exists that lives like a hermit: this statement holds especially for microbes in the soil. A network of interactions between the different components of the terrestrial ecosystem keeps the populations in a state of dynamic equilibrium. This is one of the characteristics of soils of high fertility. Information on the microbial interactions is largely confined to phenomena in models, in which two or more microorganized soil or artificial media exist. For field conditions only minimal data can be assessed. In agricultural soils, toxicants enter the terrestrial ecosystem. Depending upon their kind, they may be harmless to or eliminate the microorganisms; they may arrest growth or sporulation, or they may react with microbial products. Considerable work has been done on these aspects, but practicability of transferring results from such observations to the field depends on many variables, some of which are unknown. These variables can be briefly mentioned here, such as the diversity of conditions in the experiments, the diversity of soil types, the paucity of detail to correlate the changing pattern in the microbial population, and the lack of information on formulation products or additives which may influence the effects. The influence of fungicides on microbial populations is usually worked out on homogeneous soil samples under laboratory conditions. From such experiments, the selection of appropriate dose rates is difficult.

Fungicides are applied to soils to kill or reduce the populations of different groups of plant pathogens. Part of fungicides administered as sprays or dusts on shoots will ultimately reach the soil, as runoff or drift. Moreover, substantial quantities of subsidiary materials like solvents, carriers, diluents, synergists, and adjuvants will also reach the soil. The action of these chemicals is not always limited to the immediate objective of killing a particular pathogen. Harmless or beneficial soil organisms may be killed or temporarily reduced in numbers, the qualitative nature of the soil population may alter for varying periods of time, and chemical changes may occur within the soil. Some of the soil-inhabiting organisms play an important role in the decomposition of complex organic compounds like cellulose and lignin into humus and release simple plant nutrients in the process. Fixation of atmospheric nitrogen through microbes in soil, the main nonmonetary replenishing of soil fertility, is relied upon in formulating crop rotations. Conversion of ammonia, liberated during the degradation process to nitrite and then to nitrate, which can be readily absorbed by most plant roots, is done in soil by microbes. Similarly, *Corynebacterium* and *Flavobacterium* oxidize manganese in soil. In addition to these, a variety of other soil-inhabiting microbes regulate oxidation or solubilization of minerals releasing calcium, magnesium, phosphorus, etc. into the soil, which are essential for plant growth.

In recent years several reviews[1-7] have been put forth on the subject. A critical survey of the reviews reveals that, with few exceptions, concentrations of toxicants that result from normal rates of application for pest control have no long-lasting effect on microorganisms. Most of these reviews deal with the effects of herbicides and insecticides on soil microbial processes. In spite of the fact that fungicides are usually effective in eradicating or otherwise suppressing soilborne pathogens, very little information is available concerning the effect of fungicides on nontarget soil microorganisms and/or their activities. It would seem probable that fungicides might indeed similarly affect other soil processes. As the role of fungicides in increasing food production is well established, there is now an urgent need to elicit a full appraisal of their simultaneous influence on soil microorganisms in general, and the beneficial lot in particular. This would be essential for minimizing harmful effects, if any, through corrective measures. In this chapter, primary attention will be given in the ensuing discussion to the effects of fungicides on soil microorganisms and their activities other than the pathogen they are designed to control, and also with the effects on microbial interactions and their impact on incidence of plant diseases.

When a chemical is applied to the soil in sufficient concentration to kill plant root parasites, all or most types of soil microbes are either killed or, more commonly, reduced in numbers.[5] The actual percentage of kill depends on such factors as soil type, moisture content, temperature, soil physical condition, amount and kind of chemical applied, and method of application. After the initial kill or reduction in numbers, certain types, usually bacterial species, recover very quickly and reach numbers far in excess of those in untreated soils. In general, the greater the initial kill, the greater the subsequent peak in numbers. Fungi and other microbes become reestablished at various periods and may or may not exceed numbers in the original soils.[8] With time, which may be a year or longer, total numbers tend to approach those in the untreated soil. Conflicting reports on the effects of a few individual compounds are obvious; whether or not these effects are due to soil type difference, methods used in characterizing microbial populations, or both can only be speculated. An increase in population in fungicide-treated soil is known to be the result of a proliferation of only a few species.

II. EFFECT OF FUNGITOXICANTS ON SOIL MICROFLORA

A. Fungi

Thousands of fungi are present in a gram of cultivated soil, usually abundant near the soil surface due to better aerobic conditions.[9,10] These are responsible for the initial breakdown of complex organic matter and affect the processes of soil aggregation.[9] Saprophytic soil fungi influence the survival/pathogenicity of soilborne pathogens by competition, antagonism, or parasitism.[11,12] The fungi as a group are more susceptible and are readily killed by soil fungicides. On the basis of their effect on soil microflora, fungicides can be broadly grouped into two categories, i.e., (1) narrow- and (2) broad-spectrum fungicides. The former are effective on the target organisms. Dexon, for example, is highly specific in its action against pathogenic soil oomycetes like *Pythium, Phytophthora,* and *Aphanomyces,*[13,14] while Vitavax is selective against basidiomycetous fungi.[15,16] Broad-spectrum chemicals such as aretan, formalin, Vapam, Vorlex, etc. hit many fungi without exercising any selective action.

Chandra and Bollen[17] demonstrated that soil application of nabam (100 ppm) and Mylone (150 ppm) significantly decreased the number of fungal propagules. Corden and Young[18] studied the effect of Vapam, Metasol, Mylone, and nabam on soil fungi and found a drastic reduction in fungal population in the treated soils as compared to untreated ones. Sinha et al.[19] reported that aretan markedly reduced the number of fungal propagules in soil, particularly at 5, 10, and 20 ppm, during the early part of the experiment. On the 15th day, the fluctuation in population was nonsignificant and thereafter the fungal population gradually increased in aretan-amended soils as compared to the untreated control. The increase in fungal counts was mostly due to *Asperigillus niger, A. fumigatus, A. flavus, A. terreus,* and *Trichoderma viride* propagules.

Sinha et al.[20,21] further observed that Bavistin reduced the population of fungi in soil for 45 days and it was more effective against species of *Rhizoctonia, Penicillium, Aspergillus,* and *Trichoderma* as compared to *Helminthosporium, Curvularia,* and *Alternaria* species. Oku et al.[22] demonstrated temporary reduction in soil fungal population following thiram treatment. Agnihotri[13] demonstrated *Pythium* species to be highly sensitive to Dexon, which at 175 ppm in soil eliminated all propagules of the fungus. However, the population of *Mortierella* species increased in Dexon-treated soil. This is rather interesting as *Mortierella* is close to *Pythium* taxonomically and physiologically. Smith and Long[23] observed that phenylmercury acetate (PMA) applied in the soil at 100 ppm severely reduced the incidence of many cellulytic fungi normally present with concomitance in the increased population of *Penicillium* and *Trichoderma*. No fungicides were found immediately stimulating to soil fungal populations under the same conditions.

Several reports are on hand showing the effect of fungitoxicants on the type(s) of fungi that develop in untreated soils. Probably qualitative effects of fungicides on the soil population are of more significance than their quantitative effects. Waksman and Starkey[24] reported that fungal colonies developing on plates from fungicide-treated soils represented relatively few species as compared to those from the untreated soils. Qualitative studies revealed that certain genera like *Aspergillus, Penicillium, Fusarium*, and *Trichoderma* become abundant in most of the treated soils. Smith[25] noted that after treatment of soil with low concentrations of chloropicrin, the fungal flora which developed consisted mainly of *Trichoderma*. Tam and Clark[26] found that chloropicrin or formalin treatment of soil favored species of *Aspergillus* and *Penicillium*. Wensley[27] reported that recolonization after methyl bromide (MB) treatment was mainly by species of *Aspergillus*; McKeens[28] found *Pyronema congluens* fruiting on the surface of flats treated with MB. Katznelson and Richardson[29] noted that certain species of *Penicillium* and *Plectonaemella* increased in formalin-treated Canadian soils. *Monilia sitophila* and *Rhizopus nigricans* commonly overrun soils treated with chloropicrin.[30] Captan-treated rhizosphere soil of *Picea glauca* stimulated the growth of *Penicillium* and *Gliocladium* species.[31]

Based on the numerous published reports from different sections of the world, *Trichoderma* often becomes the most common dominant soil fungus following fungicidal treatment of the soil. It has been reported as a dominant fungal colonizer of soil treated with chloropicrin,[25] DD,[32] Vapam,[33] aretan,[19] formalin,[34] or carbon disulfide.[35] Several factors may be responsible for the ability of *T. viride* to quickly recolonize fumigated soil: (1) it is relatively more resistant to the toxic chemical than other fungi, (2) if it is initially killed, it may become reestablished while chemical in sufficient quantity remains to reduce the growth of other species, and (3) rapid proliferation rate of the fungus helps it to recolonize the soil more quickly as compared to other fungal species.

The dominance of *Trichoderma* species in fungicide-treated soil is of particular interest since some strains of the fungus are known to produce toxins and to act, under certain conditions, as antagonists to other fungi, viz., *Pythium, Rhizoctonia, Armillaria*, and *Phytophthora*.[36-38] Richardson[39] observed that thiram treatment protected plant seedlings from damping-off fungi after the chemical had decomposed in the soil. The treatment made the soil more resistant to artificial infestation with *Pythium ultimum* and prevented natural increase of damping-off organisms. It was also observed that the chemical treatment stimulated the growth of *T. viride*, and it was concluded that the beneficial effects were in part associated with the increase in number and activity of this fungus in the soil. Fumigation of soil with carbon disulfide usually controlled *Armillaria* root rot of citrus trees and other plants, although the treatment did not always kill the pathogen.[40,41] It has been noted that after fumigation, *T. viride* increased in relative numbers and activity and apparently parasitized the fungus. There are a few reports showing that the stimulation of *T. viride* in soil does not always control the detrimental activity of plant root parasites.[36] Failure to exert an antagonistic effect may be associated with the inability of certain strains to produce antibiotics.[42]

Changes in morphology of an organism due to its soil environment can alter the degree of its susceptibility to a chemical. Such morphological responses in fungi are profuse hyphae formation, or the development of rhizomorphs, pseudosclerotia, sclerotia, chlamydospores, or conidia. They vary in their resistance to toxicants. It was observed by Domsch[43] that spores are more resistant to soil fungicides than hyphae. According to Bomar,[44] pentachlorophenate was more effective against the conidia than the mycelium of *Aspergillus flavus* or *Penicillium expansum*. Morgan[45] also reported that derivatives of the experimental soil fungicides, 2-pyridinethione-1-oxide sulfide and tetrahydrothiadiazine-thione, were at least ten times more toxic to the zoospores than to the hyphae of *Phytophthora parasitica*. Conidia are not always more susceptible to a toxicant than hyphae. However, it depends on the fungicide applied, viz., the highly selective diphenyl appears to be more toxic to hyphae than it is to conidia.[46]

Sclerotium-forming fungi, in general, are more resistant to fungicides than others.[47] Tarr[48] observed that sclerotia of *Macrophomina phaseoli* were resistant to chloropicrin. Similarly, Kreutzer[30] reported that the sclerotia of *Sclerotinia sclerotiorum* and *Sclerotium rolfsii* are more resistant than their mycelia to chloropicrin and chlorobromopropene.

Soil treatment with MB controlled nonmicrosclerotial fungi, but not the microsclerotium-forming *Verticillium alboatrum*.[49] Percent kill of prewater-soaked sclerotia of *S. rolfsii* was obtained with chloropicrin, which was totally ineffective against dry ones.[50] The effect of certain physiocochemical and biological factors on the effectiveness of fungicides in soil will be discussed elsewhere.

B. Total Bacteria

The bacterial population of soil outnumbers the population of all other groups of microorganisms in both number and variety. Many kinds of bacteria — like autotrophs, heterotrophs, mesophiles, thermophiles, psychrophites, aerobes and anaerobes, cellulose digester, nitrogen fixer, etc. — are generally found in soil.[10] The populations of different groups of bacteria are generally altered by fungicide application in soil. They are reduced in number for a period, then multiply rapidly, usually to numbers exceeding those in untreated soil. Part of the rise in number may be due to decomposition of the chemical.[51] After reaching a maximum, bacterial numbers fall towards those of untreated soil.[52] The fall is sometimes very slow, taking over a year. Generally, bacterial population is initially accelerated or not affected by the application of fungicides or fumigants. Application of nabam, zineb, and maneb in soil[53] has been found to increase bacterial population. In laboratory trials 20% solution of methylisothiocyanate (MIT) applied alone at 0.5 or 0.25 mℓ in mixture of DD to soil increased the number of bacteria.[54] Similarly, Agrosan GN on cotton seeds increased the bacterial population in the rhizosphere of cotton seedlings[55] for the first 7 days. An initial increase in bacterial population in Dithane M-45 and captan-treated soil has been reported by several workers.[56-59] All concentrations (2.5, 5.0, 10.0, and 20.00 ppm) of aretan and lower concentrations of Bavistin (5 to 20 ppm) stimulated bacterial counts.[19,21] Several factors contribute to the higher numbers of microbes occurring in soils soon after a fumigation or fungicidal treatment: (1) the cell material of killed organisms provides a readily available food source for surviving microorganisms or those which become reestablished first after treatment, (2) the residual chemical may serve as a carbon or energy source for certain groups of bacteria,[60] and (3) the surviving organisms or those first to become reestablished reach higher numbers in a less competitive environment.[36]

Contrary reports exist showing reduced population of bacteria following fungicide application. Nauman[61] observed that Vapam, dazomet, and allyl alcohol when added to a soil inhibited bacterial population. Sinha et al.[33] demonstrated that Vapam considerably reduced bacterial population up to 15 days and thereafter their numbers gradually enhanced. Chandra and Bollen[17] have reported that application of Nabam (100 ppm) and Mylone (150 ppm) drastically decreased the bacterial population in the initial stages and by the 60th day the bacterial population increased significantly. Tu[62] observed a detrimental effect of Vorlex to the proliferation of soil bacteria during the first 2 weeks.

Limited work has been done on the effect of fungitoxicants on spore-forming bacteria which constitute 10% of the total bacterial population. Clark[63] and Allison[64] reported that spore-forming bacteria were quite resistant to ethylene dioxide, while McKeen[28] observed that the group was relatively resistant to MB. It has been generally observed that spore-forming bacteria quickly develop in fumigated soils.[65] Hansen and McCalla[66] found that DD favored rapid proliferation of spore formers. Agnihotri[56] has reported that the number of aerobic spore-forming bacteria in the untreated control did not differ significantly from the number in captan-treated soil up to the 7th day, after which a sharp rise in the population was noted. Agnihotri[67] further reported that in garden soil treated with thiram at 375.0, 187.5, and 93.7 ppm, the population of spore-forming bacteria increased by the 7th day, then gradually decreased. The fluctuations in population were not significant even with massive dossages of Dexon applied in soil. Similar nontoxic effects were reported by Domsch[43] for thiram, Klemner[68] for ethylene dibromide (EDB), Corden and Young[18] for Mylone, and Balasubramanian et al.[57] for Dithane M-45.

Bacterial plant pathogens, in general, are resistant to soil toxicants. Munnecke and Ferguson[69] observed that high doses of chloropicrin, MB, and sodium *N*-methyl dithiocarbamate must be used to eliminate *Agrobacterium tumefaciens*, *Corynebacterium michiganense*, and *Xanthomonas pelargonii* in infected pieces of host tissues buried in soil. Winfree et al.[70] could not control soft-rotting species of *Erwinia* and *Pseudomonas* with standard dosages of chloropicrin. Millhouse and Munnecke[71] reported that in soil fumigated with 27,000 $\mu\ell$ MB per liter of air, *Psudomonas* species were eliminated within 16 hr. *Streptomyces* spp. were greatly reduced within 64 hr and were undetectable in soil fumigated for 128 hr.

C. Azotobacter

This aerobic, nonsymbiotic bacterium is commonly present in cultivated fertile soils and fixes atmospheric nitrogen which is eventually released into the soil in the form of proteins, amino acids, and ammonia. Thus, they play a conspicuous role in nitrogen economy of soil. The *Azotobacter* species boosts plant growth through secretion of exogenous metabolites like vitamins, gibberellins, auxins, and cytokinins. *Azotobacter* usually does not flourish in acidic soils having pH below 6.0.[9]

Information is rather scanty about the effects of fungicides on the population of *Azotobacter*. In general, the bacterium seems to be sensitive to soil fumigants and long, persistant fungicides. Hoflich[72] reported that application of an even higher dose of thiram to straw-amended soil resulted in an initial decrease in number of *Azotobacter* that was followed by an increase lasting to the end of the experiment (13 weeks). Bavistin and aretan reduced *Azotobacter* count up to the 45th and 15th day, respectively, depending upon the concentration used.[19,21] Addition of carboxin to the red sandy loam soil of Hebbal and Coffee plantation soil of Coorg suppressed the population of *A. chroococcum*.[73]

Sinha et al.[33] observed that Vapam (1000 ppm) significantly reduced *Azotobacter* population by the 7th day and thereafter changes in population became nonsignificant up to the 30th day; after 45 days the bacterial numbers in Vapam- (1000 ppm) treated soil exceeded that of the control soil. Jensen[74] demonstrated a stimulatory effect of allyl alcohol on the population of *Azotobacter*. It has been reported that the fungicides which produce isothiocyanates on decomposition affect the bacterium adversely.[75]

D. Rhizobia

The nodule bacteria, a normal component of soil microflora, can survive saprophytically in the absence of a temporary host, but eventually survival depends on the availability of suitable host roots. The populations of *Rhizobia* often exceeds 10^6 in a gram of rhizosphere soil.[76] Information on the effect of fungitoxicants on *Rhizobia* is very limited.

Since nodule bacteria (*Rhizobium* sp.) are not present in the seeds of the angiosperm symbiont, seedling roots must be reinfected from the soil before nodules can be formed and nitrogen-fixing processes begin. Audus[77] explained three possible ways in which toxicants might be expected to influence this symbiosis and the nitrogen fixation which results from it. First, there might be direct action on the free-living bacterial population in the soil and this could indirectly affect the degree of infection and, hence, the amount of nodule formation. Second, the infection process itself might be influenced by a change either in the virulence of the invading bacteria or in the reaction of the plant root hairs, through which infection takes place. Both these would be actions of toxicant in the soil. Third, chemicals reaching the internal tissues of the root either from the soil or from the aerial parts of the plant could influence the development of the nodules and possibly the efficiency of the nitrogen-fixation processes going on inside. It is obvious that this is a complex situation involving a multiplicity of possible effects on two vastly different organisms and on their interaction. So far no one has attempted to study this situation as a whole even with one fungicide, although there are a few reports of fungicide effects both on bacterial growth in culture and on the degree of nodulation in legumes. In addition to these, certain reports are also on hand showing *Rhizobia* counts as affected by soil application of chemicals.

In general, fungicides are toxic to *Rhizobia* as pointed out by several workers. Ruhloff and Burton[78] reported that the nodule bacteria are quickly destroyed by chemicals used as seed protectants, e.g., Arasan, Spergan, and Phygon. Hofer[79] observed varied toxicity of fungicides towards different species of *Rhizobium*. Alfalfa *Rhizobia* were least injured by toxicants, while those of clover were most sensitive. Ceresan-M was highly toxic to clover bacteria even at 0.5 ppm. Diatloff[80] observed the toxicity of fungicides to *Rhizobia* in the following descending order: Ceresan, thiram, captan, and chloranil. Agrosan GN, thiram, and captan used for dressing seeds of groundnut adversely affected *Rhizobia* for 20, 7, and 7 days, respectively.[81] Curley and Burton[82] found thiram harmless to nodule bacteria on soybean, but demonstrated that captan and pentachloronitrobenzene (PCNB) reduced nodulation. On the contrary, Sardeshpande et al.[83] have reported that captan, arasan, and Brassicol did not adversely effect nodulation in groundnut plants. Similarly, Nene et al.[84] did not observe any adverse effect on the emergence and nodulation of soybean plants, and Ostwal and Gaur[85] reported increase yields of soybean with fungicide-treated sees. Benlate did not produce any detrimental effect on leghemoglobin content of nodules.[86] Fisher[87] made a comparative study of the effects of some protective and systemic fungicides and an eradicant fungicide on the nitrogen-fixing capacity of *R. trifolii*. Symbiotic nitrogen fixation was affected by thiram and oxycarbonxin but remained unchanged by Benlate, captan, Bavistin, etc.

All concentrations of aretan, Bavistin, and Vapam, when applied to soil, decreased the population of *Rhizobia* throughout the experimental period of 45 days.[19,21] There was no effect of D-D or metham sodium on four isolates of *R. trifolli* on white clover.[88] The isolates were tolerant to D-D and moderately tolerant to metham-sodium in vitro. There were negligible effects in pot experiments and no effects on nodulation under field conditions. Singh and Prasad[89] reported inhibition of nodulation of *Phaseolus aureus* by dazomet and 1,2-dibromo-3-chloropropane (DBCP) and by high dosages of aldicarb, carbodufan, and fensulfothion. Tewfik et al.[90] found reduced yields of *Vicia fabae* with aldicarb at various dosages applied at planting, but not with DBCP. Neither fumigant reduced yields when applied 28 days after planting. In culture media, two strains of *R. leguminosarum* were inhibited by DBCP, and another strain was unaffected by 200 ppm, but stimulated by 20 ppm. Aldicarb stimulated one of the strains which was inhibited by DBCP, and inhibited the other two.

It is, therefore, concluded that (1) various *Rhizobium* species behave differently to different concentrations of a fungicide and even within the same species there are considerable strainal variations, (2) toxicity of fungicides to *Rhizobia* can be reduced by increasing the organic matter of the soil, and (3) in general, thiram and Benlate are safe at recommended dosages for nitrogen fixation by *Rhizobium* in symbiosis with legumes.

E. Actinomycetes

Usually 1 g of cultivated soil contains several million propagules of different actinomycetes, which are commonly known as ray fungi.[91] The most predominant genera of this group are *Nocardia, Streptomyces*, and *Micromonospora*.[9] They are known to synthesize and excrete many powerful antibiotics[92] and are capable of degrading many complex substances such as cellulose and lignin and, consequently, play an important role in maintaining soil fertility. When a chemical is introduced in the soil to combat soilborne diseases, it disturbs the natural population and functions of different general of actinomycetes. The role of fungitoxicants against actinomycetes, therefore, also needs to be catalogued.

Working on deleterious effects of common soil fumigants on actinomycetes, Yatazawa et al.[93] observed that allyl alcohol at or above 1125 ℓ/ha inhibited their population. Substantially similar inhibition due to EDB,[94] chloropicrin,[95] Mancozeb,[96] metham,[97] Indar,[98] carbendazim,[21] and MB[99] has also been reported. Contrary to it, Roslycky[100] showed little initial effect on the actinomycetes population following Vorlex application at recommended rates.

It was reported by Nauman[61] that Vapam (0.15%) and formalin (0.2%) inhibited the population for a longer period at low soil temperature (12 to 15°C) than at a slightly higher temperature (20°C). Conversely, soil treatment with sodium-*N*-methyldithiocarbamate actually stimulated the population of actinomycetes.[43] This toxicant significantly reduced the population of actinomycetes for the first 7 days and thereafter the changes became nonsignificant by the 30th day. However, a significant difference occurred on the 45th day between the population of fumigated and untreated soils.[19] Dexon at 20 lb/acre did not produce any significant effects on the actinomycete population.[101] Pugashetty and Rangaswami,[55] while studying the rhizosphere microflora of cotton seedlings as influenced by pretreatment of cotton seed with Agrosan-GN, observed a reduction in the actinomycete counts in initial stages of plant growth, but not in the later part of the growth.

Agnihotri,[56] working on the effect of fungicides on soil microflora, stated that the addition of captan at 62.5, 125, and 250 ppm (active ingredient [ai])/acre to a fresh forest nursery soil in the laboratory initially decreased the population, but the counts gradually increased toward the end of the experimental period. Dexon at 87.5 to 350 ppm did not appreciably alter the number of actinomycetes in treated nursery soil except at a higher concentration (350 ppm).[13] He[67] further observed that thiram decreased actinomycete counts, whereas Balasubramanian et al.[57] demonstrated that Dithane M-45 (80 gal/acre) did not have any appreciable effect on this group of microbes. An appreciable increase in the population of actinomycetes in soil treated with aretan has been reported.[79] This effect was discernible in the early part of the experiment only.

III. EFFECTS OF FUNGICIDES ON MICROBIAL ACTIVITIES

A. Soil Respiration

The "soil respiration" measured as CO_2 evolution or O_2 consumption is the parameter that is most commonly used as indicative of the decomposition of organic matter. Liberation of CO_2 in soil also provides a reasonably good idea about the bacterial numbers, soil fertility, and consequently nitrifying capacity of the soil.[102,103] Microorganisms surviving a soil toxicant recover and then proliferate at a tremendous speed, apparently due to the absence of competitors in treated soils, raising the respiration rate even higher than in the untreated soil samples. The log period depends both on the kind and concentration of the toxicant. Broad-spectrum fungicides cause perceptible changes in soil respiration, while those with narrow spectrum (PCNB, Dexon, Vitavax) exhibit little change, indicating that soil fungi do not contribute much to CO_2 production.[2] For pesticides that are used in relatively high amounts, the decomposition of the active ingredient of the formulation product may contribute to CO_2 evolution. This has been tested by using ^{14}C-labeled products.[104] The influence of formulated products was clearly demonstrated by Saive et al.[105] Microbial respiration was inhibited when zineb formulated with product A was added to soil, but was stimulated when the fungicides were formulated with product B.

Rao[106] examined the effects of the fumigants EDB, MB, 1,3-D, and metham at field rates and at concentrations ten times greater in laboratory investigations. Carbon dioxide evolution from the decomposition of native organic matter was increased by all compounds except metham, which also retarded decomposition of added dextrose. Chandra and Bollen[17] reported that nabam and Mylone (150 ppm) depressed respiration for 28 days. After 42 days, slightly more CO_2 was evolved from nabam-treated soils, but after 56 days the treated soils produced more CO_2 than the untreated control. The use of fumigants to retard biological degradation and consequent subsidence of muck soils has been suggested. Stotzky et al.[107] demonstrated that dazomet, EDB, and several other fumigants were not effective on Rifle peat. Tu[108] observed that Vorlex increased CO_2 production at lower concentrations, but decreased it at the higher ones. The effects of fumigants (D-D, Vorlex, carbofuran, and dasonit) on the respiration of microorganisms between 15 and 30°C showed that the inhibition period observed in the early stages of incubation was negatively correlated with soil temperature.[62] The results with benomyl are conflicting. Carbon dioxide evolution from soil was only slightly influenced by benomyl even at high concentrations.[109] Even with 200 ppm of benomyl in soil treated with 0.2% cellulose or chitin, maximum CO_2 evolution occurred only a few days later than in the controls, though an increased log phase for cellulose decomposition was found. Hofer et al.[110] observed some increase in CO_2 evolution without added substrate and a small decrease with starch during 3 days incubation with benomyl. Peeples[111] observed that benomyl slightly increased the respiration rate of soil even at a lower dose (10 µg/g of soil), whereas Sinha et al.[20] reported that Bavistin (a degradation product of benomyl) significantly depressed CO_2 production for 40 days and this reduction was linked with the concentration of fungicide initially applied. It was presumed that this difference may be due to a difference in the humus content of the soil which strongly influenced the availability of benzimidazole fungicides in soil.

Radwan,[112] working with thiram, reported that this toxicant depressed CO_2 production and inhibition was directly proportional to the concentration to the toxicant applied initially. He found that after initial depression, the rate of respiration increased rapidly in treated soils, but CO_2 production decreased in the control as the period of experiment increased. This suggests a reduction in the availability of food material necessary for microbial growth and activity. Agnihotri[56] reported that captan at 62.5, 125, and 250 ppm depressed respiration for the first 6 days and no statistically significant differences appeared for the next 6 days. After 36 days, respiration of the captan-treated soils increased significantly over that of the control. The increase in CO_2 production in the latter part of the experiment was probably due to the utilization of decomposition products of captan by microorganisms. Inhibition of CO_2 production by a high concentration of mercury and Dexon has also been observed.[113,114] Helweg[115] observed that the application of 10 and 100 mg iprodione per kilogram slightly reduced CO_2 evolution from soil with added lucerne meal. Sinha[19] also reported inhibition of CO_2 production by a mercury compound, i.e., aretan. All concentrations of aretan adversely affected CO_2 production and this trend continued for 16 days, although changes became nonsignificant from the 12th day. This was probably due to the efficient decomposition of organic matter by mciroorganimss. Maximum production of CO_2 in the control soil occurred on the 4th day and thereafter it decreased, indicating depletion of food material for rapid proliferation of soil microbes. Soils treated with 0.2% Dithane M-45 stimulated soil respiration, but a concentration higher than 0.2% drastically decreased CO_2 output.[57] Captan, methylarsinic sulfide, mercuric chloride, diazoben, and PCNB were tested for their effect on oxygen consumption by soil.[116] Captan, methylarsinic sulfide, and mercuric chloride exerted only a temporary effect on soil respiration, whereas diazoben and PCNB had no measurable effect. The effect of captan on the oxidative decomposition of glucose, esculin, chitin, and tannin was also studied in soil.[117] The primary action of captan was a strong inhibition of initial oxygen uptake in soil treated with glucose, esculin, and chitin. The decomposition of the four model substrates was in no case completely inhibited. As a result, the inhibition was longer when the substrate or its breakdown products were less accessible to microbial attack.

However, even if a toxicant does not depress soil respiration as a whole, it does not necessarily mean that soil microbes have not been affected, as a toxicant may stimulate certain groups of organisms while inhibiting others. In order to assess precise action of a toxicant in soil, individual microbial count appears necessary besides an estimation of rate.

B. Ammonification

Ammonification represents the first stage in the degradation of proteins and other complex nitrogenous compounds of plant and animal origin and is carried out by the purifying bacteria in the soil. It is an essential part of the nitrogen cycle. Bacteria and fungi degrade protein at a rapid speed, while actinomycetes do it slowly.[118] Toxicants that induce physical, chemical, and/or microbiological changes in the soil are liable to affect the population of ammonifiers, as also the final organic compound(s) produced during decomposition of proteinaceous material. It could influence the speed of the process, depending upon the nature and concentration of the fungicide.

Soil fumigants and fungicides may decrease or increase the process of ammonification in soil. EDB, MB, 1,3-D, and metham slightly increased ammonia production from the native organic matter, but temporarily depressed the ammonification of added peptone.[106] Similarly, Munnecke and Ferguson[69] reported that Vapam, MB, and chloropicrin produced adverse effects on soil ammonification. On the other hand, Stark et al.[119] demonstrated that the ammonification process was not inhibited by chloropicrin. Similar results were obtained by Zayed et al.[120] with DD. Jones[121] noted that over a wide range of concentrations, EDB stimulated ammonification. Similar results were obtained by Munnecke and Ferguson[69] for metham, MB, and chloropicrin. Even a tenfold increase in the amount of ammonium nitrogen was obtained by fumigation with chloropicrin. Teuber and Poschenrieder[54] reported that a 20% solution of MIT, when applied alone at 0.5 or 0.25 mℓ in mixture with D-D to subsoil (pH 7), accelerated the process of ammonification, whereas Sinha et al.[19] reported that Vapam at lower concentrations (125 to 500 ppm) stimulated the ammonification process, and its higher concentration (1000 ppm) produced a detrimental effect for 15 days (Figure 1). A significant increase in the ammonification following D-D and Vorlex application has been demonstrated.[62,122]

Like fumigants, fungicides also exercise similar effects on the process of ammonification in soil. Captan, thiram, and a 2.5% (w/w) mercury fungicide (Verdasan) did not affect ammonification when applied at a low rate, but there was a marked increase in ammonium nitrogen with increased fungicide concentration. Balasubramaniyam and Patil[73] reported that even at the lower levels of carboxin application to red sandy loam, ammonification was not inhibited. Agnihotri[13] reported that application of Dexon stimulated the production of ammonia, and the magnitude of ammonia build-up was related to the amount of fungicide applied initially. Working with thiram, Agnihotri[67] observed a similar effect. Dubey and Rodriguez[123] reported that ammonification was inhibited by maneb and dyrene only at high concentrations of active ingredients. Application of aretan produced a detrimental effect on the ammonification process only in the initial stages of the experiment, and thereafter the rate of ammonification was stimulated over that of the control[19] (Figure 2). Contrary to this, higher concentrations of Bavistin and Indar significantly enhanced the ammonification rate throughout the experimental period.[21,98] Helweg[115] reported that application of 100 mg iprodione per kilogram caused accumulation of NH_4^+ in soil.

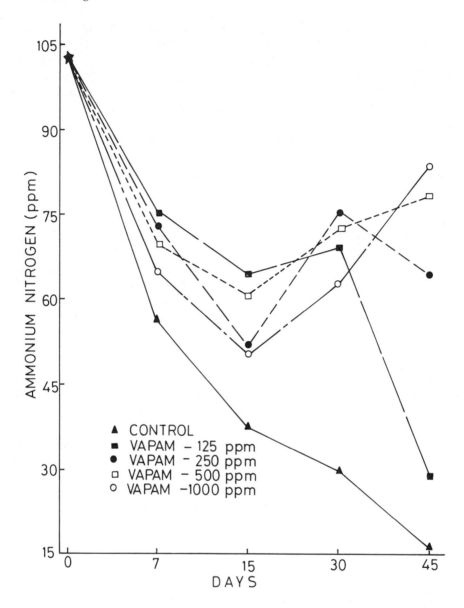

FIGURE 1.   Effect of Vapam on ammonification in soil.

Inhibition of the ammonification is an exception and when it occurs the process is rapidly restored, since the microflora that release ammonia from organic nitrogen compounds quickly recolonize the soil. Accumulation of the ammonical form of nitrogen and its beneficial effect to certain crops have been observed.[124] This is quite expected because the process of ammonification is regulated by different groups of bacteria, mostly spore formers, which proliferate both under aerobic and anaerobic conditions — with concomitant production of ammonia — and are difficult to kill.[1] The build-up of ammonia following application of fungicides is advantageous only to plants which have a liking for ammonical nitrogen, such as rice, pineapple, sugarcane, and coniferous seedlings, but is detrimental to others (which have a preference for nitrate nitrogen). In these cases it would be desirable that the seed be sown after 10 to 15 days of applying chemicals, thereby giving nitrifiers a chance to enhance their population. It is expected that by the time seedlings emerge, part of the accumulated NH_4N gets converted to NO_3-N.

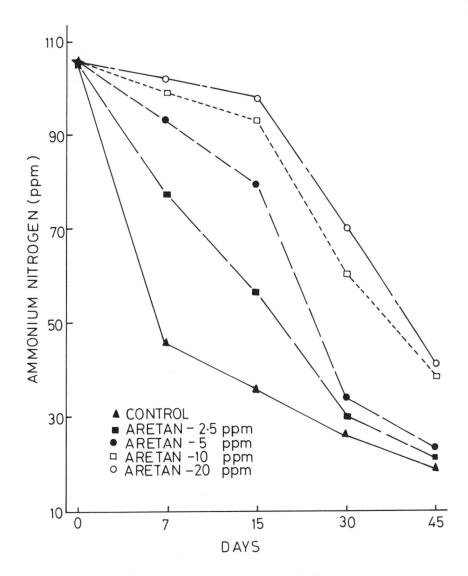

FIGURE 2. Effect of aretan on ammonification in soil.

C. Nitrification

The next stage of the nitrogen cycle is the oxidation of ammonia to nitrate in soil. It is regulated by two distinct types of bacteria: (1) oxidation of ammonia to nitrite by *Nitrosomonas* ($2 NH_3 + 2O_2 = 2 HNO_2 + 2H_2O$) and (2) oxidation of nitrite to nitrate by *Nitrobacter* ($HNO_2 + \frac{1}{2} O_2 = HNO_3$). Much attention has been paid to the effects of various fungicides on these processes, presumably since nitrate accumulation is reasonably easy to observe both in the field and in incubated or perfused soil samples in the laboratory.

Prasad et al.[125] reviewed the nitrification-retarding properties of agricultural chemicals, including both fungicides and fumigants. They noted that fungicides are effective in retarding nitrification even when applied at normal dosage.

Associated with a decrease in nitrification in such treated soils is an increase in the amount of ammonium nitrogen. The effect of dithiocarbamate (DTC) fungicides on nitrification has been studied in greater details as compared to other groups of fungicides because of their availability and efficacy. Jaques and associates[126] reported that nitrification of added ammonium sulfate was arrested by ferbam for 28 days, by manzate for 25 days, and by zineb

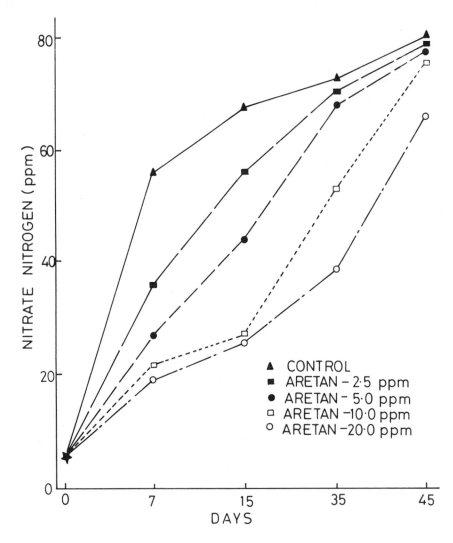

FIGURE 3.   Effect of aretan on nitrification in soil.

for 17 days. According to Caseley and Broadbent,[127] Planston arrested nitrification completely at 10 ppm, Bortan at 100 ppm, while PCNB slightly impaired nitrification at 100 ppm. The particular effect of maneb is to suppress the NH_4-oxidizing *Nitrosomonas*, but not the NO_2-oxidizing *Nitrobacter*.[123]

Mancozeb at 10 mg/kg soil decreased nitrification for 3 months.[96] Agnihotri[56] reported that captan completely suppressed nitrification for 2 to 3 weeks depending upon the concentration of the fungicide. Thiram, at 187.5 and 375 ppm, prevented nitrification for 35 days. However, at 93.7 ppm, nitrification occurred after 3 weeks, but declined sharply after 5 weeks.[67] Sinha et al.[19,21] and Sinha and Singh[98] also observed similar results for Indar, Bavistin, and aretan (Figure 3). Significantly more nitrite and nitrate ammonium was found in benomyl-treated soil than in untreated soil. Hofer et al.,[110] however, found more or less strong inhibition of nitrification in a benomyl-treated humiferous sandy soil. Van Faassen[109] observed that the oxidation of nitrite to nitrate with isolated cultures of *Nitrosomonas* and *Nitrobacter* was more sensitive to benomyl than the oxidation of ammonium to nitrite. As compared to fungicide, fumigants, in general, have a greater inhibitory effect upon nitrification, probably being more powerful biocides; due to high volatility at usual atmospheric

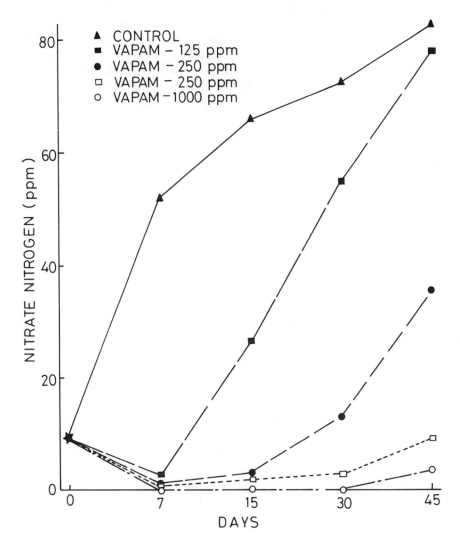

FIGURE 4. Effect of Vapam on nitrification in soil.

temperature, these come in contact with a large portion of the soil-biota sooner after application than a contact poison. In Hawaiian sugarcane soils, Koike[128] found that fumigants such as D-D, Telone, Dowfume W-85, and Vapam impaired nitrification seriously for 4 to 8 weeks; whereas other fumigants like allyl alcohol and Nemagon had less detrimental effect on nitrifying bacteria. Similarly, Sinha et al.[33] observed that Vapam had a strong, initial depressive effect on the nitrification process and the adverse effect increased as its concentration increased in soil (Figure 4). The toxicity of Vapam is thought to be due to production of a highly toxic compound, MIT. Suzuki and Watnabe[129] studied the effect of chloropicrin on nitrification at the point of injection and away from it. They found that nitrification ability was regained after 150 days at the point of injection and after 80 days in soils away from the injection point. Nitrification of ammonium sulfate and ammonium hydroxide was completely inhibited for 30 days by dazomet and nabam.[17] When added 30 days after fungicide treatment the nitrification of the two ammonium compounds was depressed by only 50%.

In a field situation, inhibition of nitrification may be considerably less than had been expected on the basis of laboratory experiments.[86] However, under special conditions, the impairment may be long-lasting, as has been observed by Lebbink and Kolenbrander[130] for soils fumigated with D-D and metham sodium for nematode control. In the Netherlands, this experiment was performed in the autumn. Because of the low soil temperature during winter, the ammonia oxidation had mostly not been resumed before the next April or even June. As has been pointed out by Goring,[131] inhibition of nitrification need not be adverse to soil fertility. It can result in preservation of nitrogen as a plant nutrient. This is because of the accumulation of NH_4^+ which is considerably less prone to leaching than NO_3^+. The contribution of this N-serve effect of the pesticides to plant nutrition has been well illustrated in the field study by Lebbink and Kolenbrander.[130] After fumigation in autumn, leaching of nitrates by winter rains was reduced, because much nitrogen was kept in the ammonia form. Accordingly, the farmers were advised to apply less nitrogen fertilizer in spring. The amount that could be saved was 20 to 40 kg ha^{-1} for sandy and peaty soil, depending on the rainfall during winter and on the fumigants used. Nitrification was less inhibited by metham sodium than by D-D. In some cases, to avoid ammonia injury or nitrate deficiency, it is advisable to use fertilizers containing nitrate nitrogen and not ammonium nitrogen. This does not apply to crops which are tolerant to ammonium nitrogen such as cotton, sugarcane, and pineapple. Overman[132] demonstrated deleterious effects of repeated application (twice a year, in total 11 times) of steam and fumigants (high rate of MB, dazomet, or Vorlex) in connection with depressed nitrification. Growth of chrysanthemums was retarded by accumulation of ammonium nitrogen and, although nematode populations were very small, yields were reduced.

Newly developed promising chemicals must be investigated for their influence on nitrification, as more potent and more specific inhibitors are required to secure the large gains in the efficiency of N fertilizers that are needed in practical agriculture.

D. Denitrification

Another bonus of fungicides with respect to the preservation of nitrogen is their effect on denitrification. Very little information is available regarding the effect of fungicides on denitrification. Bollaf and Henniger[133] estimated the effect of various toxicants on denitrification. Only fungicides, especially maneb, nabam, and captan, substantially inhibited the process. Application of maneb can cause the accumulation of nitrite and so can nabam, whereas captan inhibits denitrification without the accumulation of nitrite. Mitsui et al.[134] also demonstrated inhibition of denitrification by ziram, zineb, and ferbam. Since denitrification essentially causes loss of nitrates as plant nutrients, its inhibition is regarded as a favorable effect.

E. Availability of Nutrient Elements

Several investigators have expressed the opinion that stimulation of plant growth following fumigation of some soils is greater than can be explained on the basis of decreased parasitic activity, or that growth may be increased in the absence of plant root parasites. The phenomenon is commonly called IGR (increased growth response) and several explanations have been proposed. One is that unrecognized root parasites are present in the soil.[135] Early workers suggested that extremely small amounts of a toxic substance may stimulate rather than retard plant growth.[136] The most common view as summarized by Waksman and Starkey[52] is that the effect is due largely to increased release of plant nutrient elements, especially nitrogen, from organic and inorganic soil constituents. Soil microbes, for example, contain relatively large amounts of plant nutrient elements,[137] especially nitrogen, from organic and inorganic soil constituents. Soil microbes contain relatively large amounts of plant nutrient elements, whereas plant residue may contain less than 0.5 to about 3.0% nitrogen, microbial cells up to 5 or even 15%. Many microbial cells also contain more than ten times as much phosphorus as do plant residues. As discussed earlier, the cells of the organisms killed by a pesticidal treatment represent a ready P source of food material for the new population which quickly develops. Nutrient elements in excess of that needed by the organisms are released in available form to the plant.

Nutrient elements may also be released from decomposing fungicide residues. As the fungicides decompose, the simple products of decomposition are released, namely, CO_2, H_2O, NH_3, PO_4, SO_4, Cl, Br, As, etc., and additional simple "compounds" such as SO_4 from oxidation of CS_2Cl from D-D mixture or chloropicrin,[138,139] PO_4 from oxidation of pestox,[140] and NH_3 from cyanamide[141] may result from such decomposition. The acidifying effects of CO_2, SO_4, PO_4, and Cl will also tend to solubilize plant nutrient elements from the inorganic soil fraction. Several workers have measured increases in soluble inorganic constituents of soil following soil fumigation. At the beginning of the century Russell and Hutchinson[142] found that fumigation of the soil with CS_2 increased the phosphorus content of plants. It is an established fact that phosphates added to soil are rapidly immobilized both by polar adsorption on the soil colloids and by incorporation into organic forms through the activities of soil microorganisms. Solubilization of phosphates in soil is brought about by bacteria.[143-147] Fungitoxicants, in addition to attacking the target organism(s), may disrupt the population of phosphate-solubilizing bacteria and fungi as well. In general, fungicides, when applied at recommended rates, do not significantly alter the number of phosphate-regulating organisms, but fumigants, in general, increase soluble phosphate in the soil. Agnihotri[13] reported that application of Dexon caused fluctuations in the P content of soil, but no significant differences in the amount of P occurred between Dexon-treated and untreated soil throughout the period of experiment. In thiram-treated soil, the amount of available phosphorus enhanced significantly as compared to the control by the 7th day and the increased level continued for 35 days.

Clark[63] reported that ethylene oxide treatments substantially increased the bicarbonate-soluble phosphorus content of a soil. Martin et al.[148] observed that Telone and ethylene oxide increased bicarbonate-soluble phosphorus from 12 to 36 ppm in an alkaline soil, but did not influence solubility in an acid soil. Yatazawa et al.[93] reported that allyl alcohol treatment of soil increased both the phosphorus content of the plant as well as the extractable soil phosphorus. Rovira[149] and Rovira and Ridge[150] reported that the application of chloropicrin doubled the amount of bicarbonate extractable phosphate 28 days after fumigation. The increase in available P is likely to be beneficial to seedling growth, phosphorus being one of the essential elements for plant growth. Wainwright and Sowden[151] conducted a detailed experiment to study the influence of benomyl, captan, and thiram on the phosphate solubilization after adding insoluble phosphate. They reported that with phosphate additions of less than 100 μg/g of soil, less phosphorus was extractable from soils treated with fungicides than from untreated soils. At 20 μg P per gram of soil, however, increases in phosphate solubilization occurred.

Based on these reports it can be concluded that fungicide treatment selectively increases the proportions of microorganisms capable of solubilizing both native and added insouble phosphate.

There are considerable data in the literature showing that rhizosphere flora and organic matter in soils can bind manganese, thus making it unavailable to the crop plant.[152,153] There is also little doubt that zinc, iron, molybdenum, and copper are bound and released by soil organisms.[154] Fungicides which kill an appreciable percentage of the soil organisms capable of binding these elements or which are applied in large dosages often cause an increase in extractable salts which did not exist as chemical constituents of the original compound. Earlier reports to support this were of Darbishire and Russell[155] and Russell and Hutchinson[142] who observed that soil treatment with carbon disulfide supplied plants with more P and K than untreated soil. Aldrich and Martin[138] observed that D-D mixture, CS_2, and chloropicrin appreciably increased calcium and slightly increased potassium and magnesium in the saturation extracts of both greenhouse and field soils. In Yolo loam, for example, the saturation extract of a nontreated soil contained 1.7 meq/ℓ of calcium, while soils fumigated with chloropicrin, D-D mixture, and CS_2 contained 3.3, 3.9, and 6.8 meq/ℓ, respectively, 10 days after treatment.The effect on soluble calcium was still present 100 days later. Soluble and exchangeable manganese and to a lesser extent copper and zinc are increased in some soils when treated with fumigants. Changes in the availability of these nutrients can be detected by both plant and soil analyses.[138,148] The manganese content of spruce seedlings was increased from 0.027% in the controls to 0.064% 2 years after soil fumigation with MB.[156] In soils low in a specific micronutrient element, fumigation may stimulate plant growth by increasing the available supply.[157] Generally, increased solubility of manganese was found to occur at the time of treatment,[138] which suggests a direct chemical action of the fumigant on the soil manganese. Sherman and Fujimoto[158] indicated that fumigation of high-manganese Hawaiian soils with up to 400 lb/acre of D-D mixture or chloropicrin reduced manganese toxicity to lettuce, although the treatment may have actually increased soluble and exchangeable manganese. It was also demonstrated that fumigation increased soluble and exchangeable iron. According to Martin,[5] the Mn/Fe ratio was more important than soluble manganese with respect to manganese toxicity in soil. Increase in the solubility of soil Mn, sometimes to toxic levels, following chloropicrin treatment has been recorded by Fujimoto and Sherman,[159] after chloropicrin and formalin treatment by Daldon and Hurwitz.[160] Fungicides are also shown to have inhibitory effects. Pal et al.[161] demonstrated that benomyl at 5 μg/g (field application rate) curtailed the accumulation of iron and manganese in solution, both without and with added rice straw (0.5%). The application of benomyl, captan, thiram, and PCNB generally increased amounts of exchangeable Mn, Na, and Zn. In thiram-treated soils, there were significant increases in Na, K. and Zn. At high rates, benomyl decreased Cu, but increased Zn at the lower rates.[162] Likewise, Ray and Sethunau.   .[163] reported that oxidation of elemental sulfur in an alluvial and a lateretic soil was inhibited by benomyl at concentrations of 5 to 100 μg ai/g soil.

F. Soil Aggregation

This has not been investigated extensively. However, based on the information available, it can be said that the application of fungicides/fumigants to the soil does not measurably influence certain soil physical properties. Microbes, however, do improve soil aggregation during the decomposition of organic residues in the soil. Some species are much more effective than others.[164] It is known that if an effective species becomes dominant in the soil following a fungitoxicant treatment, improved soil aggregation may be ensured. McCalla et al.[165] observed that inoculation of the soil with an effective fungus such as *Stachybotrys atra* after fumigation improved soil binding. Martin and Aldrich[166] demonstrated that fumigation with D-D, chloropicrin, carbon disulfide, or EDB has little or no effect on soil aggregation. Where large amounts of water are used as a vehicle for a fumigant, e.g., formaldehyde, there may be loss of soil structure.[167] Some soil microorganisms are capable of producing compounds which affect soil aggregation. Dexon was observed to reduce the proportions of water-soluble aggregates in soil,[168] but there was a recovery in the aggregate with the disappearance of the fungicide. Therefore, it can be concluded that the influence of biocides on the physical structure of soil is negligible.

IV. EFFECT OF FUNGICIDES ON MICROBIAL INTERACTIONS

In the vast microbial world of the plant and in the soil, the plant pathogens are very much in the minority and, compared to their cohabitants of leaves and roots, most pathogens are weak competitors for the nutrients available. However, they escape from this competition by infecting host plants, the living tissue, and its direct surroundings — their ecological niche. Nevertheless, during certain stages of their life cycle, the pathogens are susceptible to interactions with saprophytic microflora. Through numerous interactions, different microbes in the plant environment are kept in a state of dynamic equilibrium. The use of

Table 1
MICROBIAL INTERACTIONS

Synergistic relations	Antagonistic relations
Provision of nutrients and growth factors by cross-feeding or commensalism	Competition for nutrients, oxygen, and possibly space
Production of probiotic substances	Amensalism by production of antibiotics and simple metabolites, e.g., NH_3, NO_2, C_2H_2
Inactivation of substances including pesticides toxic to the associates	Parasitism and predation

Additionally for plant pathogens

Predisposition of plant roots to infection by associated pathogens	Protection of plant roots from infection

From Bollen, G. J., *Soil-Borne Plant Pathogens,* Schippers, B. and Games, W., Eds., Academic Press, London, 1979. With permission.

fungicides may affect the microbial balance, as fungi constitute a major component. In the natural environment, suppression of sensitive species will soon result in an increase of resistant species competing for the same substrate. In addition to competing for nutrients, other mechanisms underlying the microbial balance, like production of antibiotics and mycoparasitism, may also be affected by fungicides.

As given in Table 1, interactions between species or populations may be synergistic or antagonistic.[169] A neutral relation between organisms in a micro-habitat can hardly be conceived.

A. Effect of Fungicides on Synergistic Relations

A synergistic relation is defined as one in which the growth or reproduction of one or both of the associates is stimulated by the other.[169] Sussman[170] used the term "probiosis" to describe an organism's production of stimulatory substances inducing spore germination and other specific processes in other organisms. The closer the association between the organisms, the greater the effect if one of the members is affected by a toxicant. Therefore, it is suggested that study of the toxicant's effect on soil microflora should be made critically, as it may provide information on the interactions that exist among the various members. Fungitoxicants which are subjected to continuous treatment of field soil may sometimes be degraded by soil microflora either in the process of utilizing the molecules as carbon and energy sources or by cometabolism.[171-173] According to Bollen,[169] when an organism contributes to the decrease in biocides and thus detoxifies the environment for its cohabitants, this is to be regarded as a synergistic interaction. Bollen[169] has cited several examples on the subject, particularly with herbicides. The herbicide paraquat applied as "Gramoxone W" inhibits growth of the cellulose-decomposing fungi *Chaetomium globosum* and *Trichorus* sp.[175] Sensitivity of *Mucor hiemalis* and *Zygorhynchus heterogamus* to paraquat has also been demonstrated.[174] On the other hand, paraquat is degraded by several microorganisms including *Aspergillus niger, Penicillium frequentans, Pseudomonas* sp., and particularly the yeast *Lipomyces*.[169,174] Therefore, a range of microbes detoxifies the soil for *C. globosum* and other sensitive fungi.

Use of combinations of toxicants may imply a potential hazard regarding microbial detoxification. One of the compounds may inhibit the organisms that break down the other. Under this situation persistence of the toxicant will be enhanced and also would extend the effect on soil microbes. Dobolyi et al.[176] demonstrated that the decomposition of 2,4-D sodium was slightly delayed when the herbicide had been used in combination with atrazine. This may have coincided with a shift in the fungal population.

Synergistic interactions that are involved in the etiology of plant diseases are known for associations of various organisms, especially between nematode and fungi. This has already been well demonstrated by Powell[177] when a fungus predisposes a plant to infection by an associated pathogen; application of a fungicide may decrease disease incidence even when a pathogen itself is resistant to the fungicide. Thompson et al.[178] presented sufficient data which showed that reducing the population of *M. javanica* in soil naturally infested with wilt reduced the severity of *Fusarium* wilt-root nematode complex, where *Meloidogyne* spp. predisposed the roots to infection by *Fusarium*; preplanting treatments with the nematicides EDB and DBCP caused highly significant reductions in the severity of *Fusarium* wilt of cowpea when compared to the untreated control. The *Fusarium* wilt value was reduced to 71% for the variety Grant and 41% for Chino-3 by these treatments. Root knot was reduced to 80% on Grant and 86% in Chino-3. Preplanting treatments also induced about threefold increase in yield of dry cowpeas.

Miller and McIntyre[179] demonstrated control of black shank disease of tobacco by soil treatments with oxamyl. They suggested that this might have been a direct action of oxamyl or its breakdown products on the pathogen. It is also known that root-knot nematodes interact with *Phytophthora nicotianae* in causing black shank.[177] Therefore, the possibility of disease control due to suppression of nematodes cannot be ruled out. The authors further mentioned that the chemical was only slightly effective in reducing mycelial growth of the pathogen. Bollen[169] cited in vitro tests of 29 fungi (pathogenic and nonpathogenic) against this chemical and none were found sensitive to the compound. Bollen[169] considered this nematicide-insecticide as an excellent selective agent in the study of interactions between nematodes and fungi because of its low toxicity to the latter.

B. Effect of Fungicides on Antagonistic Relations

It is often difficult to indicate the mechanism involved with regard to the effects of fungicides on antagonistic interactions. It is especially difficult to distinguish between effects on nutrient competition and those of amensalism. In amensalism, one species occupies the substrate by its ability to produce substances that are inhibitory to other organisms. In experiments where the occupation of a substrate is used as a parameter, it is impossible to indicate whether this has been achieved by reduced competition for nutrients by keeping away other organisms from the substrate. Therefore, preference should be given to the broad concept of competition advanced by Clark,[180] in which the role of antibiotics in the occupation of the substrate and nutrient competition is included. Effects of fungitoxicants on the following forms of antagonism have been demonstrated.

1. Antibiosis

Antibiosis is a condition in which one or more metabolites excreted by an organism have a harmful effect on one or more organisms.[181] Antibiotic activity is easy to demonstrate in vitro as compared to its effectiveness in natural soil. Antibiosis has certain advantages over other forms of antagonism. The toxic substances produced may diffuse in water films and water-filled pores through soil, on substrates, or through air-filled pores in the case of a volatile, and thus actual physical contact between antagonists and the subject is not necessary for the effect. Antibiosis works best where nutrients are abundant or in excess. The synthesis of antibiotics is a specific anabolic process. The effect of fungicides on it seems to be at least as specific as the process itself. Fungicides may affect the synthesis of antibiotics by fungi. Stimulation as well as suppression of the process has been reported by several workers. The effect is highly specific and inhibition of growth by a fungicide does not imply that synthesis of antibiotics is also inhibited. Bollen[169] noticed that in dual cultures, when the benomyl-sensitive fungi *Trichoderma harzianum* and *Talaromyces flavus* were grown together with insensitive organisms, big inhibition zones appeared even on plates where growth was completely inhibited by the fungicides.

Ohr and Munnecke[182] reported an example of reduced fungal antibiosis as a result of toxicant treatment. They demonstrated that *Armillaria mellea* produces antibiotic substance(s) toxic to *Trichoderma* spp., and when *A. mellea* is treated with sublethal concentrations of MB this ability is lost. They further mentioned that when growth of *A. mellea* is reduced or temporarily stopped following fumigation, antibiotic production is also reduced or eliminated and *A. mellea* becomes particularly vulnerable to attack to *Trichoderma* spp. as well as other antagonistic fungi. Earlier, Munnecke et al.[183] also obtained similar results, and in postulating how *T. viride* may become antagonistic to *A. mellea* after fumigation, the supposition was made that *A. mellea* must be weakened in the same way. The lag period observed for *A. mellea* following fungicide application may be an indication of such weakening. It is feasible that *T. viride* is able to exploit this period and exert its antagonistic action towards *A. mellea* at that time. Langerak[184] reported stimulation of the synthesis of bacteriostatic metabolites by *Cylindrocarpon destructans* by pimaricin and of fungistatic metabolites by *Penicillium janthinellum* induced by an organic mercuric fungicide.

The effect of herbicides on the synthesis of antibiotics by *Bacillus* spp. has been worked out by several workers, including Balicka and Krezel[185] and Krezel and Leszczynska.[186] The response was specific for the herbicides as well as for the synthesizing organisms. A number of herbicides inhibited bacterial growth but enhanced the production of antibiotics, while other herbicides did not affect bacterial growth but entirely suppressed antibiotic activity. Therefore, it can be concluded that an effect on growth does not imply an effect on antibiosis and vice versa. The effects of herbicides on the antibiotic activity of actinomycetes have been illustrated by Bollen.[169] It is not understood whether fungal antibiosis is affected by herbicides, but fungicides may cause similar effects to those of herbicides on bacterial antibiosis.[169]

2. Competition

Broadly, it is any interaction between two or more species populations which affects their growth or survival.[187,188] According to Clark,[180] in a narrower sense competition is a more or less active demand in excess of the immediate supply of material or condition (space etc.) on the part of two or more organisms; it is the endeavor of two or more organisms to gain the same particular thing or to gain the measure each wants from the supply of a thing when that supply is not sufficient for both.[189,190] Clark[180] concluded that "basically competition among microorganisms is for a substrate in the specific form and under the specific conditions in which that substrate is presented."

According to Domsch,[191] in the competition for nonspecific substrates, suppression of one group of organisms will result in an increase of less sensitive ones and so the "metabolic integrity" of the soil will be preserved. Reduced competition for nutrients may be the main reason for an increase of tolerant pathogens after the use of selective biostatic compounds.[169] This was probably the mechanism for the increase of *Fursarium* spp. and bacteria when quintozene was added to glucose-amended soil.[192] They reported that when PCNB was added to soil, a significant increase in seedlings damaged by fungal pathogens insensitive to PCNB often occurs. They further concluded that PCNB decreases competition for nutrients in soil by suppressing PCNB-sensitive microorganisms (actinomycetes and certain fungi) thereby permitting utilization of substrates by the tolerant fungal pathogens. Katan and Lockwood[193] observed that in soil amended with alfalfa residues, quintozene enhanced *Fusarium* spp. and *Pythium ultimum*. They reported that the proportion of fungi which was tolerant to PCNB was greater in residues recovered from soil containing PCNB than from soil without PCNB. Populations of *P. ultimum* and *Fusarium* spp. colonizing alfalfa particles increased in the presence of PCNB in soil, while those of *Rhizopus stolonifer, Penicillium oxalicum,* and *Streptomyces* decreased. These changes were correlated with the sensitivity of those microorganisms to PCNB in agar. Alfalfa hay accumulated PCNB to levels 7- to 11-fold higher than the concentration in soil. In sterile soil infested with pairs of microoganisms differing in sensitivity to PCNB, populations of the more sensitive microorganism colonizing alfalfa particles decreased and those of the more tolerant microbes increased. They further mentioned that the selective saprophytic increase in such tolerant fungal pathogens as *Pythium* and *Fusarium* may be significant in terms of increased inoculum potential. The presence of plant residues appears to enhance the effectiveness of PCNB in bringing about alterations in microbial populations by providing a site for microbial colonization and activity, and for concentration of PCNB. The accumulation of PCNB by plant debris may increase its chances of coming into direct contact, in effective concentration, with sensitive microorganisms in loci where such microbes are more likely to be abundant and in vegetative rather than in a resistant resting condition. Microbial interactions with such toxicants as PCNB might best be sought in organic microhabitats rather in the soil mass.

Several workers indicated that *S. rolfsii* in peanut is more severe when defoliation provides an organic food base.[194,195] vander Hoeven and Bollen[196] noticed an increase of sharp eye spot caused by *Rhizoctonia cerealis* after spraying of crop with the fungicide benomyl. Reinecke[197] attributed the incidence of sharp eye spot in benomyl-treated cereals to a reduced competition for host tissue at the culm base. This conclusion was based on his extensive study on the causal organisms of the foot-rot disease complex. In field trials, the numbers of culms with sharp eye-spot lesions were inversely correlated with those of culms with eye-spot lesions by *Pseudocercosporella herpotrichoides*. Competition between the pathogens is involved, as has been shown in experiments when one or both pathogens were suppressed by a chemical or the reverse was promoted by inoculation.[197] Inoculation of wheat with *R. cerealis* resulted in decrease of symptoms caused by *P. herpotrichoides*. Conversely, in-oculation of wheat with *P. herpotrichoides* suppressed incidence of sharp eye spot (as cited by vander Hoeven and Bollen).[196] In field trials, where treatment of rye with benodanil resulted in a decrease of sharp eye spot, the number of culm with eye spot caused by *P. herpotrichoides* had increased.[197] In conclusion, incidence of *R. cerealis* seems to be highly hampered by the antagonistic activity of the mciroflora present on and around the culm base, where saprophytes as well as pathogens can operate as antagonists. Suppression of either components of the microflora by environmental conditions or chemicals enhances the chance of infection by the pathogen.[196]

In paraquat-treated straw, the cellulolytic fungus *Chaetomium globosum* was replaced by tolerant species of *Fusarium*, although these fungi are less effective in degrading the straw.[175] The effect of a toxicant will not be discernible in soils which are devoid of easily available nutrients, as Farley and Lockwood failed to demonstrate effects of PCNB in unamended soil. Bollen[169] clearly demonstrated that addition of benomyl to an unamended potting soil did not affect the mycoflora, whereas in soil amended with easily utilizable carbon and nitrogen sources, some of the resistant fungi, such as Mucoraceae, increased. *Trichoderma* species, a group against which no benomyl resistance has as yet been reported, could only successfully compete for the extra substrate in the absence of the toxicant. Bollen[169] has attributed the above phenomenon only to competitions for nutrients. This is doubtful since antibiosis could also be involved. However, this has been very well discussed by Clark.[190]

3. Soil Mycostatsis

The inability of spores to germinate in soil (fungistasis) due to insufficient nutrients is known as mycostasis. Since the phenomenon of soil mycostatics as described by Baker and Snyder,[198] Hora and Baker,[199] Lockwood,[200] Smith,[201] and Watson and Ford[202] is directly or indirectly influenced by nutrient decline of microbial origin, mycostasis is likely to be affected by biocidal treatments. Biocides killed the majority of the soil microorganisms and therefore made the soil rich in nutrients. It can be expected that a decrease in mycostasis may occur in fumigated soil. Kutzera and Hoffmann[203] found only a slight reduction or even a transitory increase of mycostasis following MB treatment in greenhouse soil. A very sensitive test fungus *Phialophora cinerescens* was used in this experiment. They ascribed the mycostatic activity to bacteria, as MB treatment almost eliminated fungi and actinomycetes, but not bacteria.

Increased production of mycostatic compounds may occur in fumigant- and fungicide-treated soils.[204] With captan treatment, ethylene production was enhanced.

4. Specific Antagonism

There are many documented examples of soils that suppress root pathogens, but probably none has proved more fascinating for studies of natural biological control than soils suppressive to the wheat take-all fungus (*Gaeumannomyces graminis* var. *tritici*). Soils can become suppressive to *G. graminis* with various treatments.[205] There is the "take-all decline" phenomenon in which the suppression develops with 2 or 3 years of wheat monoculture and severe take-all; this soil becomes immune to subsequent outbreaks of take-all if cropped exclusively thereafter to wheat or barley.[206]

Bruehl,[207] Gerlagh,[208] Pope,[209] and Vojinovic[210] distinguished between general antagonism to *G. graminis* present in many soils "which moderates the potential pathogenicity of the fungus" and a specific antagonism developed during monoculture. The general antagonism was thought to be due to fungi and specific to bacteria.[209] Bacterial populations in "take-all decline" soil exceeded those in nondecline soil by 60 to 75%, actinomycetes by 12%, but there was no change in fungus population. Bacteria and actinomycetes antagonistic to *G. graminis* were three times more abundant in "take-all decline" than in nondecline soil. In this context, it can be mentioned that specific antagonism in soil is eliminated by fumigation with MB and CP.[209-211] On the contrary, general antagonism survives these treatments.[205] In field trial[149,212] *G. graminis* recolonized soil and colonized the roots of wheat growing in soil fumigated with MB or chloropicrin, but in the MB plots "take-all" was severe, while wheat in the chloropicrin plots showed no above-ground symptoms. The explanation offered is that for 4 months following chloropicrin application, the population of fluorescent *Pseudomonas* was from 500- to 1000-fold greater in the chloropicrin-fumigated plots than in the soil treated with MB, and that the populations of these antagonistic bacteria suppressed *G. graminis* on the roots. According to Cook and Rovira,[205] although the evidence strongly favors the fluorescent *Pseudomonas* as being responsible for the specific suppression of take-all, it was indicated that it is necessary for specific strains to be present in order to obtain suppression of the disease under either natural conditions or in the pot bioassay. Ridge[213] also found that the population of fluorescent *Pseudomonas* spp. in chloropicrin-treated soils exceeded by far that of the MB-treated plots.

5. Parasitism and Predation

These are the most direct microbial interactions. Nematodes are targets of various bacteria and fungi. Fungi which capture and consume nematodes are one of the most unusual and interesting groups of organisms to be found in the soil. They are widely distributed and especially abundant in surface litter and decaying organic matter. The traps by which nematophagous fungi capture their prey and have been described in detail by Drechsler,[214]

Duddington,[215] and Barron.[216] It is worthwhile to mention here that several meticulous studies have been devoted to parasitism of fungi on nematodes, but data on indirect effects of fungicides on nematode populations are scarce. Bollen[169] cited an increase of tylenchids, especially *Paratylenchus* after treatments of grass fields with prothiocarb and Captafol. The increase was doubled by both the treatments. It was concluded that these effects may be due to the suppression of nematode-destroying fungi. Bollen[169] rightly pointed out that conclusive evidence for inhibition of the nematode parasites will only be obtained when the effect of the toxicants is estimated, not only on the population of the nematodes, but also on the population of the nematode-destroying fungi.

Briggs et al.[217] observed an interesting example of differential pesticide sensitivity between a parasite and the host microbe. Of the 17 herbicides tested in vitro only 11 inhibited the growth of the obligate bacterium parasite *Bdellovibrio bacteriovorus*, while only one herbicide affected the *Pseudomonas* host.

V. EFFECTS OF FUNGICIDES ON DISEASE INCIDENCE THROUGH MICROBIAL INTERACTIONS

The most spectacular indirect evidence of soil-ecosystem alteration by treatment is based on plant response, and for the most part is phytopathological in nature. Bollen[169] critically analyzed and presented possible effects on antagonistic relationships. The indirect effect of toxicants on disease incidence may be positive or negative. Pesticides sometimes selectively enhance the antagonistic microbes or antagonism which may lead to indirect control or integrated control. However, selective inhibition may result in a change of dominant pathogens or cause boomerang effects.

A. Integrated and Indirect Control

In recent years, the increasing use of potentially hazardous fungicides in agriculture has been the subject of growing concern of both environmentalists and public health authorities. The possibilities of controlling plant pathogenic fungi by the integration of several methods have been the subject of extensive research. An integrated control which denotes the rational use of all available control measures should be considered, especially with a crop which is attacked simultaneously by numerous types and kinds of pathogens. It does offer the possibility of making up for the deficiencies of any single method. Integration of chemicals and a biotic agent for managing soilborne diseases has been considered as a novel approach, as it requires low amounts of chemicals, thereby reducing the cost of control as well as pollution hazards, with minimal interference of biological equilibrium.[218] However, the development of field technology for efficient and reliable integrated management of soilborne plant pathogens is lagging behind that for management of insects.[218]

A number of toxicants induce an increase in antagonistic activity of the microflora to pathogens. In these cases the ultimate disease control is brought about by direct inhibition of the pathogens as well as by an increased microbial antagonism. Therefore, it is generally indicated as integrated control.[219] A prerequisite for such control is that either the population of antagonists should be less sensitive to the fungicide than the pathogen or that the antagonists should be more successful than the pathogen in their colonization of plant surfaces or soil after treatment of the crop.[169] Since the microflora plays as essential role in this form of integrated control, the ultimate result of fungicide application is highly dependent on the biotic environment. An indication for the operation of integrated control is obtained when higher doses/rates are required to kill the pathogen in sterilized soil than in natural soil. This has been shown for the well-known effect of CS_2 and MB on the control of *Armillaria mellea* in orchards. This is a classical example of combined soil fumigation and biological control and may be the first instance of this type of integrated control in plant pathology. Effectiveness apparently comes from killing the more exposed mycelia and rhizomorphs by carbon disulfide weakening the rest in some undetermined manner, which enables *Trichoderma viride* to kill them. Bliss[40] found that *A. mellea* in roots was not killed immediately after soil fumigation with carbon disulfide at moderate dosage, but was killed in 24 days. The pathogen survived at least 6 years in citrus roots (2.5 cm in diameter) in nontreated soil that contained *Trichoderma*. Carbon disulfide did not kill the pathogen in roots in sterilized soil, but did when *T. viride* was present. Bliss[40] concluded that carbon disulfide could kill *Armillaria* at a high dosage, but that at a lower dosage *Trichoderma* was the effective agent. Several workers[220,221] supported Bliss's findings and concluded that the fumigant weakened *Armillaria* and that *Trichoderma* killed what the fumigant missed.

With respect to these observations, the approach to chemical control of soilborne pathogens should be as has been formulated by Baker and Cook:[219] "Rather than kill the pathogen, it may only be necessary to weaken it and make it more vulnerable to antagonism of the associated microflora." Ohr and Munnecke[222] clearly demonstrated such a weakening of a defense mechanism for *A. mellea* after treatment with sublethal doses of MB. These workers reported that antibiotic production of the pathogen decreased so that *Armillaria* failed in keeping off its antagonists. An increase of microbial antagonism as a positive side effect has been observed for fungicides of different groups, such as organomercury, DTC, benzimidazoles, quintozene, and also the antibiotic fungicide pimaricin. It is not surprising that most of the examples concern soilborne pathogens, since the role of a biotic environment is thought to be more important for pathogens in the soil than on the aerial parts of plants.

There are several examples where fungicides controlled a disease even though the material or rates were not considered directly inhibitory to the pathogen. McKeen[223] demonstrated, for example, that control of *Aphanomyces cochlioides* and *Pythium aphanidermatum* of sugar beets with thiram was partially the result of direct fungicidal action on the pathogen, but he thought that an indirect effect caused by a shift in the microbiological balance was also involved. Domsch[224] also obtained control of damping-off with captan at rates not considered directly fungicidal to the pathogen and suggested that biological control initiated by the chemical was involved. In a study on the role of antagonists in the chemical control of *Fusarium* diseases of *Narcissus* bulbs and roots, Langerak[184] demonstrated substantial evidence that increased populations of *T. viride* and *Penicillium janthinellum* markedly contributed to the ultimate disease control. He reported that after treatment of bulbs with aretan, pimaricin, and thiram, the newly developing roots were more densely colonized by *T. viride* and *P. janthinellum* during the entire growing season. These fungi were less sensitive to these fungicides than the *Fusarium* spp., and thus more or less selectively favored by the treatment. On agar plates several of these nonpathogenic rhizospheric fungi were antagonistic to the pathogenic *Fusarium* spp. Furthermore, contamination of bulbs with spore suspensions of these antagonists did provide some protection against infection by the pathogens. He further concluded that the enhanced antagonism might not have been due to increased populations of antagonists only. Aretan and thiram increased the production of antibiotics by *P. janthinellum*. These results provide the same mechanisms as have been suggested by Chinn[225] to operate in the effect of Panogen on the interaction between *Penicillium* sp. and *Cochliobolus sativus*. Although *Helminthosporium sativum* conidia were not affected directly by 5 ppm of methyl mercury dicyandiamide (MMDD), this rate controlled root rot of wheat caused by the fungus.[226] Certain microorganisms, especially *Penicillium* spp., were observed to enhance luxuriantly in the treated soil and were suggested as an explanation for the disease control obtained. It can be said that in integrated control a number of mechanisms may operate simultaneously. This has well demonstrated by Chinn[226] and Langerak.[184] The ultimate disease control was achieved by four distinct effects: a direct effect of the fungicide on the pathogen, stimulation of production of fungistatic substances by *Penicillium* spp., a significant increase in the populations of antagonists (*Penicillum* spp. and *Trichoderma* spp.), and a cross-protection mechanism. Cross protection was assumed since the subcrown internodes of wheat seedlings were more densely colonized by *Penicillium* spp. in treated plots. In daffodils, *T. viride* and *P. janthinellum* became more established on roots of bulbs treated with aretan, thiram, or pimaricin than roots of untreated bulbs. Richardson[39] observed that *Pythium ultimum* was difficult to establish in soil treated with thiram even after the fungicide was largely gone 9 weeks after treatment, and even though residual amounts of the compound were low and probably sublethal, they delayed or prevented natural build-up of the fungus in flats recropped to peas.

Vaartaja et al.[227] emphasized that for many soil fungicides, the significant control is probably in part through changes in antagonistic flora, without which the fungicide could be only partially effective. Probably the greatest role of fungicides in biological control will involve their effect on the relative competitive advantage of pathogens vs. antagonists. Many instances of control with sublethal doses of a fungicide undoubtedly involve this mechanism. The potential for selective inhibition or stimulation of organisms or groups of organisms with chemicals clearly exists, but awaits man's ingenuity to ascertain which compounds or rates to use. Clues on the indirect benefits of biocides have been found more or less by accident in tests where direct kill of the pathogen by the chemical was the objective. Far more progress would be made if researches were directed specifically at indirect control, enlisting the aid of antagonistic microorganisms. In most instances where the microflora has contributed to effective disease control by a fungicide, that supplementary role was recognized only afterwards.

Integration of chemical(s) and a biotic agent, i.e., *Trichoderma* spp., has been the subject of research during recent years. The desirable effects of combining a small amount of chemicals with a biocontrol agent have been shown by a few workers. However, few attempts have been made to integrate biological and chemical methods of control measures. Integration of biological and chemical control seems to be a very promising way of controlling pathogens with a minimal interference with biological equilibrium.[218] Curl et al.[228] observed that ineffective amounts (1 to 2 µg/g soil) of PCNB, applied together with *T. harzianum*, controlled *Rhizoctonia solani* more effectively than did *T. harzianum* alone in cotton seedling disease in the greenhouse. Similarly, Henis et al.[229] obtained greenhouse control of *R. solani* damping-off of radish by integration of PCNB (4 µg/g soil) and *T. harzianum*. There are a few other reports indicating that PCNB treatments, ineffective alone, give improved disease control when applied with *T. harzianum* in controlling *R. solani* on eggplants, bean, and tomato[230] and *Streptomyces rolfsii* on sugarbeet[231] and bean.[232,233] Lewis and Papavizas[234] have reported field control of fruit rot of cucumber caused by *R. solani* by integration of chlorothalonils with *T. harzianum* and cultural practices. Lewis and Papavizas[234] obtained field control of root rot of cucumbers and crown rot of pepper by integration of chlorothalonils and metalaxyl, respectively, with *T. harzianum*. It has been found that conidia of the antagonist do not survive in the rhizosphere of pea and bean seedlings, when given as a seed coating or soil drench, but treating the seed with metalaxyl before coating with conidia of the fungus improves their survival in the rhizosphere, possibly by reducing soil soprophytes that compete with *T. harzianum* for food base.[235] Chandra[236] reported that integration of both chemical and biological control measures has a synergistic effect on the control of damping-off in sugar beet. Seed treatment with metalaxyl alone at 0.01% was not very effective, but when used with 10.5 g m^{-1} of *T. harzianum* preparation, it provided enhanced control of damping-off of sugar beet. Therefore, the present investigation and the earlier report on control of *S. rolfsii* on sugar beet by Mukhopadhyay and Upadhyay[231] have opened up the possibility of controlling two major seedling pathogens of sugar beet by integration of biological and chemical control measures. Mukhopadhyay et al.[237] and Mukhopadhyay and Chaturvedi[238] also obtained successful control of damping-off of tobacco and eggplants by the application of *Trichoderma* preparation to soil and integrating it with metalaxyl seed treatment. Carter and Price[239] reported integrated control of *Eutypa armeniace*, an airborne, vascular pathogen of apricot. Upon invading wounds after pruning, the pathogen is antagonized by *F. lateritium*. Benomyl effectively controlled the disease, and *F. lateritium* was less sensitive to the fungicide than the pathogen. When a conidial suspension of the antagonist was added to the fungicide it colonized the wounded tissue, resulting in a control supplementary to that of the fungicide.

Examples of effective disease control due to effects on microbial interactions only are limited. The most important examples of indirect control of fungal pathogens in the field are side effects of nematicides on root-infecting fungi. These are known as complex diseases, where nematodes predispose roots to infection by *Fusarium* and other pathogens.

An interesting example of disease control of a tolerant pathogen is that of *Pythium debaryanum* on cucumber seedlings treated with an experimental fungicide (6-azauracil). Stankova-Opocenska and Dekker[240] reported that treatment of seeds with this chemical at a lower dose resulted in a significant increase in the number of bacteria in the rhizoshpere. It was concluded that the enhanced bacterial population protected the roots from infection by *P. debaryanum*. Joyner and Couch[241] observed reduced severity of Pythium blight and red leaf spot caused by *Helminthosporium erythrospilum* in *Agrostis palustris,* one of the grasses in turf treated with benzimidazole fungicides.

B. Dominance Change of Pathogens (Disease Trading)

Kreutzer[242] described it as "a situation in which a dominant pathogen is controlled by soil treatment, and a minor pathogen is elevated to major importance, thus becoming the new dominant pathogen." He termed it as "disease trading". The effect is due to differential fungicide sensitivity between the pathogens themselves or between the pathogen and their specific antagonists.[169] Baker and Cook[219] mentioned that where the chemical selectively inhibits antagonists more than the pathogen, or where competing pathogens are differentially affected by the chemical, the pathogen least affected may cause considerably more damage than without the fungicide. This side effect is known for Dexon and quintozene. One well-known example is the increase in damping-off of seedlings of several crops by *R. solani* when Dexon is used to control *Pythium* spp. or the increased seedling disease from *Pythium* and *Fusarium* spp. when PCNB is used to control *R. solani*.[243] In still other instances, control of both *R. solani* and *Pythium* has increased seedling damage caused by *Fusarium*.

Widespread use of benzimidazole compounds may favor the growth of tolerant fungi. Such fungi would either survive treatment with the chemical(s) or be the primary colonizer after treatment. Since some important pathogens are tolerant to benzimidazole fungicides, disease caused by them may increase in incidence after use of these compounds. There are a few reports indicating the appearance of rare or unknown pathogens following chemical application on the target pathogen. Smith et al.[244] reported the development of a brown patch, caused by a Basidiomycetes on turf that had been sprayed with benomyl to control *Sclerotinia homoeocarpa*. This basidiomycetes tolerated benomyl at 25 $\mu g/m\ell$ and was stimulated by concentrations of 0.5 to 1.0 $\mu g/m\ell$. Similarly, Fletcher[245] observed development of a less sensitive fungus (*Mortierella bainieri*) on commercial mushrooms where benomyl was applied to combat the common fungal pathogens.

It is an established fact that benzimidazole compounds do not have any deleterious effects on "pythiaceous fungi". However, an increase of diseases caused by these pathogens after the use of the fungicide to control other pathogens has been reported. Warren et al.[246] observed an increase in the severity of *Pythium* blight on greenhouse-grown 'Penn' cross bent grass treated with benomyl at an equivalent of 23 mg active ingredients per square meter. Williams and Ayanaba[247] showed that during a 3-year field trial with benomyl, the incidence of cowpea wet stem rot caused by *P. aphanidermatum* was significantly greater in plots treated with benzimidazole fungicides than in plots treated with nonbenzimidazole chemicals or untreated plots. Similar results have been reported by Dekker.[248]

The exacerbation of diseases caused by tolerant pathogens may not merely be attributed to a selective inhibition of the antagonistic microflora. Isolated strains of *Botrytis cinerea* resistant to benomyl were highly virulent[249] and spread readily in unsprayed plants.[250] Swinburne[251] concluded that the disease increase in benomyl-treated plants might be indirectly related to virulence of these isolates. He further observed that benomyl reduced enzyme activities in barley seedlings and pointed out that inhibition of normal host-resistant mechanisms may also be involved. Bollen[169] rightly pointed out that in order to ascertain whether an effect is attributable to suppression of microbial antagonism in soil, the effect on the population of pathogens should also be worked out, which should not only be done by disease rating, but also in the absence of a host. Still another mechanism operates through a change of the microclimate in the crop, for instance, brought about by a denser canopy, because of control of foliar pathogens. Backman et al.[252] observed that the control of peanut leaf spot caused by *Cercosposa arachidicola* and *Cercosporidium personatum* was associated with an enhanced damage cuased by *Sclerotium rolfsii*. Control of leaf spot by spraying benzimidazole fungicides and chlorothalonil promoted a dense foliage. In the subcanopy environment, a humid atmosphere is conducive to infection by *S. rolfsii*. They pointed out that, in addition to this side effect, the fungicide suppressed the population of *T. viride*, a natural antagonist of *S. rolfsii*. The effect was more pronounced in benomyl than in chlorothalonil-treated plots, probably because *S. rolfsii* is highly resistant to benzimidazoles. The change in microclimates is also responsible for the increase of severity of blight caused by *Sclerotinia minor* in peanuts treated with captafol and chlorothalonil.[253] However, unlike *S. rolfsii* this pathogen was controlled by a high rate of benomyl. In *Sclerotinia*-infested fields, peanut yield was significantly lower in captafol- and chlorothalonil-treated plots than in the untreated and benomyl-treated plots. Fokkema et al.[254] reviewed several reports concerning a decreased antagonism caused by treatment with benomyl for some tolerant aerial pathogens. He mentioned that increased infection of *Chochliobolus sativus* after benomyl treatment is due to the reduction of saprophytic mycoflora to a level which is no longer antagonistic. Therefore, while using fungicides it is necessary to be aware of susceptibility of a naturally occurring biological control mechanism which may come to light only after disturbance of the biological balance.

Increased severity of nontarget diseases with the use of benzimidazole fungicides has been reported by several workers. Jackson[255] has noted that in studies directed at the control of stripe smut on Kentucky blue grass with benomyl, there was a resultant increase in the severity of melting out (caused by *Helminthosporium vagans*). An increase in *Alternaria* decay in pears[255] and in *Alternaria* leaf spot[187] has been noted when benzimidazole fungicides were used. Joyner and Couch[241] conducted field trials using four benzimidazole chemicals. They reported that blight of *Lolium perenne* caused by *P. ultimum* was not affected, but that infection of *Agrostis palustris* was reduced, demonstrating the specificity in the indirect effect of benzimidazole fungicides.

A change in pathogen dominance may also occur between pathogens that only slightly differ in their sensitivity to the fungicide. It was demonstrated by the incidence of sharp eye spot caused by *Rhizoctonia cerealis* in benomyl-treated cereals, where *Pseudocercosporella herpotrichoides* and *Fusarium* species were suppressed.[169] In vitro, the isolates of *R. cerealis* were not resistant to the fungicide. They were, however, less sensitive than isolates of *F. culmorum* and *P. herpotrichoides*. Reinecke[197] mentioned the effect of the fungicide on the competition between *P. herpotrichoides* and *R. cerealis* during infection of the culm base.

C. Boomerang Effect

Kreutzer[242] defined it as "the disappearance of the dominant pathogen after treatment, which is soon followed by its reappearance in even greater quantities."

Unfavorable results of soil treatment with toxicants are those of disease accentuation and disease exchange. Collectively, these have been called the boomerang phenomenon.[258] Accentuation of disease can occur as a result of soil treatment for the control of soil inhabitants. Kreutzer observed rapid reinvasion of soils treated with chloropicrin and dibromopropene by *R. solani* and species of *Pythium*. Gibson[259] stated that disease accentuation from damping-off organisms resulted following the use of organic mercurials. Such effects are undoubtedly the result of the killing or inhibition of antagonists, permitting the reinvading pathogen to grow speedily through the soil without biological opposition. Smith[25] found that *T. viride* increased in soil fumigated with chloropicrin. If, however, *P. ultimum* was introduced into the treated soil, it gave greater damage than in untreated soils. Smith concluded that a likely reason for this effect was not only the direct removal of *Pythium* antagonists by the chemical, but the inhibitory effect of *T. viride* on surviving antagonists such as *Rhizoctonia*. Butler[260] has shown that *R. solani* parasitizes species of *Pythium*. When both fungi were added simultaneously to steamed soil, *R. solani* had a definite antagonistic effect on *P. ultimum*.

In experiments with herbicides, Altman and Ross[261] have shown that an effect analogous to the boomerang effect can occur. Sugar beets grown in preplant Tillam- and Pyramin-treated fields were predisposed to more disease than comparable beets grown in nontreated soil. These authors suggested that the herbicide affected the host plant by maintaining it in a juvenescent state for a prolonged period (7 to 14 days) after emergence. These beets were then more susceptible to extremely low levels of *Rhizoctonia* inocula present in the field.

A special type of boomerang effect has been noticed during recent years. Accentuation of diseases appeared in fungicide-treated crops, where resistant strains appeared in the population of the pathogens. If resistance emerges in a treated crop the pathogen population may be either unaffected or increased by inhibition of natural antagonists. The appearance of resistance has been noticed especially with the single-site inhibitors, particularly with a number of systemic fungicides. There are several reports indicating enhanced disease incidence caused by resistant strains in benzimidazole fungicide-treated crops. *Penicillium* rot of lily bulbs,[262] collar spot on turf grass caused by *Sclerotinia homeocarpa*,[262] and dry bubble of commercial mushrooms caused by *Verticillium fungicola*[264] are such examples. In these cases, suppression of antagonistic microbes has accounted for increases of disease incidence. However, the possibility of the occurrence of incidently resistant strains that are more virulent than the wild strain should not be overlooked. In the experiment on the incidence of dry bubble in mushrooms, Bollen and VanZaayen[264] noticed that the yield of plots infested with a resistant strain was lower than of those infested with a wild strain. These results suggest that the strain was more virulent than the wild strain.

Judicious application of benzimidazole fungicides and especially their use with other fungicides have made the situation relatively better, and a number of reports on boomerang effects and change in dominance associated with the use of chemicals have gone down. According to Dekker,[265] the occurrence of boomerang effects caused by resistant strains may have decreased, because the use of combinations of fungicides may suppress build-up of a resistant population. Changes of the dominant pathogen under those circumstances are less likely to occur than after the use of a single fungicide, because fewer pathogens will escape the broad-spectrum activity of fungicide combinations.[169]

REFERENCES

1. **Bollen, W. B.,** Interactions between pesticides and soil micro-organisms, *Annu. Rev. Microbiol.,* 15, 69, 1961.
2. **Domsch, K. H.,** Soil fungicides, *Annu. Rev. Phytopathol.,* 2, 293, 1964.
3. **Singh, K. and Agnihotri, V. P.,** Funbicides on soil microbes exosystems and related biochemical transformation, in *Recent Advances in the Biology of Microorganisms,* Bilgrami, K. S. and Vyas, K. M., Eds., Beshen Singh and Mahendra Pal Singh, Dehradun, India, 1980, 27.
4. **Helling, C. S., Kearney, P. C., and Alexander, M.,** Behavior of pesticides in soils, *Adv. Agron.,* 23, 147, 1971.
5. **Martin, J. P.,** Influence of pesticides residues on soil microbiological and chemical properties, *Residue Rev.,* 4, 96, 1963.
6. **Munnecke, D. E.,** Factors affecting the efficacy of fungicides in soil, *Annu. Rev. Phytopathol.,* 10, 375, 1972.
7. **Tu, C. M. and Miles, J. R.,** Interactions between insecticides and soil microbes, *Residue Rev.,* 64, 17, 1976.
8. **Martin, J. P., Bames, R. C., and Erwin, J. O.,** Influence of soil fumigation for citrus replants on the fungal population of soil, *Soil Sci. Soc. Am. Proc.,* 21, 163, 1957.
9. **Rangaswamy, G.,** *Agricultural Microbiology,* Asia Publishing House, Bombay, 1966, 413.
10. **Waksman, S. A.,** *Soil Microbiology,* John Wiley & Sons, New York, 1952, 356.
11. **Warcup, J. H.,** Effect of partial sterilization by steam or formalin on the fungus flora of an old forest nursery soil, *Trans. Br. Mycol. Soc.,* 34, 519, 1951.
12. **Weindling, R., Katznelson, H., and Beale, H. P.,** Antibiosis in relation to plant diseases, *Annu. Rev. Microbiol.,* 4, 247, 1950.
13. **Agnihotri, V. P.,** Effect of Dexon on soil microflora and their ammonification and nitrification activities, *Indian J. Exp. Biol.,* 11, 213, 1973.
14. **Alconero, R. and Hagedorn, D. J.,** The persistence of Dexon in soil and its effects on soil microflora, *Phytopathology,* 58, 34, 1968.
15. **Agnihotri, V. P., Singh, K., and Budhraja, T. R.,** Persistence and degradation of vitavax in soil and sugarcane setts and its effect on soil fungi, *Proc. Indian Natl. Sci. Acad.,* 39, 561, 1974.
16. **Edgington, L. V. and Baron, G. L.,** Fungitoxic spectrum of oxathiin compounds, *Phytopathology,* 57, 1256, 1967.
17. **Chandra, P. and Bollen, W. B.,** Effect of nabam and Mylone on nitrification soil respiration and microbial numbers in four Oregon soils, *Soil Sci.,* 92, 387, 1961.
18. **Corden, M. E. and Young, R. A.,** Changes in the soil microflora following fungicide treatments, *Soil Sci.,* 99, 272, 1965.
19. **Sinha, A. P., Agnihotri, V. P., and Singh, K.,** Persistence of aretan in soil and its effect on bioecosystems and their related biochemical activity, *Proc. Indian Natl. Sci. Acad.,* 45, 261, 1979.
20. **Sinha, A. P., Agnihotri, V. P., and Singh, K.,** Persistence of carbendazim in soil and its effect on rhizosphore fungi in sugarbeet seedlings, *Indian Phytopathol.,* 30, 21, 1980.
21. **Sinha, A. P., Agnihotri, V. P., and Singh, K.,** Effect of bavistin on soil microflora, their related physiological activities and growth of sugarbeet seedlings, *Indian J. Exp. Biol.,* 18, 489, 1980.
22. **Oku, H., Oku, K., Shiraishi, T., Sato, K., and Ouchi, S.,** Effect of fungicides, benomyl and thiram on soil microflora and some soil inhabitant fungi, *Sci. Rep. Fac. Agric. Okayame Univ.,* 54, 1, 1979.
23. **Smith, R. N. and Long, P. A.,** The effect of two fungicides, benlate and phenyl mercury acetate, on a population of cellulotytic fungi in soil and in pure culture, *Int. Biodeterior. Bull.,* 16, 119, 1980.
24. **Waksman, S. A. and Starkey, R. L.,** Microbiological analysis of soil as an index of soil fertility. VII. Carbon dioxide evolution, *Soil Sci.,* 17, 141, 1924.
25. **Smith, N. R.,** The partial sterilization of soil by chloropicrin, *Soil Sci. Soc. Am. Proc.,* 3, 188, 1939.
26. **Tam, R. K. and Clark, H. E.,** Effect of chloropicrin and other soil disinfectants on the nitrogen nutrition of the pineapple plant, *Soil Sci.,* 56, 245, 1943.
27. **Wensley, R. N.,** Microbiological studies of some selected soil fumigants, *Can. J. Bot.,* 31, 277, 1953.
28. **McKeen, C. D.,** Methylbromide as a soil fumigant for controlling soil-borne pathogens and certain organisms in vegetables seed beds, *Can. J. Bot.,* 32, 101, 1954.
29. **Katznelson, H. and Richardson, L. T.,** The microorganisms of the rhizosphere of tomato plants in relation to soil sterilization, *Can. J. Res.,* 21, 219, 1943.
30. **Kreutzer, W. A.,** Selective toxicity of chemicals to soil micro-organisms, *Annu. Rev. Phytopathol.,* 1, 101, 1943.
31. **Agnihotri, V. P.,** Solubilization of insoluble phosphates by some soil fungi isolated from nursery seed beds, *Can. J. Microbiol.,* 16, 877, 1949.
32. **Martin, J. P.,** Effect of fumigation and other soil treatments in the green house on the fungus population of old citrus soil, *Soil Sci.,* 69, 107, 1950.

33. **Sinha, A. P., Agnihotri, V. P., and Singh, K.,** Effect of soil fumigation with Vapam on the dynamics of soil microflora and their related biochemical activity, *Plant Soil.,* 53, 89, 1979.
34. **Warcup, J. H.,** Effect of partial sterilization by steam or formaline on damping-off of sitka spruce, *Trans. Br. Mycol. Soc.,* 35, 248, 1952.
35. **Evans, E.,** Survival and recolonization by fungi in soil treated with formalin or carbon disulphide, *Trans. Br. Mycol. Soc.,* 38, 355, 1955.
36. **Garrett, S. D.,** *Biology of Root Infecting Fungi,* Cambridge University Press, London, 1956, 293.
37. **Gregory, K. F., Allen, O. N., Riker, A. J., and Peterson, W. H.,** Antibiotics and antagonistic micro-organisms as control agents against damping-off of alfalfa, *Phytopathology,* 42, 613, 1952.
38. **Weindling, R.,** Studies on a lethal principle effective in the parasitic action of *Trichoderma lignorum* on *Rhizoctonia solani* and other soil fungi, *Phytopathology,* 24, 1153, 1934.
39. **Richardson, L. T.,** The persistence of thiram in soil and its relationship to the microbiological balance and damping-off control, *Can. J. Bot.,* 32, 335, 1954.
40. **Bliss, D. E.,** The destruction of *Armillaria mellea* in citrus soils, *Phytopathology,* 41, 665, 1951.
41. **Darley, E. F. and Wilbur, W. D.,** Some relationship of carbon disulphide and *Trichoderma viride* in the control of *Armillaria mellea, Phytopathology,* 44, 485, 1954.
42. **Warcup, J. H.,** Chemical and biological aspects of soil sterilization, *Soils Fert.,* 20, 1, 1957.
43. **Domsch, K. H.,** Die Prufung von Bondenfungiciden. I. Pilzsubstrat-fungicid-Kombinationen, *Plant Soil,* 10, 114, 1959.
44. **Bomar, M.,** Estimation of efficiency of fungitoxic compounds according to the inhibition of mycelium growth, *Folia Microbiol. Prague,* 7, 185, 1962.
45. **Morgan, O. D.,** Greenhouse and laboratory studies on chemical soil drenches to control *Phytophthora parasitica* var. *nicotianae,* the cause of black shank of tobacco, *Phytopathology,* 49(Abstr.), 525, 1959.
46. **Horsfall, J. G.,** *Principles of Fungicidal Action,* Chronica Botanica, Waltham, Mass., 1956, 279.
47. **Stark, F. L., Jr.,** Investigations of chloropicrin as a soil fumigant, *Cornell Univ. Exp. Stn. Mam.,* 278, 61, 1948.
48. **Tarr, S. A. J.,** Stem canker (*Macrophomina phaseoli*) of cotton seedlings in the Sudan Gezira, *Nature (London),* 178, 935, 1956.
49. **Munnecke, D. E. and Lindgren, D. L.,** Chemical measurements of methyl bromide concentration in relation to kill of fungi and nematodes in nursery soil, *Phytopathology,* 44, 605, 1954.
50. **Davey, A. E. and Leach, L. D.,** Experiments with fungicides for use against *Sclerotium rolfsii* in soils, *Hilgardia,* 13, 523, 1941.
51. **Mattews, A.,** Partial sterilization of soil by antiseptics, *J. Agric. Sci.,* 14, 1, 1924.
52. **Waksman, S. A. and Starkey, R. L.,** Partial sterilization of soil, microbiological activities and soil fertility. I, II, and IV, *Soil Sci.,* 37, 247, 1923.
53. **Eno, C. F.,** Field evaluation of insecticide residues in soils. Effect of soil applications of carbamate fungicides on the soil microflora, *Fla. Agric. Expl. Stn. Rep.,* p. 142, 1957.
54. **Teuber, M. and Poschenrieder, H.,** Investigations on the microflora of cultivated fen soil of nematicides that contain mustard oil, *Bayer. Landwirtsch. Jahrb.,* 41, 350, 1964.
55. **Pugashetty, B. K. and Rangaswami, G.,** Rhizosphere microflora of cotton seedlings as influenced by certain pretreatment of the seed, *Mysore J. Agric. Sci.,* 3, 99, 1969.
56. **Agnihotri, V. P.,** Persistence of captan and its effects on microflora, respiration and nitrification of a forest nursery soil, *Can. J. Microbiol.,* 17, 377, 1971.
57. **Balasubramanian, A., Siddaramappa, R., and Oblisami, G.,** Studies on the effect of biocides on microbiological and chemical properties of soil. I. Effect of simazine and dithane M-45 on soil microflora and certain soil enzymes, *Pesticides,* 7, 13, 1973.
58. **Cram, W. H. and Vaarteya, O.,** Rate and timing of fungicidal soil treatments, *Phytopathology,* 47, 169, 1957.
59. **Domsch, K. H.,** The effects of soil fungicides. III. Quantitative changes in soil microflora, *Z. Pflanzenbau Pflanzenschutz,* 66, 17, 1959.
60. **Audus, J. J., Ed.,** *The Physiology and Biochemistry of Herbicides,* Academic Press, London, 1964, 555.
61. **Nauman, K.,** The effects of some environmental factors on the reaction of soil microflora to pesticides, *Zentralbl. Bakteriol. Parasitenkd. Infektionskr. Skr. Hyg. Abt. 2,* 127, 379, 1972.
62. **Tu, C. M.,** The temperature dependent effect of residual nematicides on the activities of soil micro-organisms, *Can. J. Microbiol.,* 19, 855, 1973.
63. **Clark, F. E.,** Changes induced in soil by ethylene oxide sterilization, *Soil Sci.,* 70, 345, 1950.
64. **Allison, L. E.,** Vapour-phase sterilization of soil with ethylene oxide, *Soil Sci.,* 72, 341, 1951.
65. **Burges, A.,** *Micro-Organisms in the Soil,* Hutchinson, London, 1958, 188.
66. **Hanesn, E. W. and McCalla, T. M.,** Effects of fumigation in a stubble mulch system on the biological and related properties of the soil, *Bacteriol. Proc.,* 58, 70, 1958.
67. **Agnihotri, V. P.,** Thiram induced changes in soil microflora their physiological activities and control of damping-off in chillies (*Capsicum annum*), *Indian J. Exp. Biol.,* 12, 85, 1974.

68. **Klemner, H. W.,** Response of bacterial fungal and nematode populations of Hawaiian soils to fumigation and liming, *Soc. Am. Bacteriol. Proc.,* 57, 12, 1957.

69. **Munnecke, D. E. and Ferguson, J.,** Effect of soil fungicides upon soil-borne plant pathogenic bacteria and soil nitrogen, *Plant Dis. Rep.,* 44, 552, 1960.

70. **Winfree, J. P., Cox, R. S., and Harrison, D. S.,** Influence of bacterial soft rot, depth to water table, source of nitrogen and soil fumigation on production of lettuce in the everglades, *Phytopathology,* 48, 311, 1958.

71. **Millhouse, D. E. and Munnecke, D. E.,** Effects of methyl bromide dosage on microorganisms in soil before and after growth of *Nicotiana glutinosa, Phytopathology,* 71, 418, 1981.

72. **Hoflich, G.,** Effect of biocides on soil microflora and decomposing activity. IV. Influence of substances that inhibit straw decomposition on nitrification, *Zentralbl. Bakteriol. Parasitenkd. Infektionskr. Hyg. Abt.* 2, 132, 155, 1977.

73. **Balasubranyam, R. H. and Patil, R. B.,** Effect of carboxin on soil biological properties, *Indian J. Microbiol.,* 21, 303, 1981.

74. **Jensen, H. L.,** Notes on the biology of *Azotobacter, Proc. Soc. Appl. Bacteriol.,* 14, 89, 1951.

75. **Ampova, G. and Stefanov, D.,** Azotobacter in the tobacco rhizosphere after soil fumigation against root knot nematode, in *Vtori Kongres po Mikrobiologiya IV,* Chast, B'Lgarskata Akademiya na Naukite, 1971, 273.

76. **Nutman, P. S.,** The relation between nodule bacteria and the legume host in the rhizosphere and in the process of infection, in *Ecology of Soil-Borne Pathogens, Prelude to Biological Control,* Baker, K. F. and Snyder, W. C., Eds., University of California Press, Berkeley, 1965, 231.

77. **Audus, L. J.,** Herbicide behaviour in the soil. II. Interactions with soil microorganisms, in *The Physiology and Biochemistry of Herbicides,* Academic Press, London, 1964, 163.

78. **Ruhloff, M. and Burton, J. C.,** Compatibility of *Rhizobia* with seed protectants, *Soil Sci.,* 72, 289, 1951.

79. **Hofer, A. W.,** Selective action of fungicides on *Rhizobium, Soil Sci.,* 86, 282, 1958.

80. **Diatlofe, A.,** The effect of some pesticide on root nodule bacteria and subsequent nodulation, *Aust. J. Exp. Agric. Anim. Husb.,* 10, 562, 1970.

81. **Muthusamy, S.,** Effect of seed dressing and soil fungicides on the growth of *Rhizobium* of groundnut, *Pesticides,* 7, 27, 1973.

82. **Curley, R. L. and Burton, J. C.,** Compatibility of *Rhizobium japonicum* with chemical seed protectants, *Agron. Abstr.,* 132, 1974.

83. **Sardeshpande, J. S., Kulkarni, J. S., and Bagyraj, D. J.,** Influence of fungicides seed treatment on the nodulation of groundnut inoculated with *Rhizobium, Pesticides,* 7, 11, 1973.

84. **Nene, Y. L., Agarwal, V. K., and Srivastava, S. S. L.,** Influence of fungicidal seed treatment on the emergence and nodulation of soybean, *Pesticides,* 3, 26, 1969.

85. **Ostwal, K. P. and Gaur, A. C.,** Effect of seed dressing fungicides and antibiotics aureofungin on soybean inoculated with *Rhizobium japonicum, Hind. Antibiot. Bull.,* 13, 73, 1971.

86. **Mazur, A. R. and Hughes, T. D.,** Nitrogen transformations in soil as affected by the fungicides benomyl, dyrene and maneb, *Agron. J.,* 67, 755, 1975.

87. **Fisher, D. J.,** Effects of some fungicides on *Rhizobium trifolii* and its symbiotic relationship with white clover, *Pestic. Sci.,* 7, 190, 1976.

88. **Libbink, G.,** De invloed van bestrijdingsmiddelen op de functie van het bodemleven, in *Ladbouw en Plantenziekten,* special ed., Bestrijdingsmiddelen en Bodemleven, Rijksconsulentschop voor Plantenziekten, Wageningen, The Netherlands, 1972, 63.

89. **Singh, I. and Prasad, S. K.,** Effect of some nematicides on nematodes and soil micro-organisms, *Indian J. Nematol.,* 3, 109, 1973.

90. **Tewfik, M. S., Embabi, M. S., and Hamdi, Y. A.,** Efficiency of *Rhizobium leguminosarum* as affected by certain herbicides and nematicides, *Zentralbl. Bakteriol. Parasitenkd. Infektionskr. Hyg. Abt.,* 130, 725, 1975.

91. **Alexander, M.,** *Introduction to Soil Microbiology,* John Wiley & Sons, New York, 1967, 472.

92. **Waksman, S. A. and Lacheval, H. A.,** *The Actinomycetes III. The Antibiotics of Actinomycetes,* William & Wilkins, Baltimore, 1962, 430.

93. **Yatazawa, M., Persedsky, D. J., and Wilde, S. A.,** Effect of allylalcohol on microorganisms of prairie soil and growth of tree seedlings, *Soil Sci. Soc. Am. Proc.,* 24, 313, 1960.

94. **Moje, W., Martin, J. P., and Baines, R. C.,** Structural effect of some organic compounds on soil organisms and citrus seedlings grown in an old citrus soil, *J. Agric. Food. Chem.,,* 5, 32, 1957.

95. **Tajawa, H. and Tanaru, R.,** Effect of chloropicrin treatment on the soil microflora, *Bull. Hatano Tob. Exp. Stn.,* 65, 63, 1969.

96. **Doneche, B., Seguin, G., Ribereau-Gayon, P.,** Mancozeb effect on soil microorganisms and its depradation in soil, *Soil Sci.,* 135, 361, 1983.

97. **Bollen, W. B., Morrison, H. E., and Crowell, H. H.,** Effect of field treatments of insecticides on numbers of bacteria, streptomyces and molds in soils, *J. Econ. Entomol.,* 47, 302, 1954.

98. **Sinha, A. P. and Singh, A.,** Effect of Indar on soil microflora and their related nitrification and ammonification activities, *Proc. Indian Natl. Sci. Acad.,* 45, 255, 1979.
99. **Martin, J. P., Baines, R. C., and Erwin, J. O.,** Influence of soil fumigation for citrus replants on the fungus population of the soil, *Soil Sci. Soc. Am. Proc.,* 21, 163, 1957.
100. **Roslycky, E. B.,** Fungicidal activity of vorlex and accumulation of linuron in a vorlex-linuron treated soil, *Can. J. Soil Sci.,* 60, 651, 1980.
101. **Davis, F. A.,** Effects of fungicide treatment of soil on certain components of the soil microflora, *Diss. Abstr.,* 27, 666, 1966.
102. **Russell, E. J. and Appleyard, A.,** The atmosphere of the soil, its composition and cause of variation, *J. Agric. Sci.,* 7, 1, 1915.
103. **Neller, J. R.,** Studies on the correlation between the production of carbon dioxide and the accumulation of ammonia by soil organisms, *Soil Sci.,* 5, 225, 1918.
104. **Grossbard, E.,** Effect on the soil microflora, in *Herbicide, Physiology, Biochemistry, Ecology,* Vol. 2, Audus, J. L., Ed., Academic Press, London, 1976, 99.
105. **Saive, R., Eteve, G., and Brakel, J.,** The effect of four commercial fungicides on the soil microflora, *Rev. Ecol. Biol. Sol.,* 12, 557, 1975.
106. **Rao, D. P.,** Effect of Certain Fumigants on Nitrification and Other Soil Microbial Activities, Master's thesis, Oregon State University, Corvallis, 1959.
107. **Stotzky, G., Martin, W. P., and Mortensen, J. L.,** Certain effects of crop residues and fumigant applications on the decomposition of an Ohio muck soil, *Soil Sci. Soc. Am. Proc.,* 20, 392, 1956.
108. **Tu, C. M.,** Effect of 4 nematicides on activities of microorganisms in soil, *Appl. Microbiol.,* 23, 398, 1972.
109. **Van Faassen, H. G.,** Effect of the fungicide benomyl on some metabolic processes and on numbers of bacteria and actinomycetes in the soil, *Soil Biol. Biochem.,* 6, 131, 1974.
110. **Hofer, J., Beck, T., and Wallnofer, P.,** Effect of the fungicides benomyl on the microflora of the soil, *Z. Pflanzenbau Pflanzenschutz,* 78, 399, 1971.
111. **Peeples, J. L.,** Microbial activity in benomyl treated soils, *Phytopathology,* 64, 857, 1974.
112. **Radwan, M. A.,** Persistence and efffect of TMTD on soil respiration and nitrification in two nursery soils, *For. Sci.,* 11, 152, 1965.
113. **Van Faassen, H. G.,** Effect of mercury compounds on soil microbes, *Plant Soil,* 38, 485, 1973.
114. **Karnath, N. G. K. and Vasantharajan, V. N.,** Persistence and effect of Dexon on soil respiration, *Soil Biol. Biochem.,* 5, 679, 1973.
115. **Helweg, A.,** Influence of the fungicide iprodione on respiration, ammonification and nitrification in soil, *Pedobiologia,* 25, 87, 1983.
116. **Domsch, K. H.,** Der Einflub von fungiziden wirkstoffen auf die Boden atmung, *Phytopathol. Z.,* 49, 291, 1964.
117. **Domsch, K. H.,** Der Einfluss von captan auf den Abbau von glukose, Aesculin, chitin, und Tannin in Boden, *Phytopathol. Z.,* 52, 1, 1965.
118. **Waksman, S. A.,** Microbiological analysis of soil as an index of soil fertility. IV. Ammonia accumulation (ammonification), *Soil Sci.,* 15, 49, 1972.
119. **Stark, F. L., Smith, J. B., and Howard, F. L.,** Effect of chloropicrin fumigation on nitrification and ammonification in soil, *Soil Sci.,* 48, 433, 1939.
120. **Zayed, M. N., Nasser, Abd-El. M., and Malek, Abd-El. Y.,** Effect of D. D. on nitrogen transformation in soil, *Zentralbl. Bakteriol. Parasitenkd. Infektionskr. Hyg. Abt. 2,* 122, 527, 1968.
121. **Jones, L. W.,** Effects of some pesticides on microbial activities of the soil, *Utah Agric. Exp. Stn. Bull.,* 390, 17, 1956.
122. **Ebbels, D. L.,** Effect of soil fumigation on soil nitrogen and on disease incidence in winter wheat, *Ann. Appl. Biol.,* 67, 235, 1971.
123. **Dubey, H. D. and Rodriguez, R. L.,** Effect of dyrene and maneb on nitrification and ammonification and their degradation in tropical soils, *Soil Sci. Soc. Am. Proc.,* 34, 435, 1970.
124. **Sahrawat, K. L.,** Effect of pesticides on nitrification, *Pesticides,* 8, 30, 1974.
125. **Prasad, R., Rajale, G. B., and Lakhdive, B. A.,** Nitrification retarders and slow release nitrogen fertilizers, *Adv. Agron.,* 23, 337, 1971.
126. **Jaques, R. P., Robinson, J. R., and Charee, F. E.,** Effect of thiourea, ethyl urethane and some dithiocarbamate fungicides on nitrification in fox sandy loam, *Can. J. Soil Sci.,* 39, 235, 1959.
127. **Caseley, J. C. and Broadbent, F. E.,** The effect of five fungicides on soil respiration and some nitrogen transformation yolo fine sandy loam, *Bull. Environ. Contam. Toxicol.,* 3, 58, 1968.
128. **Koike, H.,** The effect of fumigants on nitrate production in soil, *Soil Sci. Soc. Am. Proc.,* 25, 204, 1961.
129. **Suzuki, T. and Watnabe, I.,** Re-establishment of nitrifying micro-organisms in field soil after chloropicrin fumigation applied in different ways. Re-establishment of nitrifying micro-organisms after soil sterilization, *J. Soil Sci. Manure (Tokyo),* 37, 579, 1967.

130. **Lebbink, G. and Kolenbrander, G. J.,** Quantitative effect of fumigation with 1,3-dichloropropene mixtures and with metham-sodium on the soil nitrogen status, *Agric. Environ.,* 1, 283, 1973.

131. **Goring, C. A. I.,** *Pesticides in the Soil,* Michigan State University, East Lansing, 1970, 51.

132. **Overman, A. J.,** Use of repeat application of soil fumigants as they affect commercial chrysanthemum production, *Proc. Fla. State Hortic. Soc.,* 81, 432, 1969.

133. **Bollaf, J. M. and Henniger, N. M.,** Influence of pesticides on denitrification in soil and with an isolated bacterium, *J. Environ. Qual.,* 5, 15, 1976.

134. **Mitsui, S., Watanabe, I., Hohma, M., and Honda, S.,** The effect of pesticides on denitrification in paddy soil, *Soil Sci. Plant Nutr.,* 10, 15, 1964.

135. **Newhall, A. G.,** Disinfestation of soil by heat, flooding, and fumigation, *Bot. Rev.,* 21, 189, 1955.

136. **Fred, E. B.,** Uber die Beschleunig ung der Lebenstatigkeit hohrer und niederer Pflanzen durch Kleine Giftmengen, *Zentralbl. Bakteriol. Bodenk.,* 31, 185, 1971.

137. **Waksman, S. A.,** *Principles of Soil Microbiology,* Bailliere, Tindall and Cox, London, 1927, 897.

138. **Aldrich, D. G. and Martin, J. P.,** Effect of fumigation on some chemical properties of soils, *Soil Sci.,* 73, 149, 1952.

139. **Hannesson, H. A., Raynor, R. N., and Crafts, A. S.,** Herbicidal use of carbon disulphide, *Univ. Calif. Agric. Exp. Stn. Bull.,* p. 693, 1945.

140. **Verona, O. and Picci, G.,** Intornoall' azione esercitata dagli insecticidi sistemici sulla mikroflora del terreno, *Agric. Ital.,* 52, 61, 1952.

141. **McCool, M. M.,** Properties and uses of calcium cyanamide, *Contrib. Boyce Thompson Inst. Pl. Res. Prof. Pap.,* 1, 226, 1933.

142. **Russell, E. J. and Hutchinson, H. B.,** The effect of partial sterilization of the soil on the production of plant food, *J. Agric. Sci.,* 3, 111, 1909.

143. **Chhonkar, P. K. and Subba Rao, N. S.,** Phosphate solubilization by fungi associated with legume root nodules, *Can. J. Microbiol.,* 13, 749, 1967.

144. **Das, A. C.,** Utilization of unsoluble phosphates by soil fungi, *J. Indian Soc. Soil Sci.,* 11, 203, 1963.

145. **Gerretsen, F. C.,** The influence of micro-organisms of the phosphate intake by the plant, *Plant Soil,* 1, 51, 1948.

146. **Katznelson, H. and Bose, B.,** Metabolic activity and phosphate dissolving capability of bacterial isolates from wheat roots, rhizosphere and nonrhizosphere soil, *Can. J. Microbiol.,* 5, 79, 1959.

147. **Louw, H. A. and Webley, D. M.,** A study of soil bacteria dissolving certain mineral phosphate fertilizers and related compounds, *J. Appl. Bacteriol.,* 22, 227, 1959.

148. **Martin, J. P., Murphy, W. S., and Brodford, G. R.,** Effect of soil fumigation on growth and chemical composition of citrus plants, *Soil Sci.,* 75, 137, 1953.

149. **Rovira, A. D.,** Studies on soil fumigation. I. Effects of ammonium, nitrate and phosphate in soil and on the growth, nutrition and yield of wheat, *Soil Biol. Biochem.,* 8, 241, 1976.

150. **Rovira, A. D. and Ridge, E. H.,** The effect of methyl bromide and chloropicrin on some chemical and biological properties of soil and on the growth and nitrition of wheat, in *Soil Disinfestation,* Mulder, D., Ed., Elsevier, New York, 1979, 231.

151. **Wainwright, M. and Sowden, F. J.,** Influence of fungicide treatment on $CaCl_2$-extractable phosphorus and phosphate solubilizing micro-organisms on soil, *Plant Soil,* 48, 335, 1977.

152. **Gerretsen, F. C.,** Manganese deficiency of oats and its relation to soil bacteria, *Ann. Bot.,* 1, 207, 1937.

153. **Heintze, S. G. and Mann, P. J. G.,** Studies on soil manganese, *J. Agric. Sci.,* 39, 80, 1949.

154. **Thornton, H. G.,** The development and present problems of soil microbiology, *J. Sci. Food Agric.,* 7, 93, 1956.

155. **Darbishire, F. V. and Russell, E. J.,** Oxidation of soils and its relation to productiveness. II. The influence of partial sterilization, *J. Agric. Sci.,* 2, 305, 1908.

156. **Ingestad, T. and Molin, N.,** Soil disinfection and nutrient status of spruce seedlings, *Physiol. Plant.,* 13, 90, 1960.

157. **Timonin, M. I.,** Microflora of the rhizosphere in relation to the manganese deficiency disease of oats, *Soil Sci. Soc. Am. Proc.,* 11, 284, 1947.

158. **Sherman, G. D. and Fujimoto, C. K.,** The effect of the use of lime, soil fumigants, and mulch on the solubility of manganese in Hawaiian soils, *Soil Sci. Soc. Am. Proc.,* 11, 206, 1946.

159. **Fujimoto, C. K. and Sherman, G. D.,** Manganese availability as influenced by steam sterilization of soils, *J. Am. Soc. Agron.,* 40, 527, 1948.

160. **Daldon, F. H. and Hurwitz, C.,** Effect of volatile disinfectants on survival of microflora in soil, *Soil Sci.,* 66, 233, 1948.

161. **Pal, S. S., Sudhakar, B., and Sethunathan, N.,** Effects of benomyl on iron, and manganese reduction and redox potential in flooded soil, *J. Soil Sci.,* 30, 155, 1975.

162. **Wainwright, M. and Pugh, G. J. F.,** The effects of fungicides on certain chemical and microbial properties of soils, *Soil Biol. Biochem.,* 6, 263, 1974.

163. **Ray, R. C. and Sethunathan, N.,** Effect of commercial formulation of hexachlorohexane and benomyl on the oxidation of elemental sulphur in soil, *Soil Biol. Biochem.,* 12, 451, 1980.

164. **Martin, J. P., Martin, W. P., Page, J. B., Raney, W. A., and DeMent, J. D.,** Soil aggregation, *Adv. Agron.,* 7, 1, 1955.

165. **McCalla, T. M., Haskins, F. A., and Curley, R. D.,** Soil aggregation by microorganisms following soil fumigation, *Soil Sci. Am. Proc.,* 23, 311, 1958.

166. **Martin, J. P. and Aldrich, D. G.,** Effect of fumigation on soil aggregation, *Soil Sci. Soc. Am. Proc.,* 16, 201, 1952.

167. **Lawrence, W. J. C.,** *Soil Sterilization,* Allen and Unwin, London, 1956, 171.

168. **Karanath, N. G. K. and Vasantharajam, V. N.,** Effect of fungicide Dexon (p-dimethyl-aminobenzenediazo sodium sulfonate) on soil aggregation, *Curr. Sci.,* 40, 394, 1971.

169. **Bollen, G. J.,** Side effect of pesticides on microbial interactions, in *Soil-Borne Plant Pathogens,* Schippers, B. and Gams, W., Eds., Academic Press, London, 1979, 451.

170. **Sussman, A. S.,** Dormancy of soil micro-organisms in relation to survival, in *Ecology of Soil-Borne Plant Pathogens, Prelude to Biological Control,* Baker, K. F. and Snyder, W. C., Eds., University of California Press, Berkeley, 1965, 99.

171. **Burns, R. G.,** *Microbial Control of Pesticide Persistence,* Beynon, K. I., Ed., Monogr. No. 17, British Corp. Prot. Council, 1976.

172. **Walker, N.,** *Soil Microbiology,* Butterworths, London, 1975, 181.

173. **Sinha, A. P.,** Effect of Pesticides on Soil Microflora and Their Related Physiological Activities with Special Reference to Sugarbeet Crop, Ph.D., thesis, University of Kanpur, India, 1976, 250.

174. **Smith, S. N., Lyon, A. J. E., and Sahid, I. B.,** The breakdown of paraquat and diquat by soil fungi, *New Phytol.,* 77, 735, 1976.

175. **Grossbard, E. and Harris, D.,** Selective action of "gramoxone W" and "round up" on *Chaetomium globosum* in relation to straw decay, *Trans. Br. Mycol. Soc.,* 69, 141, 1977.

176. **Dobolyi, Cs., Pasztor, Zs., and Kecskes, M.,** Xenobiotics and soil microbioata affected by xenobiotic interactions. III. 2, 4-D-Na and the species composition of fungi in a chernozem, *Acta Phytopathol. Acad. Sci. Hung.,* 12, 81, 1977.

177. **Powell, N. T.,** Interactions between nematodes and fungi in disease complexes, *Annu. Rev. Phytopathol.,* 9, 253, 1971.

178. **Thompson, I. J., Erwin, D. C., and Garber, M. J.,** The relationship of the root-knot nematodes, *Meloidogyne javanica,* to *Fusarium* wilt in cow pea, *Phytopathology,* 49, 602, 1959.

179. **Miller, P. M. and McIntyre, J. L.,** Oxamyl-treated soil protects tobacco against black shank, *Phytopathology,* 66, 221, 1976.

180. **Clark, F. E.,** The concept of competition in microbial ecology, in *Ecology of Soil-Borne Plant Pathogens, Prelude to Biological Control,* Baker, K. F. and Snyder, W. C., Eds., University of California Press, Berkeley, 1965, 339.

181. **Jackson, R. M.,** Antibiosis and fungistasis of soil micro-organisms, in *Ecology of Soil-Borne Plant Pathogens, Prelude to Biological Control,* Baker, K. F. and Snyder, W. C., Eds., University of California Press, Berkeley, 1965, 363.

182. **Ohr, H. D. and Munnecke, D. E.,** Effect of methyl bromide on antibiotic production by *Armillaria mellea, Trans. Br. Mycol. Soc.,* 62, 65, 1974.

183. **Munnecke, D. E., Kolbezen, M. J., and Wilbur, W. D.,** Effect of methylbromide carbon disulphide on *Armillaria* and *Trichoderma* growing on agar medium and relation to survival of *Armillaria* in soil following fumigation, *Phytopathology,* 63, 1352, 1973.

184. **Langerak, C. J.,** The role of antagonists in the chemical control of *Fusarium oxysporum* f. sp. *narcissi, Neth. J. Plant Pathol.,* Suppl. 1, 365, 1977.

185. **Balicka, N. and Krezel, Z.,** The influence of herbicides upon the antagonism between *Bacillus* sp. and *Pseudomonas phaseoli, Weed Res.,* 9, 37, 1969.

186. **Krezel, Z. and Leszczynska, D.,** The effect of herbicides on the antibiotic activity of *Bacillus subtilis, Acta Phytopathol. Acad. Sci. Hung.,* 12, 13, 1977.

187. **Elton, C.,** Competition and the structure of ecological communities, *J. Anim. Ecol.,* 15, 54, 1946.

188. **Odum, E. P.,** *Fundamentals of Ecology,* W. B. Saunders, Philadelphia, 1959, 546.

189. **Clements, F. E. and Shelford, V. E.,** *Bio-Ecology,* John Wiley & Sons, New York, 1939, 425.

190. **Milne, A.,** Definition of competition among animals in *Mechanisms in Biological Competition,* Vol. 15, Milthospe, F. L., Ed., *Symp. Soc. Exptl. Biol.,* Cambridge University Press, London, 1961, 40.

191. **Domsch, K. H.,** *Pesticides in the Soil,* Michigan State University, East Lansing, 1970, 42.

192. **Farley, J. D. and Lockwood, J. L.,** Reduced nutrient competition by soil micro-organisms as a possible mechanism for pentachloronitrobenzene-induced disease accentuation, *Phytopathology,* 59, 718, 1969.

193. **Katan, J. and Lockwood, J. L.,** Effect of pentachloronitrobenzene on colonization of alfalfa residues by fungi and streptomycetes in soil, *Phytopathology,* 60, 1578, 1970.

194. **Garren, K. H.,** The stem rot of peanuts and its control, *Va. Agric. Exp. Stn. Tech. Bull.,* 144, 59, 1959.

195. **Garren, K. H.,** Studies on *Sclerotium rolfsii* in the southeast United States in the relationship of soil micro-organisms to soil-borne plant pathogens, *South Coop. Ser. Bull. Va. Polytech. Inst. State Univ. Blacksburg,* 183, 98, 1973.

196. **vander Hoeven, E. P. and Bollen, G. J.,** Effect of benomyl on soil fungi associated with rye. I. Effect on the incidence of sharp eyespot caused by *Rhizoctonia cerealis, Neth. J. Plant Pathol.,* 86, 163, 1980.

197. **Reinecke, P.,** Untersuchungen zum Erregerspektrum des Fusskrank heitskomplexes and Getreide unter besonderer Berucksichtigung von Rhizoctonia solani Kuhn, Ph.D. thesis, University of Gottingen, 1977.

198. **Baker, K. F. and Snyder, W. C., Eds.,** *Ecology of Soil-Borne Plant Pathogens, Prelude to Biological Control,* University of California Press, Berkeley, 1975, 571.

199. **Hora, T. S. and Baker, R.,** Soil fungistasis: microflora producing volatile inhibitors, *Trans. Br. Mycol. Soc.,* 59, 491, 1972.

200. **Lockwood, J. L.,** Soil fungistasis, *Annu. Rev. Phytopathol.,* 2, 341, 1964.

201. **Smith, A. M.,** Ethylene: a cause of fungistasis, *Nature (London),* 246, 311, 1973.

202. **Watson, A. G. and Ford, E. J.,** Soil fungistasis — a reappraisal, *Annu. Rev. Phytopathol.,* 10, 327, 1972.

203. **Kutzera, J. W. and Hoffmann, G. M.,** Soil fungistasis after treatment with methyl bromide, *Z. Pflanzenkr. Pflanzenschutz,* 83, 497, 1976.

204. **Wainwright, M.,** Effects of fungicides on the microbiology and biochemistry of soil — a review, *Z. Pflanzenrnaehr. Bodenkd.,* 140, 587, 1977.

205. **Cook, R. J. and Rovira, A. D.,** The role of bacteria in the biological control of *Gaeumannomyces graminis* by suppressive soils, *Soil Biol. Biochem.,* 8, 269, 1976.

206. **Shipton, P. J.,** Take-all decline during cereal monoculture, in *Biology and Control of Soil-Borne Plant Pathogens,* Bruehl, G. W., Ed., American Phytopathological Society, St. Paul, Minn., 1975, 137.

207. **Bruehl, G. W.,** *Biology and Control of Soil-Borne Plant Pathogens,* Bruehl, G. W., Ed., American Phytopathological Society, St. Paul, Minn., 1975, 261.

208. **Gerlagh, M.,** Introduction of *Ophiobolus graminis* into new polders and its decline, *Neth. J. Plant Pathol.,* 74 (Suppl. 2), 1, 1968.

209. **Pope, A. M. S.,** The Decline Phenomenon in Take-All Disease of Wheat, Ph.D. thesis, University of Surrey, Guildford, England, 1972, 161.

210. **Vojinovic, Z. D.,** Antagonists from soil and rhizosphere to phytopathogens, *Final Tech. Rep. Inst. Soil Sci. Beograd Yugoslavia,* p. 130, 1972.

211. **Shipton, P. J., Cook, R. J., and Sitton, J. W.,** Occurrence and transfer of a biological factor in soil that suppresses take-all of wheat in eastern Washington, *Phytopathology,* 63, 511, 1973.

212. **Warcup, J. H.,** Studies on soil fumigation. IV. Effects on fungi, *Soil Biol. Biochem.,* 8, 261, 1976.

213. **Ridge, E. H.,** Studies on soil fumigation. II. Effects on bacteria, *Soil Biol. Biochem.,* 8, 249, 1976.

214. **Drechsler, C.,** Predaceous fungi, *Biol. Rev.,* 16, 265, 1941.

215. **Duddington, C. L.,** *The Friendly Fungi. A New Approach to The Eelworm Problem,* Faber and Faber, London, 1957, 188.

216. **Barron, G. L.,** The nematode-destroying fungi, *Can. Biol.,* Guelph, 1977.

217. **Briggs, W. N. and Klein, D. A.,** Herbicide effects on *Bdello vibrio bacteriovorus* parasitism of a soil pseudomonad, *Soil Biol. Biochem.,* 3, 143, 1971.

218. **Papavizas, G. C.,** Status of applied biological control of soil-borne plant pathogens, *Soil Biol. Biochem.,* 5, 709, 1973.

219. **Baker, K. F. and Cook, R. J.,** *Biological Control of Plant Pathogens,* S. Chand, New Delhi, 1979, 433.

220. **Garrett, S. D.,** Effect of a soil microflora selected by carbondisulphide fumigation on survival of *Armillaria mellea* in woody host tissues, *Can. J. Microbiol.,* 3, 135, 1957.

221. **Ohr, H. D., Munnecke, D. E., and Bricker, J. L.,** The interaction of *Armillaria mellea* and *Trichoderma* spp. as modified by methylbromide, *Phytopathology,* 63, 965, 1973.

222. **Ohr, H. D. and Munnecke, D. E.,** Effects of methylbromide on antibiotic production by *Armillaria mellea, Trans. Br. Mycol. Soc.,* 62, 65, 1974.

223. **McKeen, W. E.,** A study of sugarbeet root rot in southern Ontario, *Can. J. Res.,* 27, 284, 1949.

224. **Domsch, K. H.,** Untersuchungen zur wirkung einiger Bodenfungizide, *Mitt. Biol. Bundesanst. Land Forstwirtsch. Berlin Dahlem,* 97, 100, 1959.

225. **Chinn, S. H. F.,** Biological effects of Panogen PX in soil on root rot of wheat seedlings, in *Biology and Control of Soil-Borne Plant Pathogens,* Bruehl, G. W., Ed., American Phytopathological Society, St. Paul, Minn., 1975, 205.

226. **Chinn, S. H. F.,** Biological effect of Panogen PX in soil on common root rot and growth response of wheat seedlings, *Phytopathology,* 61, 98, 1971.

227. **Vaartaja, O., Wilner, J., Cram, W. C., Salisbury, P. J., and Crookshanks, A. W.,** Fungicide trials to control damping-off of conifers, *Plant Dis. Rep.,* 48, 12, 1964.

228. **Curl, E. A., Wiggind, E. A., and Anders, S. C.,** Interaction of *Rhizoctonia solani* and *Trichoderma* with PCNB and herbicides affecting cotton seedlings diseases, *Proc. Am. Phytopathol. Soc.,* 3, 221, 1976.

229. **Henis, Y., Ghaffar, A., and Baker, R.,** Integrated control of *Rhizoctonia solani* damping-off of radish: effect of successive planting, PCNB and *Trichoderma* on pathogen and disease, *Phytopathology,* 68, 900, 1978.

230. **Hadar, Y., Chet, I., and Henis, Y.,** Biological control of *Rhizoctonia solani* damping-off with wheat bran culture of *Trichoderma harzianum, Phytopathology,* 69, 64, 1979.

231. **Mukhopadhyay, A. N. and Upadhyay, J. P.,** Mechanism of reduction of *Sclerotium* root rot of sugarbeet through nitrogenous amendments and exploitation of *Trichoderma viride* as a potential biocontrol agent, paper presented at 3rd Int. Symp. on Plant Pathology, New Delhi, India, December 14 to 18, 1981.

232. **Chet, I., Hadar, Y., Elad, Y., Katan, J., and Henis, Y.,** Biological control of soil-borne plant pathogen by *Trichoderma harzianum,* in *Soil-Borne Plant Pathogens, Prelude to Biological Control,* Schippers, B. and Gams, W., Eds., Academic Press, London, 1979, 585.

233. **Elad, Y., Chet, I., and Katan, J.,** *Trichoderma harzianum:* a biocontrol agent effective against *Sclerotium rolfsii* and *Rhizoctonia solani, Phytopathology,* 70, 119, 1986.

234. **Lewis, J. A. and Papavizas, G. C.,** *Biological Control in Crop Production,* Papavizas, G. C., Ed., Hllanheld Osmun Publishers, Granada, 1981, 461.

235. **Papavizas, G. C.,** Survival of *Trichoderma harzianum* in soil and in pea and bean rhizosphere, *Phytopathology,* 72, 121, 1982.

236. **Chandra, I.,** Studies on Damping-Off of Sugarbeet Caused by *Pythium aphanidermatum* (Edson) Fitz, Ph.D. thesis, G. B. Pant University of Agriculture & Technology, Pantnagar, India, 1984, 131.

237. **Mukhopadhyay, A. N., Patel, G. J., and Brahmabhat, A. B.,** *Trichoderma harzianum* — a potential biocontrol agent for tobacco damping-off, *Tob. Res.,* 12, 26, 1986.

238. **Mukhopadhyay, A. N. and Chaturvedi, C.,** Biological control of damping-off by *Trichoderma harzianum,* in Proc. Annu. Meet. Soc. Mycol. Plant Pathol., Coimbatore, India, 1986.

239. **Carter, M. V. and Price, T. V.,** Biological control of *Eutypa armeniacae.* III. A comparison of chemical biological and integrated control, *Aust. J. Agric. Res.,* 26, 537, 1974.

240. **Stankova-Opocenska, E. and Dekker, J.,** Indirect effect of 6-azauracil on *Pythium debaryanum* in cucumber, *Neth. J. Plant Pathol.,* 76, 152, 1970.

241. **Joyner, B. G. and Couch, H. B.,** Relation of dosage rates, nutrition, air temperature and suscept genotype to side effects on systemic fungicides on turf grasses, *Phytopathology,* 66, 806, 1976.

242. **Kreutzer, W.,** The reinfestation of treated soil, in *Ecology of Soil-Borne Plant Pathogens,* Baker, K. F. and Snyder, W. C., Eds., University of California Press, Berkeley, 1965, 495.

243. **Garren, K. H.,** Evidence for two different pathogens of peanut pod rot, *Phytopathology,* 53, 746, 1963.

244. **Smith, A.M., Styness, B. A., and Moore, K. J.,** Benomyl stimulates growth of a basidiomycete on turf, *Plant Dis. Rep.,* 54, 774, 1970.

245. **Fletcher, J. T.,** Shaggy stipe, a new disease of cultivated mushroom caused by *Mortierella bainieri, Plant Pathol.,* 22, 25, 1973.

246. **Warren, C. G., Sanders, P., and Cole, H.,** Increased severity of *Pythium* blight on creeping bentgrass treated with benomyl, *Phytopathology,* 65, 836, 1975.

247. **Williams, R. J. and Ayanaba, A.,** Increased incidence of *Pythium* stem rot in cowpeas treated with benomyl and related fungicides, *Phytopathology,* 65, 217, 1975.

248. **Dekker, J.,** Selectivity of and resistance against systemic fungicides, *Eur. Mediterr. Plant Prot. Organ. Bull.,* 10, 47, 1973.

249. **Bollen, G. J. and Scholten, G.,** Acquired resistance to benomyl and some other systemic fungicide in a strain of *Botrytis cinerea* in cyclamen, *Neth. J. Plant Pathol.,* 77, 83, 1971.

250. **Jordan, V. W. L. and Richmond, D. V.,** The effects of benomyl on sensitive and tolerant isolates of *Botrytis cinerea* infecting strawberries, *Plant Pathol.,* 23, 81, 1974.

251. **Swinburne, T. R.,** The effect of benomyl and other fungicides on phenylalanine-ammonialyase activity in *Hordeum vulgare* and *Phaseolus vulgaris, Physiol. Plant Pathol.,* 5, 81, 1975.

252. **Backman, P. A., Rodriguez-Kabana, R., and Williams, J. C.,** The effect of peanut leaf spot fungicides on the non target pathogen, *Sclerotium rolfsii, Phytopathology,* 65, 773, 1975.

253. **Porter, D. M.,** Increased severity of *Sclerotinia* blight in peanuts treated with captafol and chlorothalonil, *Plant Dis.,* 64, 394, 1980.

254. **Fokkema, N. F., Laar, J. A. J., VandeNelis-Blomberg, A. L., and Schippers, B.,** The buffering capacity of the natural mycoflora of rye leaves to infection by *Cochliobolus sativus,* and its susceptibility to benomyl, *Neth. J. Plant Pathol.,* 81, 176, 1975.

255. **Jackson, N.,** Evaluation of some chemicals for the control of stripe smut in Kentucky blue grass turf, *Plant Dis. Rep.,* 54, 168, 1970.

256. **Spalding, D. H.,** Post harvest use of benomyl and thiabendazole to control blue mold rot development in pears, *Plant Dis. Rep.,* 54, 655, 1970.

257. **Manning, W. J. and Papia, P. M.,** Benomyl soil treatments and natural occurrence of *Alternaria* leaf spot of carnation, *Plant Dis. Rep.,* 56, 9, 1972.

258. **Kreutzer, W.,** Soil treatment, in *Plant Pathology*, Vol. 3, Horsfall, J. G. and Dimond, A. E., Eds., Academic Press, New York, 1960, 431.

259. **Gibson, I. A. S.,** An anomalous effect of soil treatment with ethyl mercury phosphate of the incidence of damping-off in pine seedlings, *Phytopathology*, 46, 181, 1956.

260. **Butler, E. E.,** *Rhizoctonia solani* as a parasite of fungi, *Mycologia*, 49, 354, 1957.

261. **Altman, J. and Ross, M.,** Plant pathogens as a possible factor in unexpected preplant herbicide damage in sugarbeets, *Plant Dis. Rep.*, 51, 86, 1967.

262. **Rattink, H. and Beuzenberg, M.,** Problemen rond gebruik van Benlate in lelieteelt, *Bloembollencultuur*, 83, 398, 1972.

263. **Warren, C. G., Sanders, P., and Cole, H.,** *Sclerotinia homoeocarpa* tolerance to benzimidazole configuration fungicides, *Phytopathology*, 64, 1139, 1974.

264. **Bollen, G. J. and VanZaayen, A.,** Resistance to benzimidazole fungicides in pathogenic strains of *Verticillium fungicola*, *Neth. J. Plant Pathol.*, 81, 157, 1975.

265. **Dekker, J.,** Resistance, in *Systemic Fungicides*, Marsh, R. W., Ed., Longman, New York, 1977, 176.

Chapter 11

NONTARGET EFFECTS OF PESTICIDES ON SOILBORNE PLANT PATHOGENS

I. INTRODUCTION

The application of pesticides to the soil environment usually can be expected to result in a diverse array of effects on target and nontarget organisms.[1] Only when the side effects become apparent are questions raised regarding the adequacy of a pesticide for generalized field use, the environmental and economic importance of the side effect, and the desirability of testing for such effects prior to use of the material. There are many pesticides having a wide range of biological activities other than those specified by the manufacturer. However, groups of toxicants are named according to their intended use. Therefore, insecticides are formulated to use against insects, fungicides are designed to control pathogenic fungi, nematodes are intended to provide a certain spectrum of activity against undesirable nematodes, and herbicides are manufactured to control noncrop (weed) plants. Within this commonly accepted terminology, unintended or nontarget activities of certain toxicants on pathogens/diseases will be discussed in this chapter.

The toxicants used represent a wide diversity of chemicals varying in chemical, physical, and toxicological properties as well as persistence, degradability, and a range of other properties. With the prospect of continued use of pesticides in agriculture and the possibilities of adverse or beneficial interactions, there is a need to develop a thorough understanding of the total effects of the these important compounds in our environment.

The increasing use of pesticides for the control of undesirable vegetation, insect pests, and diseases has a tremendous impact on crop production and it appears that their use will continue to expand in the future. The biological activity of any pesticide is usually not restricted to the target organism, but extends to nontarget organisms as well. Therefore, inhibitory and stimulatory effects of beneficial or harmful nontarget organisms in the environment are possible. It seems, however, that less emphasis has been given on the indirect effects of pesticides as compared to the direct ones.

Pesticides may affect crop plants in addition to targets (weeds/pathogens/insects) whether directly or indirectly by their effect on other organisms which may lead to harmful or beneficial effects. Crop plants form various kinds of relationships with other organisms, e.g., pathogens. Plant disease is the final result of a compatible interaction which occurs under suitable conditions between a particular pathogen and host. Other living components connected with disease are the surrounding microflora and fauna which may affect either the pathogen, the host, or both by their antagonistic or synergistic action. Pesticides in their original form or their degradation products may interact in different ways with any one of the organisms involved in the disease, at one or more points in the chain of events leading to disease development (pathogenesis). The final result may be an increase, decrease, or no change in disease severity or its incidence. The extreme case may be a severe outbreak of a "new" disease which was of negligible importance before the application of the herbicide, whereas the other extreme would be the complete elimination of an existing severe disease. The effect on disease might be immediately during the same growing season, or it might be a long-term effect due to build-up, especially where persistent chemicals are involved. Changes in disease were also noticed following the uses of toxicants, particularly insecticides and fungicides.[1] Soil pathogens are more likely to be affected, since most of the pesticides reach the soil sooner or later irrespective of the method of application used. This occurs upon direct application to the soil, as seed treatment, as drift from treated aerial parts of the plants, or through the decomposition of treated plant tissues after their incorporation in soil. When in the soil, the pesticides are in continuous contact with soilborne pathogens which survive there and invade the host through its subterranean parts. Furthermore, pesticides, particularly herbicides used as preemergence treatment, are usually applied at higher rates as compared to when used for foliar treatments.

Pesticides affect disease indirectly in addition to their effect on the host, the pathogen, and the surrounding microorganisms. The eradication of weeds and plant residues by herbicides may in turn affect disease incidence, because many weeds serve as hosts or symptomless carriers of many agricultural crop pathogens. Plant residues often harbor pathogens: weeds also affect the microclimate and, therefore, disease development.[2-5]

This chapter deals with the effects that pesticides have on the increase or decrease of disease incidence, particularly soilborne plant pathogens. Those toxicants are included which are used mainly or exclusively in soil against weeds, insect pests, or diseases. The effects of fungicides on other soil microbes and their related biochemical activities have already been well covered in Chapter 10.

II. FUNGICIDES

Determination of the short- and long-term effects of fungicides is essential to avoid (1) disastrous environmental disruptions, such as increase of nontarget diseases and (2) destruction of antagonistic microorganisms which may act to hold nontarget disease organisms in check.

A. Direct Nontarget Effects of Nonsystemic Fungicides

Nonsystemic fungicides have an inherent toxicity for a broad spectrum of microorganisms. Although they might be expected to have a greater impact on the soil microenvironment, they have not been implicated in several undesirable effects on nontarget soilborne plant pathogens. Little information is available in the literature on the subject.

For the first time, a sudden change of a relatively minor disease of tomato, i.e., gray mold (*Botrytis cinerea*), to a disease of great economic importance under field conditions has been observed.[6] The workers were of the opinion that an increase of a gray mold may have been caused by the increased use of some ethylenebisdithiocarbamate (EDBC) sprays to control foliar diseases in the early 1950s. Ethylmercury phosphate, examined at low rates as a soil sterilant, enhanced damping-off in pine seedlings (*Pinus patula* and *P. radiata*).[7] Gibson concluded that the nontarget enhancement of the disease was brought about by proliferation of the pathogen in the soil environment under attenuated competition. The fungicide neither directly stimulated growth of the pathogens involved (*Pythium ultimum* and *Rhizoctonia solani*) nor damaged pine seedlings. Pentachloronitrobenzene (PCNB) reduced damping-off of the two pine species incited by *R. solani* and enhanced postemergence damping-off caused by a *Pythium* species.[8] The nontarget effect of PCNB was presumed to be due to the suppression of the natural antagonist *Penicillium paxilli,* as this chemical was not toxic to *Pythium.* Rich and Miller[9] obtained slightly more complex nontarget effects for PCNB. They reported increase in strawberry wilt (*Verticillium alboatrum*) which was attributed to an increase of the nematode (*Pratylenchus penetrans*) by PCNB, which, in turn, enhanced wilt in tandem. Powell[10] has demonstrated synergism between nematodes and fungi in the development of soilborne plant pathogens.

Another example on the subject is the increase of *Sclerotinia* blight (*S. sclerotiorum*) disease in the U.S. The disease was first reported in 1971 from Virginia, which later became a serious problem with peanuts (*Arachis hypogea*). Application of chlorothalonil significantly enhanced the blight on peanuts.[11] Porter[12] demonstrated that chlorothalonil enhanced *Sclerotinia* blight and decreased peanut pod yields in two field locations in Virginia heavily infested with *S. sclerotiorum.*

Based on this information, there is no doubt that nonsystemic fungicides may sometimes bring about undesirable nontarget effects, which results in increase of disease that has not been a problem previously. It is not unusual for the nonsystemics, as they have inherent toxicity, to minimize or eliminate beneficial nontarget microbes in soil. Fungicides have the broad ability to reduce or eliminate not only the target pathogen and the nontarget antagonists, but also nontarget, hitherto secondary, pathogens.[13]

B. Direct Nontarget Effects of Systemic Fungicides

Systemic fungicides are known for their chemotherapeutic activity for reducing or controlling plant diseases otherwise uncontrolled by nonsystemics.[13] Of the several systemic fungicides, benzimidazoles are known for their side effects on nontarget soil microorganisms. Soilborne nontarget fungi nonsensitive to benzimidazoles formerly of no pathogenic significance were observed to be enhanced in numbers.[6] Extensive use of benzimidazoles for longer periods of time as compared to other systemics led to this situation. There are many reports showing that application of benzimidazole fungicides applied to control certain plant diseases increased nontarget soilborne pathogens of minor importance, making them economically important. Benomyl controlled peanut leaf spot (*Cercospora arachidicola* and *Cercosporidium personatum*), but increased stem rot (*Sclerotium rolfsii*).[14] Benomyl increased the severity of blight caused by *P. aphanidermatum* and a *Pythium* species, when it was used to control turf grass diseases caused by *Fusarium, Rhizoctonia,* and *Sclerotinia.*[15] Jackson[16] applied benomyl and thiabendazole in the fall to a 3-year-old turf of 'Merion' Kentucky bluegrass which reduced smut (*Ustilago striiformis*), but increased melting out (*Drechslera poae*). Experiments on cotton seed (*Gossypium hirsutum*) treatments to control seedling disease caused by several soilborne pathogens revealed that benomyl consistently increased damping-off when *Pythium* spp. were present in the soil.[6] They conducted an experiment with nonsterilized soil with a natural population of *Pythium* spp. (five to ten propagules per 10 g soil). The soil was infested separately with *R. solani* or *Thielaviopsis basicola* 3 weeks before planting. For mixed pathogen experiments, the *R. solani-* or *T. basicola*-infested soils were mixed in proportions so as to provide a 1:1 ratio. The experimental fungicide BAS 389 (effective for *R. solani*) was applied to the cotton seed with the organic solvent infusion technique (OSIT). Another chemical, CGA 48988 (effective against *Pythium* spp.), was added in combination with BAS 389 with OSIT. Benomyl was incorporated directly on the seed after the solvent evaporated. Seeds were planted and the containers were kept at 27°C for 1 week and then at 20°C for 5 weeks to encourage *T. basicola* infection. It was observed that benomyl significantly enhanced damping-off of cotton when *Pythium* spp. were not eliminated from the soil by CGA 48988. The enhanced disease appreciation with benomyl application was apparently the result of infections by *Pythium* spp. Williams and Ayanaba[17] obtained a positive relationship between a field increase in the incidence of *Pythium* stem rot of cowpea (*Vigna linquiculata*) and the application of benomyl, thiobendazole, methyl-2-benzimidazole carbamate (MBC), and several nonbenzimidazole fungicides. In six field trials in Nigeria, benomyl and MBC produced the highest nontarget increase in stem rot. They concluded that increase in stem rot is probably due to an enhanced activity of *Pythium* brought about by the suppression of nontarget antagonistic microorganisms in the agro-ecosystem.

III. HERBICIDES

A. Herbicide Effect on Pathogen Physiology and Survival

There are several published reports to assess the influence of herbicides on plant pathogens. However, most of this work has been done in culture in the laboratory. These laboratory results are contradictory. In spite of voluminous research work in vitro and in vivo, meaningful correlations or extrapolations between laboratory, greenhouse, and field results have rarely been observed. The response of fungi to herbicides greatly depends on dosage, quality of the toxicant, and pH and nutritive value of the assay medium.

Growth effects in culture media have been studied to the greatest extent. In a review article by Altman and Campbell,[18] 21 of the 52 listed herbicides affected growth of pathogenic fungi. In early screening tests, it was demonstrated that 25 herbicides recommended in crop production stimulated growth of *Rhizoctonia solani* in vitro.[18] Growth was measured linearly and by comparing colony density. In all instances, *Rhizoctonia* grew better in media supplemented with herbicides at 1, 100, and 100 ppm than in unsupplemented media. Some growth occurred at 10,000 ppm with 12 out of 25 herbicides tested. Complete inhibition of growth occurred with two of the herbicides at 10,000 ppm. In these tests it was not determined whether the herbicides stimulated fungal growth or the fungus utilized the herbicides as a source of energy. Sinha et al.[19] demonstrated that growth of *Sclerotium rolfsii* was completely arrested by Diuron at 20 ppm and by alachlor at 160 ppm, while growth of other pathogenic fungi (*Pythium aphanidermatum, R. bataticola,* and *Fusarium moniliforme*) continued up to 640 ppm of these herbicides. The sporulation of most fungi was also affected, particularly at higher doses of these herbicides. In *P. aphanidermatum,* a high concentration of these herbicides (80 ppm) induced conspicuous vacuolization in the mycelium and appreciably reduced its branching. The fungus did not produce either sporangia or oospores in the presence of the two toxicants. *S. rolfsii* produced abundant reddish-brown sclerotia of mustard size in the control plates, but not in the presence of alachlor or Diuron. Trifluralin and prometryne increased production of conidia and chlamydospores of *Fusarium oxysporum* f. sp. *vasinfectum*[20,21] and oospores of *Aphanomyces euteiches,*[6] but the increase was observed in sterile soil and culture media, respectively. Similarly, mecoprop enhanced production of perithecia and microconidia of *Gaeumannomyces graminis.*[22] However, Diuron, EPTC, and fluometuron inhibited production of sclerotia of *Sclerotinia sclerotiorum* and *S. rolfsii* in sterile soil[6] and dinitramine and trifluralin inhibited zoospores production by *A. euteiche* in replacement culture.[23] Regarding the effect of herbicides on propagule germination, Chopra et al.[21] observed that prometryne at 2 μg/mℓ decreased germination of chlamydospores of *F. oxysporum* f. sp. *vasinfectum.* Zoospore germination of *A. euteiche* in maltose-peptone broth and germ tube development was suppressed by 0.06 and 0.12 μg/mℓ of dinitramine and trifluralin, respectively.[23]

There are several reports showing side effects of herbicides on sporulation and propagule germination and survival in soil. This information may provide a substantial base to analyze nontarget disease changes and delineate mechanisms of action of herbicides. Percich and Lockwood[24] observed that atrazine enhances the population of *F. solani* f. sp. *pisi* and *F. culmorum* as well as microconidial germination and chlamydospore formation. These observations are substantially more useful in interpreting nontarget effects on disease development and survival as compared to in vitro studies on similar processes. In another report by Wyse et al.,[25] the same chemical increased the populations of *F. solani* f. sp. *phaseoli* in soil and stimulated germination of microconidia, germ tube development, and subsequent chlamydospore formation. The increase in inoculum density by atrazine and stimulation of conversion of conidia to chlamydospores may result in enhanced disease severity and also increase the ability of these fungi to survive in soil.[6] Similarly, Tang et al.[26] noted that trifluralin enhanced chlamydospores production by the cotton wilt organism in soil and stimulated their germination. Nontarget effects were observed not only in soil, but in cotton rhizosphere as well.[6] Chlamydospores of cotton wilt pathogen germinated 100% better in the rhizosphere soil from plants grown in trifluralin-treated than in nontreated soil.

Little work has been done on saprophytic activity. Garren[27] pointed out that Dinoseb, a fungistatic herbicide, decreased stem rot of peanut (*S. rolfsii*). The reduction could have been the result of direct fungistatic activity of the chemical on the pathogen and its ability to reduce weeds and, therefore, deprive the pathogen of an important source of organic matter (food bases) in the peanut stem rot development. In another report, herbicides (diphenamid and trifluralin) added to *R. solani*-infested soil increased saprophytic activity,[6] whereas Prometryne or fluometuron decreased saprophytic acitivity,[28] and cycloate at 16 and 32 μg/g reduced colonization of a substrate by *R. solani* in natural soil.[29] With a low concentration of herbicides (4 or 8 μg/g soil) no reduction of saprophytism occurred. Papavizas and Lewis[6] conducted extensive experiments on the saprophytic activity of *R. solani* as affected by 17 herbicides generally incorporated in bean, cotton, and soybean culture. These herbicides were added to *R. solani*-infested soil at recommended rates, and saprophytic activity was determined at intervals up to 15 weeks after the addition of the chemical in soil. None of the herbicides tested consistently increased saprophytic activity of *R. solani*. Alachlor, cycloate, and diphenamid reduced activity considerably at some, but not all, rates.

B. Increase of Plant Diseases due to Herbicides

Increased incidences of soilborne disease caused by the application of herbicides have been reported primarily in greenhouse and to a lesser extent in field studies. The phenomenon of disease increase is not restricted to a specific group of herbicides, pathogens, or crops.[18] Most of the studies were carried out in the greenhouse in which the inoculum level of the pathogen and the environmental conditions could be controlled. It is much more difficult to control these factors in field studies, as more than one pathogen might be involved in certain diseases (e.g., damping-off complex), and therefore it is essential to study the interaction with each pathogen. Diseases caused by *R. solani* are also increased by herbicides in the field. Preplant applications of trifluralin at label rates caused stunted cotton seedling development and enhanced damping-off caused by *R. solani*.[30] In Israel, trifluralin also increased seedling disease of cotton caused by *R. solani* in field plots.[31] Altman and Ross[32] reported an interaction in field studies involving herbicides used for sugar beet weed control. The herbicides involved were pebulate and pentachloroaniline (PCA, Pyrazon) and the pathogen was *R. solani*. In all cases where combinations of either pebulate and *R. solani* or PCA and *R. solani* were present, disease incidence increased over that caused by *R. solani* alone. It was later suggested that these herbicides predisposed the plants to infection by the pathogen.[18] Diseases of pepper (*Capsicum frutescens*) and tomato caused by *R. solani* were enhanced as a result of dephenamid addition to soil at 10 μg/g.[33] The toxicants, however, had no effect on the *Pythium* disease of pepper. The following factors may be responsible for the frequent reports on increase in *Rhizoctonia* disease:[34] it is easy to detect the disease and to estimate the loss of stand, the fungus is widespread in many soils, and preemergence applications of herbicides are widely used in several crops. Other possibilities for the increase in *Rhizoctonia* disease is that this organism is more tolerant to the herbicides than its antagonists, or that herbicides increase the susceptibility of the seedlings to this particular disease. A particular herbicide-pathogen combination might produce various results as far as the disease is concerned.[34] For example, application of trifluralin did not always result in an increase in *Rhizoctonia* disease in cotton.[35] Standifer et al.[36] observed a lower stand of cotton seedlings by increasing the rates of trifluralin in one field planted early in the spring, but not in the second one sown later under warmer conditions. They attributed the differences in results to the different prevailing temperatures. Field observations by Roming and Sasser[37] indicated increased disease due to *R. solani* on snapbean when trifluralin or Dinoseb was applied to soil. Salt[38] obtained conflicting results regarding the effect of mecoprop (MCPP) on take-all disease of wheat (*Ophiobolus graminis*).

In greenhouse studies, Sumner and Glaze[39] treated soils (artificially or naturally infested with species of *Fusarium, Pythium, Rhizoctonia,* or *Colletotrichum*) with field rates of DCPA, Nitrofen, or trifluralin. In natural soils in the greenhouse, the three herbicides enhanced root diseases on turnips. In artificially infested soils, only Nitrofen caused this effect. Plant stands and foliage weights were decreased in infested soils by DCPA and Nitrofen. Cycloate, applied to steamed soil in the greenhouse infested with *R. solani,* significantly enhanced sugar beet damping-off compared to an inoculated nontreated control.[40] Dinoseb and trifluralin at field rates enhanced damping-off of snapbeans in *P. myriotylum*-infested soils.[41] Herbicides have also been demonstrated to increase *Fusarium* diseases in the field. Application of EPTC, Chloramben, Dinoseb, or fluorodifen in the field increased root rot of navy bean caused by *F. solani* f. sp. *phaseoli.*[42] Atrazine incorporated in a field in Michigan at label rates enhanced pea root rot severity caused by *F. solani* f. *pisi* three times over that in the control.[24] At 5 µg/g the herbicide enhanced the incidence of corn seedling blight caused by *F. culmorum* twofold over the control. Mussa and Russell[43] found that bentazon, trifluralin and triallate in sand at 10 µg/g significantly enhanced root rot and damping-off of beans caused by *F. solani.* f. *phaseoli.*

An extensive study on the effect of herbicides on take-all disease of wheat and barley caused by *Gaeumannomyces graminis* has been conducted in fields in Sweden.[22,44] Disease of spring and winter wheat caused by this pathogen increased after field applications of mecoprop or ioxynil. Other herbicides did not affect the disease. In studies on take-all of barley, disease was greater in plots treated with the mecoprop than in control plots or those treated with 2,4-D or MCPA.

Black root rot of two crops caused by *Thielaviopsis basicola* was also increased by some herbicides. Chloromben, applied at recommended rates, increased black root rot of soybean (*Glycine max*) caused by *T. basicola* and reduced plant stand, height, and yield.[45]

Herbicides like chloramben, EPTC, and Prometryne added to infested soil increased the severity of cotton damping-off and root rot (*T. basicola*).[6] The interaction between herbicide and pathogen may be injurious to the plant by suppressing its growth. Chandler and Santelmann[46] observed that trifluralin or Prometryne in combination with *R. solani* reduced the weight of cotton plants in growth chambers.

C. Mechanisms Involved in Increase of Diseases

The increase in disease incidence due to herbicides is the result of the positive or negative effects that the herbicide might have on each of the living organisms involved: the pathogen, the host, and the surrounding microorganisms. The sum of all these effects determines whether and to what extent the disease will be increased. A disease may be increased if the herbicide is toxic to the pathogen causing disease, and at the same time it reduces host resistance or the activity of antagonists of the pathogen.[34]

Herbicides may increase disease incidence by stimulating the growth and reproduction of pathogens, thereby increasing their population density. There are several reports on this aspect for which studies have been carried out in liquid, solid culture media, or in sterile soil. The stimulation of growth or reproduction of pathogens by herbicides in culture has been demonstrated.[34] However, the effects varied with the doses of the chemical. At high concentrations growth was inhibited, whereas at low concentrations it was stimulated. In culture media the growth of *F. oxysporum* f. sp. *lini* (maleic hydrazide, MH), *Fusarium* sp. (atrazine), *F. oxysporum* f. sp. *vasinfectum* (EPTC, paraquat, trifluralin), and *S. rolfsii* (EPTC) was enhanced by the chemical(s) given in parentheses. In sterilized soil, stimulation was recorded with *F. oxysporum* f. sp. *vasinfectum* (Prometryne, atrazine), *R. solani* (fluometuron, Prometryne), and *S. rolfsii* (atrazine, trifluralin, EPTC).[34]

Sikka et al.[47] hypothesized that the stimulatory effect of a small quantity of atrazine was due to better utilization of sugar from the media and not due to supplementary carbon and nitrogen. Richardson[48] has given a different explanation for the stimulatory effect, i.e., herbicides may neutralize or arrest the formation of self-inhibitors produced by the fungus. There are some fungi which can utilize herbicides as energy sources.[49] This is probably of little importance in natural soil, because the rates of herbicide incorporated to soil at the commercially used doses of application amount to less than 0.1% of the organic content of the soil.

Katan and Eshel[34] rightly pointed out that the studies on the effect of herbicides on soil fungi performed in sterilized soil might be more valuable for subsequent application as compared to these in culture media, since this system bears greater resemblance to the natural environment. The effects of different herbicides on certain soil pathogens were also determined by estimating CO_2 production in soil, nutrient utilization, enzyme activity, spore production, and several other processes. The most significant findings of these studies were that in most cases at particular rates, the various herbicides had a stimulating effect on the tested pathogen in the soil. A rise in carbon dioxide production in herbicide-treated soil may be due to the higher rate of growth of the pathogen, but may also result from some metabolic disturbances.[50] Other parameters used to evaluate the effects of herbicide in soil were saccharase activity and spore production of fungi,[21] which were also found to be stimulated. Saccharase activity is in direct correlation with the dry weight of mycelium.[51] Due to physicochemical forces the effective biological activity of a given concentration of most chemicals is much lower in soil than the activity of the same concentration in culture. The explanation for the stimulatory effect might be attributed to the fact that the concentration of the herbicide fell within its stimulatory range when used in soil.[34] The possible increase of different soilborne plant pathogens by the herbicides used is too dangerous to be overlooked and, hence, needs a thorough investigation.

A pathogen incites disease by means of sequential metabolic processes involving enzymatic activity and toxic production. Environmental factors, such as chemical stimuli or nutrients, might enhance these processes and hence increase the virulence of the pathogen.[52] There are reports indicating that many pesticides affect metabolism of pathogens and, thus, they may also enhance their virulence. The mechanism which might be responsible for the increase of disease caused by an herbicide has not received much attention. However, there are only a few conclusive reports on the subject. A technique for testing this possibility with a given pathogen is to grow the pathogen in a liquid medium, supplemented with the test herbicide. After a desired period of incubation, the inoculum is removed, washed free of the herbicide, and compared for pathogenicity with inoculum grown on an herbicide-free medium.[34] The pathogenicity of *Helminthosporium sativum* grown on a medium supplemented with 2,4-D was somewhat higher than the control.[53] No increase in virulence of *R. solani* grown on diphenamid-supplemented medium was noticed.[54] Some herbicides affect nutrient level in the host; they may also indirectly affect the virulence of the pathogen in addition to other influences.[34]

Herbicides may increase the susceptibility of hosts or even break their resistance, by interfering with one or more stages of the plant's defense mechanism. The possibility that after the use of herbicides the increase in susceptibility of the host is responsible for an increase in disease incidence has been demonstrated by several workers. Herbicides are known to induce abnormal growths of cells and tissues and, therefore, may provide conditions for easier penetration of the pathogens. Trifluralin, which is known as an inhibitor of root growth, increased damping-off disease.[34] Anderson and Griffin[55] demonstrated increased inhibition of root and top growth in both alfalfa and tomato when infestation with root knot nematode was in the presence of trifluralin. Abnormal growth of the diseased plant is usually accompanied by changes in the content of growth-regulating substances in the tissues.[56]

Many studies attribute greater susceptibility to changes in the nutrient content of host tissues following soil application of herbicides. *Fusarium* wilt was more severe in tomato plants with a lower phosphorus content following treatment with MH.[57] MH also increased susceptibility of flax to *Fusarium*.[58]

It is an established fact that many plant pathogenic fungi survive in the soil in an inactive form mostly as resting structures, in the absence of a host,[59] and also that carbohydrates and amino acids in root exudates of the host incite propagule germination, which is then followed by penetration of the host. Root exudates have a direct effect on disease incidence. Herbicides which enhance root exudates will, in turn, increase the degree of infection by soilborne diseases. Altman[60] postulated the increased susceptibility to *Rhizoctonia* to the greater amount of glucose exudate at the soil-plant interface when sugar beet plants were grown in herbicide-treated soil. Similarily, increase of root rot of corn in picloram-treated soil was associated with increased carbohydrate exudation.[61]

Incomplete selectivity of herbicides to plant species may result in different kinds and degrees of phytotoxicity and stunting of the crop. Damping-off diseases are mainly connected with the young seedling stage of the plant. It has been observed that herbicides increase these diseases by retarding plant growth and exposing them to infection for a longer time. However, weakened plants are not necessarily more predisposed to attack by pathogens.[34] It was shown that certain types of injury to the roots of tomato plants resulted in less severe *Fusarium* wilt.[62] Davis and Dimond[63] have shown that 2,4-D reduced growth of tomato plants and at the same time reduced *Fusarium* wilt. Several techniques have been used to demonstrate the effect of herbicides on plant susceptibility to pathogens, but in many of them other possible mechanisms for disease increase were not excluded. Use of sterile soil eliminates the effect of soil microorganisms, but not that of direct stimulation of pathogen growth or virulence. Katan and Eshel[54] followed a technique by treating plants with herbicides and later transplanting them to herbicide-free soil inoculated with the pathogen for studying the effect on susceptibility of the plants to soilborne diseases. They have demonstrated that diphenamid enhanced *Rhizoctonia* disease in pepper, but had no effect on the susceptibility of the seedlings. Cotton seedlings become more susceptible to *R. solani* infection when pretreated with trifluralin.[31]

Soilborne pathogens exist in the soil in active or in passive forms and are much influenced by the dense population of the microbes which exist in natural soil. The quantity, quality, and activity of these organisms in the soil are important to determine the inoculum density of the pathogens and, consequently, the disease incidence. These factors also determine the survival of the pathogens in the soil in the absence of their host.[34] Soil organisms which are antagonistic to pathogens are very common in soil. This phenomenon of antagonism is the main reason for the frequently observed lower pathogenicity of pathogens in natural soil than with a sterile one. Several different possible mechanisms of antagonism may exist in soil:[64] (1) competition for limited amounts of nutrients, oxygen, space, or other common requirements; (2) the release of toxic products (antibioticity) which inhibits the growth of the pathogen; and (3) direct parasitism or predation. Herbicide(s) incorporated into soil might be injurious to a certain pathogen per se and yet be beneficial to it in the soil environment. This occurs when the herbicide reduces the antagonists to a greater extent as compared to the pathogen. Herbicides might result in the disturbance of the biological equilibrium from the following adverse effects on the antagonists of the pathogen: (1) reduction of their number (2) decrease in their capacity to produce antibiotics or lytic enzymes and (3) decrease in their capacity to compete with pathogens for nutrients. Paraquat sprayed on potato haulm altered the outcome of competition between *F. culmorum*, a known cereal pathogen, and the known antagonist *Trichoderma viride* in favor of the former,[65] perhaps due to the higher sensitivity of *T. viride* to paraquat. Chopra et al.[21] reported that Prometryn affected the magnitude of antibiosis of certain antagonists of *F. oxysporum* f. sp. *vasinfectum* using the baiting technique. Neubauer and Avizohar-Hershenzen[31] observed an increase in the saprophytic activity of *R. solani* in trifluralin-treated soil. Since this chemical is inhibitory to the pathogen, its stimulating effect on saprophytism was attributed to a shift in the biological equilibrium. Enhanced saprophytic activity of *R. solani* was observed with diphenamid.[54]

It has been suggested by many authors that normal rates of most herbicides have no pronounced adverse effect on the soil microflora. They reported that a normal dose of Karmex did not have any deleterious effect on soil microbial population. However, Fink et al.[67] observed that atrazine and Simazine greatly decreased the number of certain species of *Aspergillus* and *Penicillium*. A total number count of the microbes may not, however, reveal the changes in certain specific elements which may be vital for the maintenance of soil fertility or for biological control of plant diseases.[34] The influence of herbicides on microbes in the presence of nutrients, which are important during the germination and the penetration into host tissues, might be significant to study. According to Lai and Semeniuk,[61] soil pathogens invade roots through the rhizosphere in which there is a pronounced microbial activity. Soil applications of herbicides which lead to changes in microbial composition of the rhizosphere may well affect the inoculum of the pathogen and, consequently, disease incidence.

D. Decrease of Plant Disease due to Herbicides

Herbicide incorporation may result in a decrease in the incidence of various plant diseases. It might be due to the effect the herbicide has on the pathogen, the host, or the surrounding microorganisms. Owing to its potential usefulness, this type of effect deserves much attention. Huber et al.[68] obtained better development and higher yield from winter wheat treated with diuron, which was attributed to the reduction (at 50% or more) in root rot, while trifluralin was found to increase the disease in some instances. Harvey et al.[69] demonstrated reduction in root rot (*Aphanomyces euteiches*) on shelled peas with peas with trifluralin at 0.56 kg/ha followed by propachlor at 0.56 kg/ha application. They concluded that trifluralin was responsible for the protection observed. In greenhouse tests, Dinitramine and trifluralin also reduced *Aphanomyces* root rot and enhanced pea yields in Minnesota fields.[6] Trifluralin, in combination with Dinoseb, significantly reduced pea root rot caused by a complex of fungi (*A. euterches, F. oxysporum, F. solani, R. solani, Pythium* sp.) and enhanced yields. Trifluralin alone, as well as profluralin, reduced disease in the field. As early as 1953, it was thought that the larger peanut plants in herbicide-treated soil were due to an inhibitory effect of the herbicides on soilborne plant pathogens. After several years, it was demonstrated that there was 60% fewer dead plants in *S. rolfsii*-infested Dinoseb-treated fields then in infested fields with no herbicide.[70] Garren[27] noticed that Dinoseb, at field rates, decreased peanut stem rot and enhanced yield 2 out of 3 years. Papavizas and Lewis[6] demonstrated that Dinoseb added at normal rates to soil naturally infested with *P. aphanidermatum, P. myriotylum,* and *P. ultimum* 1 week before planting greatly decreased damping-off and blight. They further observed that Dinoseb application immediately before emergence or 1 to 3 days after emergence was not as effective as the preplant applications. In vitro studies showed that Dinoseb was fungistatic to all three species of *Pythium* with ED_{50} values obtained at 2 to 7 μg/mℓ.

Herbicides not only reduced soilborne diseases of leguminous plants (bean, pea, peanut), but also some serious diseases of cereals. Diuron, applied to fields infested with *Pseudocercosporella herpotrichoides* at 1.2 kg/ha, decreased the incidence and severity of root rot as compared to control plots.[68] Buczacki[71] found that trifluralin mixed with soil before sowing of cabbage lowered the incidence of club root (*P. brassicae*). Benfluralin and isopropalin also reduced club root, while nitralin and Dinitramine did not. Huber et al.[68] showed that diuron at 1.12 kg/ha reduced severity of root rot of winter wheat caused by *Cercosporella herpotrichoides*. Growth of *C. herpotrichoides* was not inhibited below 100 ppm Diuron on cornmeal dextrose agar. Initial host penetration by the pathogen was not affected, but host resistance seems to be enhanced by the herbicide. Cole and Batson[72] demonstrated that diphenamid added at 6.72 kg/ha to a steamed sand and silty clay loam decreased preemergence damping-off and enhanced postemergence damping-off caused by *P. aphanidermatum* and *R. solani*, which resulted in overall increased tomato stand. On artificial medium-containing herbicides, growth of these pathogens was also reduced. Dinoseb and trifluralin at field rates reduced damping-off of snapbeans in *P. irregularie*-infested soil.[40] Increased growth of onions in dacthal-treated plots may be due to a fungicidal action of dacthal.[73] Paul and Schonbeck[74] tested the influence of diallate on several disease pathogens in vitro. Diallate added to agar-reduced mycelial growth of *F. avenaceum, F. culmorum, F. graminium,* and *F. moniliforme.* Diallate incorporated into soil led to reduction of root rots on maize *F. moniliforme* and on wheat *F. avenaceum* and *F. culmorum.* In hydroculture, diallate reduced root rot on maize. In maize roots treated with diallate, *F. moniliforme* was restricted to the cell layers and did not enter the stele. In roots without the herbicide, the fungus did penetrate the stele. Hyphae in roots exposed to diallate were often partially deformed and the cytoplasm seemed to be strongly granulated. Lipids, globular and spherical bodies, were less abundant, but *P. glucosidase* activity was about 25% greater in herbicide-treated plants than in control plants. Brandes and Heitefuss[75] worked on the physiological and biochemical alterations in wheat during the period of influence by Simazine and monolinuron. In Simazine-treated plants, total nitrogen content increased almost 30%, the amino acids threonine and valine enhanced slightly, and asparagine content was conspicuously higher. Total sugar was less abundant with Simazine and monolinuron treatment. Quantities of glucose, pectose, and saccharase were about one half those in nontreated plants. In herbicide-treated plants, DMBO-glucoside content increased by 33%, while DMBO-glucan content increased by approximately 50%. Brandes and Heitefuss[75] concluded that all the above alterations in toto were related to the disease decrease of *C. herpotrichoides* during the period of primary herbicidal action in wheat.

E. Mechanism Involved in Decrease of Diseases

According to Katan and Eshel,[34] the following three mechanisms may be associated with decrease of diseases due to herbicides: (1) direct toxic effects on the pathogen, (2) resistance of the host, and (3) relationships with microorganisms. Most studies on the effect of herbicides on plant pathogens have been made in culture, which showed various degrees of inhibition. Based on these studies an herbicide is usually regarded as potential fungicide. Various workers have generally used different concentrations of chemicals, and based on the data "nontoxic" and "toxic" effects are reported. In some cases an herbicide was considered "nontoxic" when it did not inhibit the tested pathogen at concentrations as low as 10 ppm, whereas in other cases, an herbicide was considered to be "toxic" when it partially curtailed the growth of a pathogen at concentrations as high as 500 ppm or more.

There are several pathogens (*Fusarium, Rhizoctonia,* and *Sclerotium*) which are inhibited even at concentrations of 10 ppm in in vitro studies.[34] Paraquat inhibited *S. rolfsii* in sterilized soil.[76] The fungitoxicity of herbicides might be affected by secondary factors. Richardson[77] reported that Dinoseb was very toxic to *F. oxysporum* at a concentration of 2.5 ppm at pH 3.5; at pH 7.5 it was not toxic even at 10 ppm. The wetting agent in the commercial formulation enhanced the fungitoxicity of paraquat.[78] *Botrytis* was more inhibited by 500 ppm of 2,4-D than was *Fusarium;*[79] 2,4 S-T was more toxic to several pathogenic fungi than 2,4-D.[80,81] Some pathogens, however, were more sensitive to 2,4-D than those mentioned above. *Pythium* was completely inhibited by 250 ppm of sodium and amine salts of 2,4-D,[82] and *Actinomyces scabies* by 50 ppm of the methyl ester.[83] Pathogens differ greatly in their sensitivity to the same herbicide. In several studies *Fusarium* was observed to be less sensitive to herbicides than other pathogens. *Rhizopus stolonifer* was fully inhibited by paraquat at 10 ppm, while *F. culmorum* was only partially inhibited at 500 ppm.[78] *B. cinerea* was more sensitive to bromoxynil than *F. nivale.*[84] *R. solani* was more sensitive to four dinitroanilines than was *Fusarium.*[85] *Rhizoctonia* and *S. rolfsii* were more sensitive to 11 herbicides than *S. bataticola.*[86] Herbicides may also suppress formation of propagation or reproduction units of pathogens and, therefore, may decrease their population. Sporulation of several pathogens was reduced by herbicides in vitro and even on the host. Sclerotia production of *S. rolfsii* was inhibited by various herbicides such as atrazine and fluometuron.[87,88]

According to Katan and Eshel,[34] few studies on herbicidal inhibition of fungi deal with the mode of action involved and its possible similarity to that in plants. It appears that the inhibitory effects of the chemical result from interference with fundamental physiological processes of the fungus which are similar to those of other oganisms. Physiological disturbances are therefore to be expected in fungi, especially when the herbicide is a mitotoxic poison, an uncoupler, or an agent which affects essential metabolic processes such as proteins and nucleic acid synthesis. Even herbicides whose toxicity is specific to processes occurring in higher plants such as photosynthesis (atrazine and substituted ureas) were found to be toxic to fungi. In such cases, fungitoxicity perhaps results from the secondary effects which are of minor importance in higher plants. It has been demonstrated that the dynamic rate of glucose catabolism by *Monilinia fructicola* was reduced by atrazine, Simazine, and fluometuron. Atrazine also suppressed the hexase monophosphate shunt (pentose) pathway.[89] EPTC caused accumulation of oxalic acid in *S. rolfsii* cultures with no increase in mycelium production. Rodriguez-Kabana and co-workers[51] suggested that this could result from a blockage in the tricarboxylic acid (TCA) cycle, which shuttles glucose into enhanced oxalic acid production.

Various reports as mentioned above suggest that herbicidal suppression of growth and reproduction formation of resting structures, and sporulation of pathogen are a possible mechanism for the decrease in disease. Katan and Eshel[34] rightly pointed out that the fungitoxicity of an herbicide in vitro cannot be used as a sole measure for its capacity to control the pathogen at its site of action. Physical and chemical forces may reduce its effective concentrations under natural conditions, making it much less toxic. They concluded that the question still remaining open is which of the several herbicides demonstrated to be fungitoxic in vitro can also act as fungicides in practice?

There are several reports showing that herbicides may reduce diseases by mechanisms other than toxicity to the pathogen. It has been reported that Prophan (IPC) and TCA reduced *Fusarium* wilt of tomato, but were not toxic to the fungus in culture.[18] Richardson[77] suggested that changes in metabolism of the host might have affected disease development. Similarly, Davis and Dimond[63] reported that 2,4-D reduced *Fusarium* wilt in tomato. They concluded that growth regulators probably reduced disease by inducing changes in host metabolism which regulate the gowth of the parasite and/or the elaboration of toxins. Miller and Ahrens[90] reported that Simazine reduced *Rhizoctonia* infestation of roots of taxus, though the growth of the fungus in culture was little affected. Diuron reduced foot rot in winter wheat, but did not suppress penetration of the fungus or its growth in culture, nor did it affect population counts of soil fungi, bacteria, and actinomycetes. Therefore, alteration in host resistance is again suggested as being responsible for decrease of the disease.[91]

There are reports indicating stimulation of antagonists by the application of herbicides, which suppress soil pathogens. Curl et al.[87] observed that in sterile soil treated with Simazine and amended with sources of carbon and nitrogen, the pathogen *S. rolfsii* was inhibited while the antagonist *T. viride* was stimulated. Growth of *T. viride* was also stimulated by fluormeturon,[92] atrazine,[93] and Simazine.[94] At some rates atrazine stimulated the inhibitory effect of certain antagonists to *S. rolfsii*.[95] Kaufmans[96] observation was that Linuron and Diuran decreased the total number of *Fusarium* and stimulated fungi, known to be antagonists of this pathogen, a shift which might contribute to its control. Von Yegen and Heitefuss[97] observed an increased population of actinomycetes by application of TCA antagonistic to *Pythium*.

Herbicides are also known to suppress the pathogens for a prolonged period. Miller and Ahrens[90] observed that Simazine reduced *Rhizoctonia* on *Taxus* for a period of 2 years. This kind of information is of high importance as the same can be used in reducing the build-up of soilborne pathogens.

IV. INSECTICIDES AND NEMATICIDES

Richardson[77] observed that the insecticides isodrin (an isomer of aldrin) and lindane enhanced tomato wilt caused by *Fusarium oxysporum* f. sp. *lycopersici,* whereas aldrin, endrin, and DDT reduced wilt. Aldrin also reduced barley seedling blight; none of these materials was toxic to the pathogens in culture. He concluded that the insecticides may alter the metabolism of the host, thereby decreasing or increasing resistance. These experiments were conducted in sand culture. Similarly, Mussa and Russell[43] observed that the insecticides nicotine, triazophos, and demeton S-methyl enhanced preemergence damping-off and root rot of bean in sterile sand caused by *F. solani* f. sp. *phaseoli.* According to these workers, applications of these insecticides to soil might aggravate foot and root rot. However, it is very difficult to draw conclusions on what may happen in the field from data obtained in sterile soil. Hacskalyo and Stewart[98] observed that phorate reduced *Rhizoctonia solani* damage to cotton, and later Erwin et al.[99] observed that seed treatment of cotton with phorate resulted in an increased number of plants per unit area of soil infested with the pathogen. However, this can lead to predisposition of the treated seed to attack by *Pythium* spp.[100] Sinha et al.[101] noticed that the growth of *R. solani* appreciably decreased as the concentration of the chemical increased in the medium (the ED_{50} value 58.0 ppm). Two phosphorothioate nematicides (ethoprop and fensulfothion) commonly used in peanuts may be toxic to *Sclerotium rolfsii* and other fungi.[102,103] Ethoprop was shown to be a toxic to *S. rolfsii* and *R. solani* on PDA, but did not significantly affect growth of the antagonistic species of *Trichoderma* or saprophytic fungi in the genera *Rhizopus* and *Aspergillus.* Ethoprop suppressed growth of *R. solani* and *S. rolfsii* in soil and enhanced invasion of *S. rolfsii* colonies by antagonistic *Trichoderma* spp. In Georgia, ethoprop enhanced the root rot and damping-off complex of snapbeans in field plots infested with *P. myriotylum, P. irregulare,* or *F. roseum* more than the other pesticides or combinations tested.[41] Sumner and Glaze[39] also noticed that ethoprop and ethoprop combined with the herbicide dimethyl tetrachloroterephthalate (DCPA) significantly enhanced root rot of turnip (*Brassica rapa* subsp. *rapifera*) caused by *R. solani, P. irregulare, F. solani,* and *F. oxysporum* and decreased plant stand and foliage weight. They concluded that the nontarget effects of the mixture on root rots of turnip were due to ethoprop alone, as DCPA alone brought about less detrimental side effects than ethoprop. The activity of ethoprop against *S. rolfsii* was equivalent to that of PCNB and field incorporations to peanuts at blooming time consistently reduced disease damage. The discrepancy on the nontarget effects of ethoprop (beneficial effects on peanuts, undesirable effects on snapbeans and turnip) may be explained by the fact that ethoprop experiments in the two states dealt with different soils, environmental conditions, and particularly different host-pathogen interactions. Fensulfothion was also observed to suppress growth of *S. rolfsii* and *R. solani* on PDA, but in contrast to ethoprop it did not stop mycelial development of *S. rolfsii* in soil and reduced production of sclerotial initials only. Fensulfothion did not affect the rate of development of *Trichoderma* spp. on colonies of *S. rolfsii* in the soil plate. In the field, fensulfothion incorporated at blooming time reduced damage by *S. rolfsii* during the early part of the season. However, this reduction was not apparent at harvest time. Thompson[104] also observed reduced incidence of *S. rolfsii* in peanut fields by fensulfothion application. Insecticides and nematicides could decrease or increase soilborne plant diseases by various mechanisms not clearly understood or elucidated. These toxicants may be directly fungicidal or fungistatic to a given soilborne pathogen in vitro or in the field over prolonged periods of time. Direct fungistatic activity has been reported for aldicarb on *R. solani,*[105] for Phorate on *R. solani* and *S. rolfsii,*[101] for 1,2-dibromo-3-chloropropane (DBCP) on pythiaceous fungi,[106,107] for fensulfothion on *R. solani* and *S. rolfsii,*[102] for dasonit on *S. rolfsii* and *P. aphanidermatum,*[101] and for lindane on *R. solani, R. bataticola, S. rolfsii,* and *P. aphanidermatum.* Ethoprop greatly inhibited growth of *S. rolfsii* in vitro and eliminated production of sclerotia.[103] Nematicides and insecticides may bring about nontarget effects on soilborne diseases by increasing or decreasing inoculum density and saprophytic activity of a given pathogen. Inoculum density of *Sclerotinia sclerotiorum* was enhanced by D-D mixture added to soil for nematode control in lettuce.[108] Enhanced D-D rates progressively enhanced stipe production from sclerotia, thus increasing ascospore inoculum. A decrease in inoculum density has been reported for DBCP on *Pythium* number in soil[107] and by ethoprop on *Pythium* populations in turnip seedling roots.[39] A decrease in saprophytic colonization has been demonstrated for aldicarb on *R. solani.*[105] Toxicants may indirectly affect soilborne pathogens by increasing or destroying nontarget antagonistic microorganisms. Rodriguez-Kabana et al.[103] observed the ethoprop stimulated growth and proliferation of *Trichoderma* spp. and *Aspergillus* spp. in soil and indirectly increased invasion of *S. rolfsii* colonies by *Trichoderma.* This latter organism (*Trichoderma*) seems to be sensitive not only to ethoprop, but also to fensulfothion.[102] The mycoparasitic action of *Trichoderma*, coupled with the fungistatic ability of ethoprop on *S. rolfsii*, may explain reduction in peanut stem blight by these toxicants.

Toxicants may indirectly affect soilborne diseases by the remaining predisposing factors (insects and nematodes). Erwin[109] has well reviewed nontarget effects of fumigants or vascular wilts by removing the predisposing insects or nematodes. Nematicides and insecticides may also enhance root diseases by being phytotoxic to a potential host, a side effect that appeared to be the case in aldicarb in the damping-off problem of sugarbeet caused by *R. solani*[105] and that of heptachlor in the barley seedling blight problem.[110] From these limited studies on insecticides, it does not appear that these chemicals bring about extensive changes in the soil ecosystem as far as soilborne plant disease development is concerned. More research on this group of pesticides is needed, however, in order to reach any conclusion.

V. PESTICIDES-SOIL-ROOT INTERFACE: PATHOGENS IN THE RHIZOSPHERE

The behavior of plant pathogens and associated microorganisms at the soil-root interface is mediated by factors which manipulate host physiology and the quantitative and qualitative nature of root exudates.[111] There are toxicants which are known to incite changes in root zones by which pathogens are affected, and changes in mycorrhizal fungi can also be expected to influence the mineral nutrition of plants with a resulting potential for predisposition to disease.

Little attention has been devoted to the pesticide effects on pathogens, especially in the rhizosphere or at the root surface. At this zone a pathogen concentrates sufficient inoculum supplemented with a readily available energy source (primary exudates) necessary for infection.[111] Those root exudates can be altered by the application of some toxicants to leaves.[112] Picloram and 2,4,5-T are released from roots after foliar threatments and the quantities exuded are large enough to affect contiguous plants.[113] Even smaller quantities would be sufficient to alter microbial activities. When roots of yellow poplar seedlings were dipped in a 150-μg/mℓ solution of benomyl and transplanted into pots of nursery soil for 10 days, a rhizosphere effect was created. The effect was evident by an increase in microbial populations, particularly species of *Trichoderma* and *Streptomyces.* Bensen[114] studied trifluralin and another cotton herbicide fluometuron for their effects in the rhizosphere of seedlings. Chlamydospores were added into the rhizosphere of cotton cultivars in cups of nonsterilized soil, then recovered 12 hr later, and stained preparations were examined for germination. Whereas fluometuron had little effect on chalmydospore germination, trifluralin at 10 μg/ g induced significantly more abundant germination in the rhizosphere than in root-free soil. However, a direct correlation between spore germination and the quantitative nature of root exudation was not established. Fluometuron increased root development of cotton seedlings and trifluralin adversely affected root systems, which suggest that stress by the latter probably induced a greater quantity of exudation that affected spore germination.

Brown[115] suggested that herbicide-induced attenuation of root exudation by a noncrop plant (*Cassia obtusifolia*) may affect *F. oxysporum* f. sp. *vasinfectum* in the rhizosphere. Three-day-old plants, cultured in tubes with glass beads and sterile water and then foliar treated with the herbicide linuron, released 19.8% more root exudates than plants that were untreated. Chlamydospore germination, buried on millipore filters in nonsterilized rhizosphere soil from herbicide-stressed weed plants, was suppressed below the natural fungistatic value recorded for soil from untreated plants. It is generally assumed that the nutrients from root exudates serve to negate the fungistatic properties of soil; the foregoing study appears to imply that enhanced exudation by this herbicide-stressed weed plant may have stimulated microflora that contributed to a more intense fungistatic condition. In Georgia, Sumner[116] reported that alterations in rhizosphere ecology by toxicants can have very practical implications. The effects of herbicide DCPA treatments on root diseases of turnip in intensive cropping systems were studied under field conditions. Root disease severity enhanced and root growth as well as yield of turnip leaves decreased in soils that received herbicide treatments. Populations of *Pythium* spp. increased in rhizosphere soil at the same time. Herbicide applications suppressed populations of *F. solani* in the rhizosphere of fall crops, but favored an increase in *Fusarium* in spring crops. According to Rodriguez-Kabana and Curl,[111] this study clearly demonstrates that the effects of herbicide applications on plant pathogens can vary considerably with season and various cropping systems.

VI. PESTICIDE-MYCORRHIZAE INTERACTION

Mycorrhizae are necessary components of most plant systems. The role of mycorrhizae-forming fungi in plant nutrients, uptake, water transport, and the biological control of some root diseases is well demonstrated.[117,118] However, there is comprehensive information available in the literature on direct and indirect effects of pesticides on those fungi. Virtually any chemical will kill an organism exposed to excessive concentrations. Glucose in optimum concentration in a culture of an extomycorrhizal fungus will promote its growth, but glucose in excess will reduce growth or even lyse the fungus. As with other soil microbes, the endophytes that lead to vesicular-arbuscular infection are greatly influenced by the soil environment, including pesticides. The results of axenic culture experiments, even at low pesticide concentrations, unfortunately do not necessarily correlate with the performance of the same fungi as mycobionts with hosts when exposed to the same pesticides in the glasshouse or field.[119,120] Such experiments can have interpretive value when fungal growth is either not affected or is enhanced at pesticide concentrations likely to be encountered under field application conditions.[121] Several interactions of mycorrhizal fungi, their hosts, and the environment must be recognized in considering the impact of pesticides. Any herbicides that severely damage the host will almost certainly damage the mycorrhizae and this the mycorrhizal fungus.[111] Moderate damage to the host, in contrast, does not necessarily suppress mycorrhizae formation or curtail the mycorrhizal fungi.[122] The influence of alachlor, trifluralin, and diazinon on the development of endogenous mycorrhizae in soybeans was investigated. It was observed that these herbicides applied at commercial rates (2 kg/ha) did not significantly affect rot colonization by mycorrhizal fungi.[30] Trifluralin and diazinon at 4 kg/ha also had no significant effect on mycorrhizal development, but a reduction in root colonization by these fungi was observed in 25- and 60-day-old soybean plants grown in soil treated with 4-kg/ha alachlor. Iyer and Wilde[123] attributed the reduction in forest productivity to the eradication or impairment of mycorrhizae-forming fungi in the rhizosphere by pesticides. Data on herbicide-mycorrhizae association are conflicting and have been derived from in vitro studies. It was noticed that several herbicides affected mycelium production of ectotrophic mycorrhiza-forming fungi, but only at concentrations of 10 μg/ml or more, unreasonable concentrations for field use. Some herbicides may even stimulate mycelial growth of mycorrhiza-forming fungi in vitro.[124] Dasilva et al.[125] studied mycorrhizae-forming *Sullus* and *Rhizopogon* and showed that six pesticides significantly affected the growth and metabolism of these fungi in vitro. Low concentrations of phenoxyacetic acid herbicides stimulated growth of *S. variegatus* and *S. luteus*. High concentrations curtailed growth of these two species and *R. roseolus*. The herbicides amitrole and paraquat both reduced growth of these fungi at 1 and 5 μg/ml, whereas the insecticide malathion stimulated their growth. Toxicants may be expected to affect growth and metabolism of mycorrhiza-forming fungi.

Other toxicants besides herbicides may influence mycorrhiza-forming fungi. Jalali and Domsch[126] demonstrated that foliar applications with conventional and systemic fungitoxicants restricted the development of endotrophic mycorrhizal fungi on wheat roots. Since foliar-applied pesticides may not be translocated intact to the roots, Jalali and Domsch[126] postulated that the side effect on mycorrhizae was brought about by changes in the spectrum of wheat root exudates as a result of the stress created by the pesticides. In further tests the systemic fungicides triforine and tridemorph applied to wheat foliage changed the pattern of amino acid exudation. Menge et al.[127] observed that soaking inoculum (hyphae, vesicles, arbuscles, and chlamydospores) of *Glomus fasciculatum* in suspensions of PCNB, ethazole, or DBCP did not impair its viability; the highest concentrations of the pesticides tested were 4000, 80, and 240 μg/ml active ingredient (a.i.), respectively. The authors also observed that when Sudan grass, inoculated with the fungus, was drenched separately with PCNB (1000 μg/ml) at the time of inoculation the fungicide restricted spore production by 70% after 104 days, whereas ethazole and DBCP did not significantly alter spore production. When the pesticides were applied 60 days after inoculation, PCNB inhibited spore production by 90%, while ethazole and DBCP increased spore production by 76 and 63%, respectively. Pfleger and Stewart[128] noticed the effect of fungicides on *Glomus* spp. and *Pisium sativum* (cv. Green Giant 539) and observed that in-furrow applications of pyroxychlor (1.1 kg/ha ai) significantly reduced numbers of chlamydospores from *G. fasciculatus*. Thiram, dichloram, captan, PCNB, and Lanstan mixed with soil restricted mycorrhizal development of *Endogone fasciculata* (*G. fasciculatus*) on corn.[129] These authors also demonstrated that the preplant compounds dazomet, etham sodium, and Vorlex had similar effects. Of the five fungicides, captan was least and PCNB the most harmful to the fungus; among the fumigants, Vorlex was somewhat less injurious than dazomet and etham sodium. Root volume and branch root development were severely limited by thiram, dichloram, and PCNB. The three materials also caused some distortion of root hairs, and thiram reduced the number of root hairs that formed. Powell et al.,[130] working with feeder root necrosis of pecans (*Carya illinoensis*) caused by *Pythium* spp. and nematodes in Georgia, observed that application of DBCP and four fungicides reduced the disease without reducing *Pythium* populations in soil. There was no apparent overall decrease in nematode populations. Since applications of DBCP and the fungicides enhanced the mycorrhizal fungus, *Scleroderma bovista*, they concluded that there was a possible side effect of the pesticides on feeder root necrosis through the mycorrhiza-forming fungus.

Stunting and chlorosis of citrus seedlings in nurseries fumigated with methylbromide (MB) in California and Florida were thought to be caused by soil toxicity resulting from fumigations. Henderson and Stone[131] demonstrated that stunting and chlorosis were caused by the complete destruction of *Endogone* spp. that enter into endomycorrhizal associations with citrus. Earlier this "disease" was believed to be caused by soil toxicity. Stunted plants from nurseries grew normally when they were inoculated with *Endogone*.

The effects of nonfumigant nematicides on vesicular arbuscular mycorrhizae (VAM) have not been studied extensively. However, influence of these compounds on VAM has been noticed. Backman and Clark[132] examined the effects of several soil toxicants on VAM in 'Florunner' peanuts in field experiments and noticed that the nematicide carbofuran caused a significant decrease in VAM in peanut roots 30 days after planting. The effect of carbofuran in the greenhouse was also noticed. Mycorrhizae development in the peanut feeder roots returned to normal levels when field plots were reexamined 120 days after planting. According to Rodriguez-Kabana and Curl,[111] this type of compound is more likely to disturb the normal pattern of mycorrhizae established in plants. The in vitro inhibitory effects of pesticides on mycorrhizae may not tell us exactly what side effects may occur in nature. However, in vitro studies will provide background information for field experiments on pesticide mycorrhizae associations.[6] In view of the importance of mycorrhizal associations in plant growth and the relative lack of knowledge concerning pesticide-mycorrhizae interactions, further research should include an evaluation of potentially beneficial and detrimental interactions whenever pesticides are incorporated directly to soil or are translocated from the plant into the soil after a foliar application.

REFERENCES

1. **Munnecke, D. E.,** Fungicides in the environment, in *Fungicides, an Advanced Treatise,* Vol. 7, Torgeson, D. C., Ed., Academic Press, New York, 1967, 510.
2. **Gooding, G. V. and Lucas, G. G.,** Tobacco stalk and root destruction with herbicides and their effect on tobacco mosaic virus, *Plant Dis. Rep.,* 53, 174, 1969.
3. **Kavangh, T.,** Influence of herbicides on plant disease. I. Temperate fruit and crops, *Sci. Proc. R. Dublin Soc.,* 132, 1979, 1969.
4. **Heathcote, G. D.,** Weeds herbicides and plant virus diseases, *Proc. Br. Weed Control Conf.,* 10, 934, 1970.
5. **Franklin, M. T.,** Inter-relationship of nematodes, weeds, herbicides, and crops, *Proc. Br. Weed Control Conf.,* 10, 927, 1970.
6. **Papavizas, G. C. and Lewis, J. A.,** Side effects of pesticides on soil-borne plant pathogens, in *Soil-Borne Plant Pathogens,* Schippers, B. and Gams, W., Eds., Academic Press, London, 1979, 483.
7. **Gibson, I. A. S.,** Trials of fungicides for the control of damping-off in pine seedlings, *Phytopathology,* 46, 181, 1956.
8. **Gibson, I. A. S., Ledger, M., and Boehm, E.,** An anomalous effect of PCNB on the incidence of damping-off caused by *Pythium* sp., *Phytopathology,* 51, 531, 1961.
9. **Rich, S. and Miller, P. M.,** Verticillium wilt of strawberries made worse by soil fungicides that stimulate Meddow nematode populations, *Plant Dis. Rep.,* 48, 246, 1964.
10. **Powell, N. T.,** Interactions between nematodes and fungi in disease complexes, *Annu. Rev. Phytopathol.,* 9. 253, 1971.
11. **Beute, M. K., Porter, D. M., and Hadley, B. A.,** Sclerotinia blight of peanut in North Carolina and Virginia and its chemical control, *Plant Dis. Rep.,* 59, 697, 1975.
12. **Porter, D. M.,** The effect of chlorothalonil and benomyl on the severity of *Sclerotinia* blight of peanuts, *Plant Dis. Rep.,* 61, 995, 1977.
13. **Erwin, D. C.,** Systemic fungicides: disease control, translocation, and mode of action, *Annu. Rev. Phytopathol.,* 11, 389, 1973.
14. **Backman, P. A., Rodriguez-Kabana, R., and Williams, J. C.,** The effect of peanut leaf spot fungicides on the non target pathogen, *Sclerotium rolfsii, Phytopathology,* 65, 773, 1975.
15. **Warren, C. G., Sanders, P. L., and Cole, H., Jr.,** Increased severity of *Pythium* blight associated with use of Benzimidazole fungicides on creeping bent grass, *Plant Dis. Rep.,* 60, 932, 1976.
16. **Jackson, N.,** Evaluation of some chemicals for control of stripe smut in Kentucky blue grass turf, *Plant Dis. Rep.,* 54, 168, 1970.
17. **Williams, R. J. and Ayanaba, A.,** Increased incidence of *Pythium* stem rot in cowpeas treated with benomyl and related fungicides, *Phytopathology,* 65, 217, 1975.
18. **Altman, J. and Campbell, C. L.,** Effect of herbicides on plant diseases, *Annu. Rev. Phytopathol.,* 15, 361, 1977.
19. **Sinha, A. P., Agnihotri, V. P., and Singh, K.,** Effect of alachlor and diuran on growth, pigmentation and morphological abnormalities of some rhizosphere fungi of sugarbeet, *Pestology,* 3, 16, 1979.

20. **Curl, E. A. and Rodriguez-Kabana, R.,** The Relation of Soil Micro-Organism to Soil-Borne Plant Pathogens, Papavizas, G. C., Ed., Southern Coop. Ser. Bull. 183, Virginia Polytechnic Institute and State University, Blacksburg, 1974, 39.

21. **Chopra, B. K. and Curl, E. A., and Rodriguez-Kabana, R.,** Influence of prometryne in soil on growth related activities of *Fusarium oxysporum* f. sp. *vasinfectium, Phytopathology,* 60, 717, 1970.

22. **Nilsson, H. E.,** Influence of the herbicides, mecoprop on *Gaeumannomyces graminis* and take-all of spring wheat, *Swed. J. Agric. Res.,* 3, 105, 1973.

23. **Grau, C. R.,** Effect of Dinitramine and Trifluralin on growth reproduction and infectivity of *Aphanomyces enteiches, Phytopathology,* 67, 551, 1977.

24. **Percich, J. A. and Lockwood, J. L.,** Influence of Atrazine on the severity of *Fusarium* root rot in pea and corn, *Phytopathology,* 65, 154, 1975.

25. **Wyse, D. L., Meggitt, W. F., and Penner, D.,** Herbicide-root rot interaction in navybean, *Weed Sci.,* 24, 16, 1976.

26. **Tang, A., Curl, E. A., and Rodriguez-Kabana, R.,** Effect of Trifluralin on inoculum density and spore germination of *Fusarium oxysporum* f. sp. *vasinfectum* in soil, *Phytopathology,* 60, 1082, 1970.

27. **Garren, K. H.,** An evaluation of role of dinoseb in "non-dirting" control for peanut stem rot, *Plant Dis. Rep.,* 43, 665, 1959.

28. **Curl, E. A. and Wiggins, E. A.,** Herbicide effects on the competitive colonization of substrate by *R. solani, Proc. Am. Phytopathol. Soc.,* 2 (Abstr.), 132, 1975.

29. **Champbell, C. L. and Altman, J.,** Pesticide-plant disease interactions: effect of cycloate on growth of *Rhizoctonia solani, Phytopathology,* 67, 557, 1977.

30. **Pincard, J. A. and Standifer, L. C.,** An apparent interaction between cotton herbicidal injury and seedling blight, *Plant Dis. Rep.,* 50, 172, 1966.

31. **Neubauer, R. and Avizohar-Hershensen, Z.,** Effect of the herbicide, Trifluralin, on Rhizoctonia disease in cotton, *Phytopathology,* 63, 651, 1973.

32. **Altman, J. and Ross, M.,** Plant pathogens as a possible factor in unexpected preplant herbicide damage in sugarbeets, *Plant Dis. Rep.,* 51, 86, 1967.

33. **Katan, J. and Eshel, Y.,** Effect of the herbicide Diphenamid on damping off disease of pepper and tomato, *Phytopathology,* 64, 1186, 1974.

34. **Katan, J. and Eshel, Y.,** Interactions between herbicides and plant pathogens, *Res. Rev.,* 45, 145, 1973.

35. **Anderson, L. P.,** Histological and Cytological Responses of Cotton to Trifluralin and Interactions with the Damping-Off Pathogens, Ph.D. thesis, University of Georgia, Athens, 1968.

36. **Standifer, L. C., Jr., Melviue, D. R., and Phillipps, S. A.,** A possible interaction between herbicidal injury and the incidence of seedling disease in cotton plantings, *Proc. South. Weed Conf.,* 19, 126, 1966.

37. **Roming, W. R. and Sasser, M.,** Herbicide predisposition of snap beans to *Rhizoctonia solani, Phytopathology,* 62(Abstr.), 785, 1972.

38. **Salt, G. A.,** Effect of herbicides on take-all in wheat, *Rep. Rothamstead Exp. Stn.,* p. 113, 1961.

39. **Sumner, D. R. and Glaze, N. C.,** Interactions of herbicides and nematicides with root diseases of turnip grown for leafy greens, *Phytopathology,* 68, 123, 1978.

40. **Altman, J. and Campbell, C. L.,** Pesticide plant disease interactions: effect of cycloate on sugar beet damping-off induced by *Rhizoctonia solani, Phytopathology,* 67, 1163, 1977.

41. **Sumner, D. R.,** Interactions of herbicides and nematicides with root diseases of snapbean and southern pea, *Phytopathology,* 64, 1353, 1974.

42. **Wyse, D. L., Meggitt, W. F., and Penner, D.,** Effect of herbicides on the development of root rot on navy bean, *Weed Sci.,* 24, 11, 1976.

43. **Mussa, A. E. A. and Russell, P. E.,** The influence of pesticides and herbicides on the growth and virulence of *Fusarium solani* f. sp. *phaseoli, J. Agric. Sci. Camb.,* 88, 705, 1977.

44. **Nilsson, H. E.,** Influence of herbicides on take all and eye spot disease of winter wheat in a field trial, *Swed. J. Agric. Res.,* 3, 115, 1973.

45. **Lee, M. and Lockwood, J. L.,** Enhanced severity of *Thielaviopsis basicola* root rot induced in soybean by the herbicide chloramben, *Phytopathology,* 67, 1360, 1977.

46. **Chandler, J. M. and Santelmann, P. W.,** Interactions of four herbicides with *Rhizoctonia solani* on seedling cotton, *Weed Sci.,* 16, 453, 1968.

47. **Sikka, H. C., Couch, R. W., Davis, D. E., and Funderburk, H. H., Jr.,** Effect of atrazine on the growth and reproduction of soil fungi, *Proc. South Weed Conf.,* 18, 616, 1965.

48. **Richardson, L. T.,** Effect of atrazine on growth response of soil fungi, *Can. J. Plant Sci.,* 50, 594, 1970.

49. **Guillemat, J., Charpentier, M., Tardieux, P., and Pochon, J.,** Interaction entre une chloro-amino triazine herbicide et la microflore fongique et bacterienne du sol, *Ann. Epiphyte,* 2, 261, 1960.

50. **Rodriguez-Kabana, R. and Curl, E. A.,** Effect of atrazine on growth of *Fusarium oxysporum* f. sp. *vasinfectum, Phytopathology,* 60, 65, 1970.

51. **Rodriguez-Kabana, R., Curl, E. A., and Peeples, J. L.,** Growth response of *Sclerotium rolfsii* to the herbicide EPTC in liquid culture and soil, *Phytopathology,* 60, 431, 1970.

52. **Weinhold, A. R., Dodman, R. L., and Bowman, T.,** Influence of exogenous nutrition on virulence of *Rhizoctonia solani, Phytopathology,* 62, 278, 1972.
53. **Hsia, Y.-T. and Christensen, J. J.,** Effect of 2,4-D on seedling blight of wheat caused by *Helminthosporium sativum, Phytopathology,* 41, 1011, 1951.
54. **Katan, J. and Eshel, Y.,** Increase in damping-off incidence of pepper caused by diphenamid, *Weed Sci. Soc. Am. Abstr.,* p. 100, 1972.
55. **Anderson, J. L. and Griffin, G. D.,** Interaction of DCPA and trifluralin with seedling infection by root-knot nematode, *Weed Sci. Soc. Am. Abstr.,* p. 5, 1972.
56. **Wood, R. K. S.,** *Physiological Plant Pathology,* Blackwell Scientific, Oxford, 1967.
57. **Waggoner, D. E. and Dimond, A. E.,** Effect of stunting agents, *Fusarium lycopersici* and maleic hydrazide, upon phosphorus distribution in tomato, *Phytopathology,* 42, 22, 1952.
58. **Nair, P. N.,** Effect of maleic hydrazide, thiourea, and 2,4-dinitrophenol on resistance to flax wilt, *Phytopathology,* 48, 288, 1958.
59. **Garett, S. D.,** *Pathogenic Root Infecting Fungi,* Cambridge University Press, New York, 1970.
60. **Altman, J.,** Predisposition of sugarbeets to *Rhizoctonia* damping off with herbicides, *Phytopathology,* 59, 1015, 1969.
61. **Lai, M. T. and Semeniuk, G.,** Picloram-induced increase of carbohydrate exudation from corn seedlings, *Phytopathology,* 60, 563, 1970.
62. **Keyworth, W. G. and Dimond, A. E.,** Root injury as a factor in the assessment of chemotherapeutants, *Phytopathology,* 42, 311, 1952.
63. **Davis, D. and Dimond, A. E.,** Inducing disease resistance with plant growth regulators, *Phytopathology,* 43, 137, 1953.
64. **Alexander, M.,** *Introduction to Soil Microbiology,* John Wiley & Sons, New York, 1961.
65. **Wilkinson, V. and Lucas, R. L.,** Influence of herbicides on the competition ability of fungi to colonize plant tissues, *New Phytol.,* 68, 701, 1969.
66. **Sinha, A. P., Agnihotri, V. P., and Singh, K.,** Population dynamics of microbes in a herbicide (Karmex) amended sugarcane soil, *Indian Sugar Crops J.,* 3, 1, 1979.
67. **Fink, R. J., Fletchall, O. H., and Calvert, O. H.,** Relation of triazine residues to fungal and bacterial colonies, *Weed Sci.,* 16, 104, 1968.
68. **Huber, D. M., Seely, C. I., and Watson, R. D.,** Effects of the herbicide diuron on foot rot of winter wheat, *Plant Dis. Rep.,* 50, 852, 1966.
69. **Harvey, R. G., Hagedorn, D. J., and Deloughery, R. L.,** Influence of herbicides on root rot in processing peas, *Crop Sci.,* 15, 67, 1975.
70. **Chappell, W. E. and Miller, L. T.,** The effect of certain herbicides on plant pathogens, *Plant Dis. Rep.,* 40, 52, 1957.
71. **Buczacki, S. T.,** Effects of trifluralin and related dinitroaniline herbicides on club root in Brassicae, *Ann. Appl. Biol.,* 75, 25, 1973.
72. **Cole, A. W. and Batson, W. E.,** Effects of diphenamid on *Rhizoctonia solani, Pythium aphanidermatum,* and damping-off of tomato, *Phytopathology,* 65, 431, 1975.
73. **Altman, J., Ross, M., and Sunik, J.,** Soil fumigation for onion disease and weed control, *Colo. Agric. Exp. Stn. Prog. Rep.,* p. 144, 1965.
74. **Paul, V. and Schonbeck, F.,** Untersuchungen über der Einfluss des Herbizide Diallat auf emige Getreidekrankheiten, *Phytopathol. Z.,* 85, 1289, 1976.
75. **Brandes, W. and Heitefuss, R.,** Nebenwirkung von Herbiziden auf *Erysiphe graminis* and *Cercosporella herpotri, choides* an weizen. II. Physiologische und biochemische ursachen des veranderten Befalls der Pflanze, *Phytopathol. Z.,* 72, 34, 1971.
76. **Rodriguez-Kabana, R. and Curl, E. A.,** Effect of paraquat on growth of *Sclerotium rolfsii* in liquid culture and soil, *Phytopathology,* 57, 911, 1967.
77. **Richardson, L. T.,** Effect of insecticides and herbicides applied to soil on the development of plant disease. II. Early blight and *Fusarium* wilt of tomato, *Can. J. Plant Sci.,* 39, 30, 1959.
78. **Wilkinson, V. and Lucas, R. L.,** Effects of herbicides on the growth of soil fungi, *New Phytol.,* 68, 709, 1968.
79. **Mostafa, M. A. K. and Gayed, S. K.,** A comparative study of the effect of cotton dust and 2,4-D on three pathogenic fungi, *Mycopathol. Mycol. Appl.,* 13, 198, 1960.
80. **Erickson, L. C., Dewolf, T. A., and Brannaman, B. L.,** Growth of some citrus fruit pathogens as affected by 2,4-D and 2,4,5-T, *Bot. Gaz.,* 120, 31, 1958.
81. **Tourneau, D. Le. and Buer, L.,** The toxicity of some chlorinated phenols and aryloxyalkane carboxylic to *Verticillium albo-atrum, Phytopathology,* 51, 128, 1961.
82. **Bever, W. M. and Slife, F. W.,** Effect of 2,4-D in culture medium on the growth of three pathogenic fungi, *Phytopathology,* 38, 1038, 1948.

83. **Michaelson, M. E., Schaal, L. A., and Fults, J. L.,** Some effects of 2,4-dicholrophenoxyacetic acid, its salts, and esters on several physiologic strains of the potato scab organism *Actinomyces scabies* (Thaxt.) Guss., *Soil Sci. Soc. Am. Proc.,* 13, 267, 1949.

84. **Smith, M. E. and Fletcher, W. W.,** 3,5-Dihalogeno-4-hydroxybenzonitriles and soil micro-organisms, *Hortic. Res.,* 4, 60, 1964.

85. **Eshel, Y. and Katan, J.,** Effect of dinitroanilines on solanaceous vegetables and soil fungi, *Weed Sci.,* 20, 243, 1972.

86. **Bain, D. C.,** Effect of various herbicides on some soil fungi in culture, *Plant Dis. Rep.,* 45, 814, 1961.

87. **Curl, E. A., Rodriguez-Kabana, R., and Funderburk, H. H., Jr.,** Influence of atrazine and varied carbon and nitrogen amendments on growth of *Sclerotium rolfsii* and *Trichoderma viride* in soil, *Phytopathology,* 58, 323, 1968.

88. **Bozarth, G. A. and Tweedy, B. G.,** Effect of pesticides on growth and sclerotial production of *Sclerotium rolfsii, Phytopathology,* 61, 1140, 1971.

89. **Tweedy, B. G. and Loeppky, C.,** The use of ^{14}C-labeled glucose, glucuronate, and acetate to study the effect of atrazine, simazine, and fluometuron on glucose catabolism in related plant pathogenic fungi, *Phytopathology,* 58, 1522, 1968.

90. **Miller, P. M. and Ahrens, J. F.,** Effect of an herbicide, a nematocide and a fungicide on *Rhizoctonia* infestation of Taxus, *Phytopathology,* 54, 901, 1964.

91. **Huber, D. M., Seely, C. I., and Watson, R. D.,** Effects of nonfungicidal chemical control of foot rot of winter wheat, *Phytopathology,* 58, 1054, 1968.

92. **Bozarth, G. A., Funderburk, H. H., Jr., and Curl, E. A.,** Interaction of fluometuron and soil microorganisms, *Weed Sci. Soc. Am. Abstr.,* p. 236, 1969.

93. **Rodriguez-Kabana, R. and Curl, E. A.,** Effect of atrazine on growth response of *Sclerotium rolfsii* and *Trichoderma viride, Can. J. Microbiol.,* 13, 1343, 1967.

94. **Eno, C. F.,** The effect of simazime and atrazine on certain of the soil microflora and their metabolic process, *Soil Sci. Soc. Fla. Proc.,* 22, 49, 1962.

95. **Curl, E. A. and Funderburk, H. H., Jr.,** Some effects of atrazine on *Sclerotium rolfsii* and inhibitory soil microorganisms, *Phytopathology,* 55, 497, 1965.

96. **Kaufman, D. D.,** Effect of S-triazine and phenylurea herbicides on soil fungi in corn and soybean cropped soil. *Phytopathology,* 54, 897, 1964.

97. **Von Yegen, O. and Heitefuss, R.,** Nebenwirkungen von Natriumtrichloracetat (NaTA) auf den Wurzelbrand der Riben und das antiphytopathogene Potential des Bodens, *Zucker,* 3, 723, 1970.

98. **Hacskaylo, J. and Steward, R. N.,** Efficacy of phorate as a fungicide, *Phytopathology,* 52, 371, 1962.

99. **Erwin, D. C., Reynolds, H. T., and Garber, M. J.,** Effect of seed treatment of cotton with thimet, a systemic insecticide, on seedling disease in the field, *Plant Dis. Rep.,* 43, 558, 1959.

100. **Erwin, D. C., Reynolds, H. T., and Garber, M. J.,** Predisposition to *Pythium* seedling disease and an activated charcoal-fungicide interactions as factors influencing emergence of cotton seed treated with phorate, *J. Econ. Entomol.,* 54, 855, 1961.

101. **Sinha, A. P., Agnihotri, V. P., and Singh, K.,** Effect of insecticides on the growth and sporulation of rhizosphere fungi of sugarbeet seedlings, *Indian Phytopathol.,* 33, 96, 1980.

102. **Rodriguez-Kabana, R., Backman, P. A., Karr, G. W., Jr., and King, P. S.,** Effects of the nematicide fensulfothion on soil-borne pathogens, *Plant Dis. Rep.,* 60, 521, 1976.

103. **Rodriguez-Kabana, R., Backman, P. A., and King, P. S.,** Antifungal activity of the nematicides ethoprop, *Plant Dis. Rep.,* 60, 255, 1976.

104. **Thompson, S. S.,** Control of southern stem rot of peanuts with PCNB plus fensulfothion, *Peanut Sci.,* 5, 49, 1978.

105. **Tisserat, N., Altman, J., and Campbell, C. L.,** The influence of Aldicarb on growth of *R. solani* Kuhn and sugarbeet seedling damping-off, *Proc. Am. Phytopathol. Soc.,* 3(Abstr.), 220, 1976.

106. **Brodie, B. B.,** Use of 1,2-dibromo 3,chloropropane as a fungicide against *Pythium ultimum, Phytopathology,* 51, 798, 1961.

107. **Bumbieris, M.,** Effect of DBCP on pythiaceous fungi, *Plant Dis. Rep.,* 54, 622, 1970.

108. **Partyka, R. E. and Mai, W. F.,** Nematicides in relation to *Sclerotial* germination in *Sclerotinia sclerotiorum, Phytopathology,* 48, 519, 1958.

109. **Erwin, D. C.,** Control of vascular pathogens, in *Antifungal Compounds,* Vol. 1, Siegel, M. R. and Sisler, H. D., Eds., Marcel Dekker, New York, 1977, 163.

110. **Richardson, L. T.,** Effect of insecticides and herbicides applied to soil on the development of plant disease. I. The seedling disease of barley caused by *Helminthosporium sativum* P. K. & B., *Can. J. Plant Sci.,* 37, 196, 1957.

111. **Rodriguez-Kabana, R. and Curl, E. A.,** Nontarget effects of pesticides on soil-borne pathogens and diseases, *Annu. Rev. Phytopathol.,* 18, 311, 1980.

112. **Hale, M. G., Foy, C. L., and Shay, F. J.,** Factors affecting root exudation, *Adv. Agron.,* 23, 89, 1971.

113. **Grovannetti, M. and Riess, S.,** Effects of soil applications of systemic fungicides on bulb formation in onions, *Plant Soil,* 57, 463, 1980.

114. **Bensen, G. I.,** Effect of Fluormeturon, Trifluralin, and the Rhizosphere of Cotton Seedlings on the Inoculum Potential of *Fusarium oxysporum* f. sp. *vasinfectum,* Ph.D. thesis, Auburn University, Auburn, Ala., 1976, 108.

115. **Brown, S. L.,** Rhizosphere Effect of Herbicide-Stressed Sickle-Pod (*Cassia obtusifolia*) on *Fusarium oxysporum* f. sp. *vasinfectum,* M. S. thesis, Auburn University, Auburn, Ala., 1979, 74.

116. **Sumner, D. R., Glaze, N. C., Dowler, C. C., and Johnson, A. W.,** Herbicide treatments and root diseases of turnip in intensive cropping systems, *Plant Dis. Rep.,* 63, 801, 1979.

117. **Marx, D. H.,** Ectomycorrhizae as biological deterrents 3558 to pathogenic root infections, *Annu. Rev. Phytopathol.,* 10, 429, 1972.

118. **Mosse, B.,** Advances in the study of vesicular-arbuscular mycorrhiza, *Annu. Rev. Phytopathol.,* 11, 171, 1973.

119. **Cudlin, P., Mejstrik, V., and Skoupy, J.,** Effect of pesticides on ectomycorrhizae on *Pinus sylvestris* seedlings, *Plant Soil,* 71, 53, 1983.

120. **Marx, D. H. and Rowan, S. J.,** Fungicides influence growth and development of specific ectomycorrhizae on loblolly pine seedlings, *For. Sci.,* 27, 167, 1981.

121. **Trappe, J. M., Molina, R., and Michael, C.,** Reaction of mycorrhizal fungi and mycorrhiza formation to pesticides, *Annu. Rev. Phytopathol.,* 22, 331, 1984.

122. **Nemee, S.,** Effect of 11 fungicides on endomycorrhizal development in sour orange, *Can. J. Bot.,* 58, 522, 1986.

123. **Iyer, J. G. and Wilde, S. A.,** Effect of vapam biocide on the growth of pine seedlings, *J. For.,* 63, 703, 1965.

124. **Uhlig, S. K.,** Untersuchungen uber die wechselwirkung Zeischen-chlor-bisathylamino-s-triazin (Simazin) and mykorrhizabildenden Pilzen, *Wiss. Z. Tech. Univ. Dresden,* 15, 639, 1966.

125. **Dasilva, E. J., Henrikson, L. E., and Udris, M.,** Growth responses of mycorrhizal *Boletus* and *Rhizopogon* species to pesticides, *Trans. Br. Mycol. Soc.,* 68, 434, 1977.

126. **Jalali, B. L. and Domsch, K. H.,** Effect of systemic fungitoxicants on the development of endotrophic mycorrhiza, in *Endomycorrhizas,* Sanders, F. E., Mosse, B. and Tinker, P. B., Eds., Academic Press, New York, 1975, 619.

127. **Menge, J. A., Minassian, V., and Johnson, L. J. V.,** The effects of heat treatment and three pesticides upon inoculum of the vesicular arbuscular mycorrhizae fungus *Glomus fasciculatus, Proc. Am. Phytopathol. Soc.,* 3(Abstr.), 274, 1976.

128. **Pfleger, F. L. and Stewart, E. L.,** The influence of fungicides on endemic populations of *Glomus* species in association with field grown peas, *Proc. Am. Phytopathol. Soc.,* 3, 274, 1976.

129. **Nesheim, O. N. and Linn, M. B.,** Deleterious effect of certain fungitoxicants on the formation of mycorrhizae on corn by *Endogone fasciculate* and on corn root development, *Phytopathology,* 59, 297, 1969.

130. **Powell, W. M., Hendrix, F. F., Jr., and Marx, D. H.,** Chemical control of feeder root necrosis of pecans caused by *Pythium* species and nematodes, *Plant Dis. Rep.,* 52, 577, 1968.

131. **Henderson, G. S. and Stone, E. L.,** Interactions of phosphorus availability, mycorrhizae and soil fumigation on coniferous seedlings, *Soil Sci. Soc. Am. Proc.,* 34, 314, 1970.

132. **Backman, P. A. and Clark, E. M.,** Effect of carbofuran and other pesticides on vesicular-arbuscular mycorrhizae in peanuts, *Nematropica,* 7, 13, 1977.

Chapter 12

FUNGICIDE RESISTANCE

I. INTRODUCTION

All around the world, agricultural crops are under constant attack by pests and diseases. In several situations, the use of resistant varieties and other cultural practices proves inadequate for the control of pests and diseases and use of chemical control measures becomes necessary. In modern agriculture, chemicals play a predominant role in the fight against hunger, especially in developing countries. Unfortunately, however, the use of chemicals has its own limitations. Living organisms possess the ability to adapt to changing environmental conditions and the causal organism becomes resistant to the chemicals. The ability of organisms to adapt to changing environmental conditions and to survive under often adverse conditions is one of the most fundamental properties of living matter.[1] This adaptation may also occur to changes brought about by man. The adaptation of bacteria to antibiotics is well known and becomes apparent shortly after the application of these biocides for control of bacterial diseases in human beings. In recent years, especially with the introduction of systemic fungicides with selective action on fungi, several cases of fungicide resistance have been reported.[2-5]

Organisms which adapt to a toxicant by a reduction of their sensitivity are called "resistant" or "tolerant". Such a decrease in sensitivity may be caused by genetic or nongenetic changes in the fungal cell. The latter type of change is not stable and usually disappears rapidly in the absence of the toxicant. The FAO Panel of Experts on Pest Resistance to Pesticides (1979) has recommended that the term "resistance" should apply to these hereditable changes in sensitivity in fungi and bacteria, and the word "tolerance" should not be used in this sense, as it is ambiguous.[1] The sudden appearance of resistance in a pathogen population may, when it is not recognized at an early stage, result in failure of disease control and, consequently, serious crop losses.[2-5] Under such situations, the use of an originally effective compound has to be restricted or even abandoned. Farmers are put in a difficult situation if no other substitute fungicides are available. Resistance certainly has the potential to increase the farmer's cost of production and decrease his profits, unless he can pass on the increased costs to the consumers.

II. OCCURRENCE OF RESISTANCE

Discovery of the bordeaux mixture by Millardet in 1882 was the beginning of the commercial application of fungicides to fight plant diseases. No problem of resistance to this fungicide has been reported, and it occupied a prime place for more than 50 years. The same is true for organic mercury compounds and dithiocarbamates (DTC) which were introduced around 1914 and 1930, respectively. These compounds provide protection at the surface of the plant. After World War II, a search started for development of fungicides which could penetrate the plant and eradicate pathogens from the site of infection. Compounds which are taken up by the plant and transported throughout the plant system are called systemic fungicides.

A. Protective Fungicides
These fungicides have rarely given serious problems with respect to resistance, and there are only a few reports on development of resistance in practice.[6] Failure to control *Pyrenophora avenae* on oats by seed treatment with organic mercuric compounds has been reported

from Scotland.[7] Cole et al.[8] and Nicholson et al.[9] reported development of resistance in *Sclerotinia homoeocarpa* on turf grass to cadmium-containing compounds and to dyrene, respectively, in the U.S. Failure in effectiveness of dodine for apple scab control was observed in the U.S. 10 years after its first use.[10] Isolates of *Venturia inaequalis* from orchards with inadequate scab control were less sensitive to dodine as compared with the isolates from those orchards where adequate scab control was obtained.[10,11] Several workers have reported development of resistance for fungicides belonging to the aromatic hydrocarbon group, viz., pentachloronitro-benzene (PCNB),[12] dicloram,[13,14] diphenyl,[15] and hexachlorobenzene (HCB).[16] Three excellent reviews on the development of resistance for fungicides in general, and for inorganic and organic fungicides have been written by Georgopoulos,[17] Ashida,[18] and Georgopoulos and Zaracovitis.[19]

B. Systemic Fungicides

Various research groups and the chemical industry played important roles in the search for systemic fungicides, testing thousands of available and newly synthesized chemicals. Various fungicides that have systemic action were discovered in this way. Most of them were not suitable for practical use, however, since they gave phytotoxic side effects. It was difficult to find compounds selective enough to kill the fungus without harming the crop. It also became clear that there might be other problems. While working with 6-azauracil, an experimental systemic fungicide, Dekker[20] found strains of *Cladosporium cucumerinum* that suddenly had become resistant to the chemical. Soon after the application of systemic fungicides was started in practice, reports appeared about increasing failure of some of these compounds to provide disease control. Benomyl resistance in cucumber powdery mildew was reported within 1 year after introduction of this fungicide.[21] Large-scale failures of disease control due to the development of benzimidazole resistance by target fungi[3,4,21,23] have been reported frequently in the last few years, and undoubtedly there must be additional cases which have not appeared in the literature.[17]

Development of resistance in *Sphaerotheca fuliginea* to dimethirimol is another example. The fungicidal suspension applied to the soil around the base of cucumber plants in greenhouses gave protection for several weeks. However, failure in satisfactory disease control was observed within 2 years, and Bent et al.[22] reported development of dimethirimol-resistant strains from those areas. Resistance to benzimidazole fungicides in strains of *Botrytis cinerea*[23] and *Verticillium fungicola*[24] has also been reported. Ergosterol-biosynthesis inhibitors are broad-spectrum fungicides, toxic to representatives of Ascomycetes, Deuteromycetes, and Basidiomycetes. So far, practical application of fungicides that inhibit ergosterol biosynthesis has not led to any confirmed case of resistance. Recently, however, Walmsley-Woodward et al.[25] reported increased levels of resistance of *Erysiphe graminis* f.sp. *hordei* to tridemorph (Calixin) in both greenhouse and field experiments. In tridemorph-resistant strains, pathogenicity and resistance were found to be negatively correlated.[26,27]

The acylalanines have become in a short time increasingly important in the control of several important plant pathogenic fungi belonging to the order Peronosporales. Damaging pathogens such as downy mildews of a number of crops, potato late blight, blue mold of tobacco, and soilborne *Pythium* and *Phytophthora* spp. are among the main target fungi. During the 1979 to 1980 growing season, cucumber growers from the southern Greek mainland, Crete, and Israel reported loss of effectiveness of metalaxyl in controlling cucumber downy mildew caused by *Pseudoperonospora cubensis*.[5,28,29] Their results showed that the fungicide was almost completely inactive against isolates from these areas. Inoculation of detached leaves floating on fungicide solutions demonstrated that a considerable difference existed in minimal inhibitory concentrations for sensitive and resistant isolates. Resistance appeared to be very stable. Even maintaining the organism for 20 generations over a period of 7 months on untreated plants did not result in any loss of resistance.

Davidse[30] obtained metalaxyl-resistant isolates of *Phytophthora megasperma* f. sp. *medicaginis*, the causal organism of root rot of alfalfa. Adaptation and mass selection from zoospores yielded 19 strains with a relatively low degree of resistance, all of which were less virulent than the original isolate in an alfalfa seedling assay. Poor performance of metalaxyl in controlling potato late blight (*P. infestans*) was reported by farmers in the Netherlands in 1980. Sensitivity of several isolates of *P. infestans* to metalaxyl was compared by leaf-disc assay.[2] Isolates originating from fields where metalaxyl failed were usually highly resistant, and up to a 1000-fold difference in sensitivity was noticed between the isolates. Resistant isolates were also resistant to furalaxyl, milfuram, and Galben.

III. CROSS RESISTANCE

"Cross resistance" means resistance to two or more toxicants mediated by the same genetic factor.[17] The term cross resistance is used when a change in one genetic factor results in resistance to different fungicides. It usually occurs only between compounds with a similar mechanism of action. When a fungicide fails because of the development of resistance by the target organism, in practice it is very important to know whether the effectiveness of other fungicides has been affected.[31] If mutation affects sensitivity to one toxicant and not to another, the two compounds are said to be uncorrelated in terms of cross resistance. A positive correlation exists when the mutant is less sensitive or less resistant to both toxicants than the wild-type strain. If the mutant is more resistant than the wild type to one chemical and more sensitive than the wild type to the other, one speaks of negatively correlated cross resistance.

A positive correlation between the carbendazim and thiabendazole appears to be the rule.[32] However, a low percentage of mutants of *Aspergillus nidulans* and *A. Niger*, obtained on benomyl medium, is not less sensitive to thiabendazole than the wild type, and somewhat higher percentages of those isolated on thiabendazole are more sensitive to benomyl (and to carbendazim) than the wild type. The mechanism responsible for the existing correlations is very well understood, since sensitivity differences in vivo are reflected in differences in the affinity of tubulin for the fungicides.[33,34]

In several fungi, mutants obtained by selection on one of the hydrocarbon fungicides are resistant to all members of this group of fungicides. The degree of resistance appears to be very high and it is impossible to achieve a concentration inhibitory to the resistance mutants.[19,35] A positive correlation between aromatic hydrocarbon fungicides and the dicarboximides (vinclozolin) has been recognized.[36] On the basis of their chemical structures, this positive correlation would be rather unexpected.

A great variety of cross-resistance relationships exist among members of the carboxamide group of fungicides. Georgopoulos and Ziogas[37] reported resistance in *Ustilago maydis*. The strain is also resistant to many carboxin analogues, including oxycarboxin and pyracarboloid. The phenomenon, which is called "negatively correlated cross resistance", has been reported in organophosphates, viz., phosphoramidate and phosphorothiolate fungicides in *Pyricularia oryzae*.[38,39] Ergosterol biosynthesis-inhibiting fungicides, viz., triforine, fenarimol, imazalil, tridemorph, etc., are mostly positively correlated for cross resistance.[40,41]

IV. DETECTION AND MEASUREMENT

A disease-control failure must be carefully studied before fungicide resistance can be confirmed as the cause. The factors that can lead to poor disease control with fungicides are as follows: improper fungicide or formulation; inadequate or faulty equipment; improper dosage, volume, or timing; incompatibility with other pesticides; severe epidemic conditions; human error; and finally resistance. After all other factors (such as faulty application on unusual weather) have been eliminated, resistance may be suspected. Fungicide resistance may be detected and measured in various ways, depending on the fungus-fungicide combination. However, the general principles are the same. As with other organisms, the recognition of resistant strains of fungi must be made by comparison with data obtained with sensitive strains. Therefore, "base-line sensitivity" for the fungus-fungicide combination in question must be established by experiment with wild-type strains (sensitive to the fungicide) or by using data from the literature.

Two important parameters must be measured after detection of resistance: the extent of resistance or the proportion of the population that no longer exhibits the sensitivity of the wild strains and the degree of resistance or the magnitude of the difference in the sensitivity. Plating spores or other propagules on agar media containing a fungicide concentration that completely prevents the growth of the wild type is the usual method for detecting resistance and its subsequent monitoring. Germination of spores in fungicide solutions or on media containing fungicide is also used to detect and measure resistance. The rate of increase of colony diameter on treated agar medium appears to be the most frequently used criterion for measuring fungitoxicity. The rate of dry weight increase in liquid media containing the fungicide is used less frequently for measuring fungitoxicity because its measurement requires more time.

Resistance can also be detected and measured on living, appropriately treated plants or plant parts. This method is practically indispensable in the case of obligate parasites like powdery mildews and downy mildews.

V. MECHANISMS OF RESISTANCE

The mechanisms whereby fungi become resistant to fungicides are diverse.[17] In some instances the mechanism may involve the site of action, giving an insight into the mode of action, but in other instances it may be unrelated. Some chemicals are toxic because they interact with cell constituents in a way that adversely affects important cellular processes. The ability of an organism to develop resistance to such a chemical is due to its cellular components being able to exist in more than one form without preventing survival under normal conditions. To understand a change of cellular components from one form to another, it is important to recognize whether there are genetic determinants (genes) involved and of what kind; and what cellular components (enzymes or others) are coded by these genes and in what way they have changed.[43] If a change in sensitivity does not involve a change in chromosomal or extrachromosomal DNA, it is thought to be the result of phenotypic adaptation. Such adaptation may take place in culture after long exposure to the fungicides. Phenotypic resistance is usually rapidly lost upon transfer to a toxicant-free medium. Whether phenotypic adaptation to fungicides is possible under field conditions is not known. Instability of resistance of field isolates may indicate phenotypic adaptation, but it should not be confused with regression of resistance in the field, which may be due to reduced fitness of genetic mutants.

A. Genetics

Most genetic studies on resistance to fungicides have revealed the involvement of chromosomal genes. As a rule, several genes control sensitivity to the same fungicides or group of fungicides. In many of the reported cases of acquired resistance to a fungicide, the phenomenon has not been analyzed genetically. In others, the genetic nature of sensitivity differences may have been established, but the study has been restricted to one or a very few resistant mutants; thus, the picture of genetic control is far from complete. However, the few more detailed studies available indicate that the genetics of fungicide resistance is complex. The easiest way to recognize genetic resistance is to use heterothallic ascomycete whose perfect stage is produced readily and that gives easily analyzable progeny within a few days. These advantages are offered by the nonpathogen *Neurospora crassa*,[44] and more or less by some pathogens, *Nectria haematococea*[45] and *Venturia inaequalis*.[46]

The example in which the action of a known gene that controls sensitivity to agricultural fungicides has been completely elucidated is that of the ben-A gene for resistance to benzimidazoles in *Aspergillus nidulans*.[47] This gene codes for B-tubulin, one of the subunits of the tubulin molecules. Mutation of this gene affects the electrophoretic properties of B-tubulin and at the same time the ability of the protein to bind carbendazim, which is inversely correlated to carbendazim resistance. Chromosomal genes for resistance to benzimidazoles have also been recognized in several others fungi,[48,49] although their actions have not been studied at the biochemical level. Another instance in which a gene for resistance has been identified and its action at the biochemical level has become known is the Oxr-1 gene for resistance to oxathiin carboxanilides in *Ustilago maydis*.[50] Of the five genes for resistance to aromatic hydrocarbon fungicides (quintozene, hexachlorobenzene, dicloran, and chloroneb),[51] two have been recognized in *A. nidulans*.[52] The same mechanism must control resistance to dicarboximides, which are classified with the aromatic hydrocarbon group on the basis of cross resistance and the ability to increase the frequency of mitotic segregation.[53] Involvement of three chromosomal loci has clearly been demonstrated with kasugamycin resistance in *Pyricularia oryzae*.[54] All kasugamycin-resistant progeny were without exception, resistant to Blasticidin-S. Both kasugamycin and Blasticidin-S act by inhibition of the protein-synthesizing system in *P. oryzae*.

B. Biochemistry

The fungi possess the potential for a wide range of metabolic variations resulting in resistance to those fungicides which interfere at specific sites with the fungal metabolism.[6] For the study of biochemical mechanisms it is usually necessary to examine the effect of the fungicide on the wild type and mutant at the cellular and subcellular level. The mechanisms by which resistance to a toxic compound may develop are modification of the sensitive site, by-pass of the site due to operation of an alternative pathway, decreased uptake or increased secretion of the toxicant, detoxification, or decreased conversion of nontoxic into a toxic compound.[43] Of these mechanisms, the first two are important for the development of resistance to mainly site-specific fungicides, the remaining three to specific and multisite fungicides. With resistance to the same toxicant, different mechanisms may operate in different organisms, or in different mutants of the same organism.

VI. COUNTERMEASURES FOR AVOIDING RESISTANCE

During a century of large-scale use of conventional fungicides, such as compounds based on copper, few problems with fungicide resistance have been encountered. However, since the introduction of systemic fungicides some 20 years ago, disease control failure due to fungicide resistance is reported with increasing frequency. Genetic changes in a pathogen resulting in fungicide resistance occur much more readily with fungicides acting primarily at one particular site in the metabolism of the fungal cell than with fungicides that interfere at many sites with metabolic processes.[6,17] All systemic fungicides appear to be single-site inhibitors, while the majority of conventional fungicides are multisite inhibitors. Among the systemic fungicides, some do encounter resistance problems more readily than others[55] (Table 1).

Experiences with multisite and specific site compounds indicate that the resistance problem may be tackled if use of specific site compounds is minimized or supplements use of the multisite compounds. Before a new fungicide is introduced commercially, knowledge about the risks of fungicide resistance is very desirable. Information about the potential of pathogens

Table 1
**RISK OF FAILURE OF DISEASE
CONTROL IN PRACTICE DUE TO
DEVELOPMENT OF RESISTANCE TO
SPECIFIC-SITE FUNGICIDES**

Site-specific fungicides	Risk of failure
Antibiotics	
Kasugamycin, streptomycin, pimaricin	Moderate to high
	Very low
Acylalanines	
Metalaxyl, furalaxyl	High
Benzimidazoles	
Benomyl, carbendazim, thiabendazole	High
Carboxamides	
Carboxin, oxycarboxin	Low to moderate
Dicarboximides	
Iprodione, procymidone, vinclozolin	Moderate
Hydroxypyrimidines	
Ethirimol	Low to moderate
Dimethirimol	High
Imidazoles	
Imazalil	Low
Morpholines	
Dedemorph, tridemorph, fenpropimorph	Very low
Organic phosphorus compounds	
Pyrazophos	Low
Edifenphos, kitazin	Moderate
Piperazines	
Triforine	Very low
Pyrimidines	
Fenarimol, nuarimol	Low
Thiophanates	
Thiophanate-methyl	High
Triazoles	
Triadimefon, triadimenol, bitertanol	Low

Note: This is a rough indication only, since risk not only
depends on type of fungicide, but also on type of
disease and use strategy for fungicide.

Adapted from Dekker.[55]

to become resistant to such fungicides is of prime importance. In case it is possible to choose different types of chemicals for the control of a particular disease, those with the lower risk should be preferred. To avoid unduly high selection pressure, the amount of fungicide should not exceed the minimum necessary for adequate disease control. Combined or alternate use of fungicides with different mechanisms helps to a great extent in delaying development of resistance. Kable and Jeffery[56] have developed a mathematical model for such use. Use of a mixture of fungicides with negatively correlated cross resistance also helps check build-up of the fungicide-resistant population. It is possible to prolong the useful life of fungicides by strategies of fungicide treatment that delay or avoid the development of resistance.[55] Fortunately, however, the problems of resistance in soil fungicides are far less as compared to the same fungicides when used as foliage spray.[57]

REFERENCES

1. **Dekker, J. and Georgopoulos, S. G.,** *Fungicide Resistance in Crop Protection,* Pudoc, Centre for Agricultural Publishing and Documentation, Wageningen, The Netherlands 1982, 265.
2. **Davidse, L. C., Looyen, D., and Turkensteen, L. G.,** Occurrence of metalaxyl-resistant strains of *Phytophthora infestans* in Dutch potato field and van dar Wal, D, *Neth. J. Plant Pathol.,* 87, 85, 1981.
3. **Giannopolitis, C. N.,** Occurrence of the strains of *Cercospora beticola* resistant to triphenyl tin fungicide in Greece, *Plant Dis. Rep.,* 62, 205, 1078.
4. **Georgopoulos, S. G. and Dovas, C.,** A serious outbreak of strains of *Cercospora beticola* resistant to benzimidazole fungicides in northern Greece, *Plant Dis. Rep.,* 57, 321, 1973.
5. **Georgopoulos, S. G. and Grigoriou, A. C.,** Metalaxyl, resistant strains of *Pseudoperonospora cubensis* in cucumber greenhouses of southern Greece, *Plant Dis.,* 65, 729, 1981.
6. **Dekker, J.,** Resistance, in *Systemic Fungicides,* 2nd ed., Marsh, R. W., Ed., Longman Group, London, 1977, 401.
7. **Noble, M., Maggarvie, Q. D., Hams, A. F., and Leafe, L. L.,** Resistance to mercury of *Pyrenophora avenae* in Scottish seed oats, *Plant Pathol.,* 15, 23, 1966.
8. **Cole, H., Taylor, B., and Dutch, J.,** Evidence of differing tolerances to fungicides among isolates of *Sclerotinia homoeocarpa, Phytopathology,* 58, 683, 1968.
9. **Nicholson, J. F., Meyer, W. A., Sinclair, J. B., and Buller, J. D.,** Turf isolates of *Sclerotinia homoeocarpa* tolerant to dyrene, *Phytopathol. Z.,* 72, 169, 1971.
10. **Szkolnik, M. and Gilpatrick, J. D.,** Apparent resistance of *Venturia inaequalis* to dodine in New York apple orchards, *Plant Dis. Rep.,* 53, 861, 1969.
11. **Gilpatrick, J. D. and Blowers, D. R.,** Ascospore tolerance to dodine in relation to orchard control of apple scab, *Phytopathology,* 64, 649, 1974.
12. **Shatla, M. N. and Sinclair, J. B.,** Tolerance to PCNB and pathogenicity correlated in naturally occurring isolates of *Rhizoctonia solani Phytopathology,* 53, 1047, 1963.
13. **Locke, S. B.,** Botran tolerance of *Sclerotium cepivorum* isolates from fields with different Botran treatment histories, *Phytopathology,* 59, 13, 1969.
14. **Webster, R. K., Orawam, J. M., and Bose, E.,** Tolerance of *Botrytis cinerea* to 2, 6-dichloro-4-nitroaniline, *Phytopathology,* 60, 1489, 1970.
15. **Harding, P. R.,** Assaying for diphenyl resistance in *Penicillium digitatum* in California lemon packing houses, *Plant Dis. Rep.,* 48, 43, 1964.
16. **Kuiper, J.,** Failure of hexachlorobenzene to control common bunt of wheat, *Nature (London),* 206, 1219, 1965.
17. **Georgopoulos, S. G.,** Development of fungal resistance to fungicides, in *Antifungal Compounds,* Vol. 2, Siegel, M. R. and Sisler, H. D., Eds., Marcel Dekker, New York, 1977, 674.
18. **Ashida, J.,** Adaptation of fungi to metal toxicants, *Phytopathology,* 3, 153, 1965.
19. **Georgopoulos, S. G. and Zaracovitis, C.,** Tolerance of fungi to organic fungicides, *Annu. Rev. Phytopathol.,* 5, 109, 1967.
20. **Dekker, J.,** Conversion of 6-azauracil in sensitive and resistant strains of *Cladosporium cucumerinum,* in *Wirkungsmechanismen von Fungiziden and Antibiotica,* in *Proc. Symp. Reinhardsbrunn, D. D. R.,* May 1966, 1967, 333.
21. **Schroeder, W. T. and Provvidenti, R.,** Resistance to benomyl in powdery mildew of cucurbits, *Plant Dis. Rep.,* 53, 271, 1969.
22. **Bent, K. J., Cole, A. M., Turner, J. A. W., and Woolner, M.,** Resistance of cucumber powdery mildew to dimethirimol., *Proc. 6th Br. Insectic. Fungic. Conf.,* 1, 274, 1971.
23. **Bollen, G. J. and Scholten, G.,** Acquired resistance to benomyl and some other systemic fungicides in a strain of *Botrytis cinerea* in cyclamen, *Neth. J. Plant Pathol.,* 77, 83, 1971.
24. **Bollen, G. J. and van Zaayen, A.,** Resistance to benzimidazole fungicides, in pathogenic strains of *Verticillium fungicola, Neth. J. Plant Pathol.,* 81, 157, 1975.
25. **Walmsley-Woodward, D. J., Laws, F. A., and Whittington, W. J.,** Studies on the tolerance of *Erysiphe graminis* f.sp. *hordei* to systemic fungicides, *Ann. Appl. Biol.,* 92, 199, 1979.
26. **Walmsley-Woodward, D. J., Laws, F. A., and Whittington, W. J.,** The characteristics of isolates of *Erysiphe graminis* f.sp. *hordei* varying in response to tridemorph and ethirimol, *Ann. Appl. Biol.,* 92, 211, 1979.
27. **Walmsley-Woodward, D. J., Laws, F. A., and Whittington, W. J.,** Comparison between isolates of *Erysiphe graminis* f.sp. *hordei* tolerant and sensitive to the fungicide tridemorph, *Ann. Appl. Biol.,* 94, 305, 1980.
28. **Reuveni, M., Eyal, H., and Cohen, Y.,** Development of resistance to metalaxyl in *Pseudoperonospora cubensis, Plant Dis.,* 64, 1108, 1980.
29. **Katan, T. and Bashi, E.,** Resistance to metalaxyl in isolates of *Pseudoperonospora cubensis,* the downy mildew pathogen of cucurbits, *Plant Dis.,* 65, 798, 1981.

30. **Davidse, L. C.,** Resistance to acylalanine fungicides in *Phytophthora megasperma* f.sp. *medicaginis, Neth. J. Plant Pathol.,* 87, 11, 1981.

31. **Georgopoulos, S. G.,** Cross resistance, in *Fungicide Resistance in Crop Protection,* Dekker, J. and Georgopoulos, S. G., Eds., Pudoc, Centre for Agricultural Publishing and Documentation, Wageningen, The Netherlands, 1982, 265.

32. **Van Tuyl, J. M.,** Genetics of fungal resistance to systemic fungicides, *Meded. Landbouwhogesch. Wageningen,* 77, 1977, 137.

33. **Davidse, L. C. and Flach, W.,** Differential binding of methyl benzimidazole-2-yl-carbamate to fungal tubulin as a mechanism of resistance to this antimitotic agent in mutant strains of *Aspergillus nidulans, J. Cell Biol.,* 72, 174, 1977.

34. **Davidse, L. C. and Flach, W.,** Interaction of thiabendazole with fungal tubulin, *Biochem. Biophys. Acta,* 543, 82, 1978.

35. **Tillman, R. W. and Sisler, H. D.,** Effect of chloroneb on the metabolism and growth of *Ustilago maydis, Phytopathology,* 63, 219, 1973.

36. **Sztejnberg, A. and Jones, A. L.,** Tolerance of the brown rot fungus *Monilinia fructicola* to iprodione, vinchlozolin, and procymidone fungicides, *Phytopathol. News,* 12, 187, 1978.

37. **Georgopoulos, S. G. and Ziogas, B. N.,** A new class of carboxin resistant mutants of *Ustilago maydis, Neth. J. Plant Pathol.,* 83, 235, 1977.

38. **Vesugi, Y., Katagiri, M., and Noda, O.,** Negatively correlated cross-resistance and synergism between phosphoramidate and phosphorothiolates in their fungicidal actions on rice blast fungi, *Agric. Biol. Chem.,* 38, 907, 1974.

39. **Katagiri, M., Uesugi, Y., and Umehara, Y.,** Development of resistance to organophosphorus fungicides in *Pyricularia oryzae* in the field, *J. Pestic. Sci.,* 5, 417, 1980.

40. **Sherald, J. L., Ragsdale, N. N., and Sisler, H. D.,** Similarities between the systemic fungicides triforine and triarimol, *Pestic. Sci.,* 4, 719, 1973.

41. **Fuchs, A., de Ruig, S. P., van Tuyl, J. M., and de Vries, F. W.,** Resistance to triforine: a nonexistent problem?, *Neth. J. Plant Pathol.,* 83, 189, 1977.

42. **Georgopoulos, S. G.** Detection and measurement of fungicide resistance, in *Fungicide Resistance in Crop Protection,* Dekker, J. and Georgopoulos, S. G., Eds., Pudoc, Centre for Agricultural Publishing Documentation, Wageningen, The Netherlands, 1982, 265.

43. **Georgopoulos, S. G.,** Genetical and bio-chemical background of fungicide resistance, in *Fungicide Resistance in Crop Protection,* Dekker, J. and Georgopoulos, S. G., Eds., Pudoc, Centre for Agricultural Publishing Documentation, Wageningen, The Netherlands 1982, 265.

44. **Borck, K. and Braymer, H. D.,** The genetic analysis of resistance to benomyl in *Neurospora crassa, J. Gen. Microbiol.,* 85, 51, 1974.

45. **Kappas, A. and Georgopoulos, S. G.,** Genetic analysis of dodine resistance in *Nectria haematococca, Genetics,* 66, 617, 1970.

46. **Polach, F. J.,** Genetic control of dodine tolerance in *Venturia inaequalis, Phytopathology,* 63, 1189, 1973.

47. **Sheir-Neiss, G., Lai, M. H., and Morris, N. R.,** Identification of a gene — tubulin in *Aspergillus nidulans, Cell,* 15, 639, 1978.

48. **Brasier, C. M. and Gibbs, J. N.,** MBC tolerance in aggressive and nonaggressive isolates of *Ceratocystis ulmi, Ann. Appl. Biol.,* 80, 231, 1975.

49. **Shabi, E. and Ben-Yephet, Y.,** Tolerance of *Venturia pirina* to benzamidazole fungicides, *Plant Dis. Rep.,* 60, 451, 1976.

50. **White, G. A., Thorn, G. D., and Georgopoulos, S. G.,** Oxathiin carboxamides highly active against carboxin-resistant succinic dehydrogenase complexes from carboxin selected mutants of *Ustilago maydis* and *Aspergillus nidulans, Pestic. Biochem. Physiol.,* 9, 165, 1978.

51. **Georgopoulos, S. G. and Panopoulos, N. J.,** The relative mutability of the cnb loci in *Hypomyces, Can. J. Gen. Cytol.,* 8, 347, 1966.

52. **Therlfall, R. J.,** The genetics and biochemistry of mutants of *Aspergillus nidulans* resistant to chlorinated nitrobenzenes, *J. Gen. Microbiol.,* 52, 35, 1968.

53. **Georgopoulos, S. G., Sarris, M., and Ziogas, B. N.,** Mitotic instability in *Aspergillus nidulans* caused by the fungicides iprodione, procymidone, and vinclozolin, *Pestic. Sci.,* 10, 389, 1979.

54. **Taga, M., Nakagawa, H., Tsuma, M., and Ueyama, A.,** Identification of three different loci controlling Kasugamycin resistance in *Pyricularia oryzae, Phytopathology,* 69, 463, 1978.

55. **Dekker, J.,** Countermeasures for avoiding fungicide resistance, in *Fungicide Resistance in Crop Protection,* Dekker, J. and Georgopoulos, S. G., Eds., Pudoc, Centre for Agricultural Publishing and Documentation, Wageningen, The Netherlands, 1982, 265.

56. **Kable, P. F. and Jeffery, H.,** Selection of tolerance in organisms exposed to sprays of biocide mixture: a theoretical model, *Phytopathology,* 70, 8, 1980.

57. **Mukhopadhyay, A. N. and Singh, U. S.,** Recent thoughts in plant disease control. II. Fungal resistance to fungicides, *Pesticides,* 19, 23, 1985.

Chapter 13

EVALUATION OF SOIL FUNGICIDES

I. INTRODUCTION

The value and importance of evaluating fungicides as being basic to selection for commercial development have appropriately been recognized. Prevost[1] was the first to demonstrate the efficacy of chemicals against fungal spores by means of laboratory tests. He had also correlated the laboratory results with field studies and was fully aware of the important factors of cost effectiveness, ease of handling and application, and toxicology. Since the beginning of planned research for new or improved fungicides, studies have been made in the laboratory with the aim that from the many chemicals which could be evaluated by these simpler methods, a few might be selected for the time-consuming and costly field trials. The bioassay techniques have been divided into basic and practical types.[2] Basic methods are in vitro techniques which are used to determine whether the test chemical has any inherent fungitoxic property, whereas practical methods are designed for the characterization of practical usefulness of the compound.

The primary object of bioassay is to determine the response of the organism exposed to the fungicide under conditions which minimize the influence of all but one of the factors affecting the response. The variable factors are usually the quantity of the compound to which the organism is exposed and time of exposure. Variations in other factors may arise through differences in the resistance of the individual organisms in their exposure to the fungicide and in the environmental conditions.

The science of fungicide research depends on the art of careful testing methodology and also on the reexamination of the methods in use. These should be based on four criteria: suitability, reliability, efficiency, and economy.

II. BASIC FUNGITOXICITY TESTS

The standard laboratory tests using the petri plate, glass slide, or test tube dilution methods are based primarily on the measurements of growth, germination, and respiration of fungi. Of these, the first two techniques are widely used to measure intrinsic fungitoxicity. However, Zentmyer[3] and Kendrick and Middleton[4] point out the inadequacy of these methods of evaluation of soil fungicides and the need for direct tests using soil. The degree of toxicity of a chemical to an organism in pure culture is markedly different from that shown against the same organism in its natural habitat, i.e., soil. The first involves only a fungicide-organism interaction; the second, fungicide-organism-soil interaction. However, spore germination and agar plate tests are still commonly used to measure fungitoxity, and after sorting out the effective chemicals from a large number of compounds, the selected chemicals may then be processed for specific characterization depending on their practical utility. The major advantages of these methods are their efficiency and economy. The tests can be performed with the limited facility in a short period of time, and require a smaller quantity of chemical. Latham and Linn[5] rightly pointed out that a combination of evaluation procedures is required for the characterization of fungicide for use in soil. The techniques are briefly reviewed below.

A. Spore Germination
The germination of fungus spores in the presence of the chemical has been particularly favored by plant pathologists and is probably the most extensively studied and widely used

of the fungicide test methods. Prevost[1], one of the fathers of modern plant pathology, demonstrated that copper sulfate solutions inhibited germination of wheat bunt spores. Later the spore germination method was also used by Carleton[6] and Swingle.[7]

1. Slide Germination

The technique was first introduced by Reddick and Wallace.[8] It consists of spraying the glass slides with the test chemicals and allowing them to dry. One drop of the test-spore suspension in water is placed on the treated dry slide and after a suitable incubation in a moist chamber the observation is recorded on spore germination. This technique was modified by Montgomery and Moore[9] by pipetting the fungicide drops on the slide and allowing them to dry before fungus spore drops are added. The method has been subjected to critical analyses, which have resulted in marked improvement, and it is now considered a precise technique. The errors associated with spore germination tests of fungicides arise from two sources: biological and mechanical. Rigid control and standardization of the techniques of producing and germinating the spores[10-13] have greatly lessened the biological errors, while further development in precision apparatuses for applying sprays and dusts has reduced the mechanical variation.[11,14-16]

A standard method[17] for evaluating protectant chemicals was suggested in 1943 by the Committee on Standardization of Fungicide Tests of the American Phytopathological Society. The method involves application of fungicide to chemically clean glass slides by means of a precision technique such as a settling tower or horizontal sprayer. The former is suitable for sprays or dust, the latter for sprays only. The deposition of the chemical is regulated to give a series of dosages varying in geometric progression. The slides are allowed to dry and then are placed on moist chambers. Fungus spores obtained under controlled conditions regarding species, strains, medium, age, temperature, concentrations, and stimulants, if desired, are suspended in distilled water and pipetted on the sprayed or dusted slides and held at a temperature in a moist chamber suitable for germination. In the case of water-soluble chemicals, the spores are added directly to the solution and aliquot, then pipetted onto untreated slides. After a specified time, the slides are placed under a low-power microscope and the spores are examined for germination. The percentage of spores inhibited from germination based on a specific count are plotted on logarithmic probability paper, and the LD_{90} or LD_{95} is obtained for comparing the relative merit of the various chemicals.

The flexibility of the test is such that its variations are nearly as numerous as the investigators who use it.

a. Slides

The glass slides used are the standard (3 × 1 in.) size. Suggestions for solving the difficulties associated with variations in drop size include the use of etched circles,[9] raised cover glass mounted on the slides,[18] ordinary slides coated with cellulose nitrate,[19-24] cavity slides or microbeakers,[25,26] and petri dishes,[27] and transparent materials such as plastic have been substituted for glass slides.[28]

b. Application of Chemicals

The chemicals may be applied with the help of precision apparatuses such as the settling tower,[16,29] horizontal sprayer,[14,15,19,30,31] vertical sprayer,[32] or pendulum sprayer.[33] However, in spite of the complexity of these techniques of obtaining uniform and reproducible amounts of the fungicide on glass slides, the results frequently have been variable.[34] Other methods of applying chemicals include the use of macropipette,[9] a solid glass rod with round top,[18] or test tube dusters.[24] However, until techniques can be demonstrated to be equal or better in precision than the settling tower or horizontal sprayer, they cannot be advocated for a standard technique.[17]

Settling towers — Standard settling towers function on the principle of filling the tower uniformly with the spray or dust and then allowing it to settle on the glass slides. Fungicide suspensions kept in a beaker and consistently stirred are sprayed up into the tower through an atomizer nozzle located near the base of the tower. When the tower is filled with a fine suspension, spraying is stopped and then a preliminary settling period is allowed for the heavy drops to fall. The slides are then introduced at the bottom. After a given period of exposure, the slides are withdrawn and the tower evacuated with a forced air draft. The procedure is then repeated and a series of chemical deposits are obtained on the slides according to the number of times they were exposed. It is absolutely necessary to standardize factors such as dimension of towers, spraying pressure, time of spraying, time of preliminary settling, time of exposure, and positional effect, if any, of the slides. A settling tower of known precision has been described by McCallan and Wilcoxon.[29] Every individual settling tower constructed must be calibrated; in addition, a special calibration is necessary for chemicals of unusual physical nature. Settling towers for dusts are modified as a given amount of chemical is placed in a "gun" and shot upward into the tower by sudden release of air pressure. Details about the apparatus have been reported.[16]

Horizontal sprayer — A horizontal sprayer is a stationary apparatus in which the chemical is sprayed horizontally through an atomizer nozzle onto a facing glass slide held at a distance of 20 to 30 in. The duration of spraying is accurately controlled by a cutoff valve or stop cork and the amount of chemical is determined by the time of spray. Details of the technique with illustrations have been described.[14,15,29]

c. Dose Ratio

Dosage, i.e., settling tower exposure, spraying time, or concentration of soluble materials, should be varied in geometric progression. The standard ratios of $\sqrt{2}$ (i.e., 1.414), 2, 4, $\sqrt{10}$ (i.e., 3.16), or 10, are recommended.[11,35] It is also suggested that sufficient dosages should be run so that at least three finite germination responses will be obtained.[11]

d. Standardization of the Test Fungus

Several species of fungi have been extensively used and gave satisfactory results.[17] Any species may be used provided it satisfies the requirements of reproducibility of results, ease of counting, and production of spores.[11] It has been seen that the percent of germination response varies with the dose/spore and there is a direct relation between the log, LD_{50} and density of spore suspension.[11,15] Therefore, it is necessary to control the number of spores in suspension. The final concentration should be adjusted to 50,000 spores per milliliter and this suspension will give about 35 spores/low power field (15 × ocular, 16 mm obj.).[11]

The spore suspension is applied to the slides by means of a 1- or 2-cc pipette and two pairs of drops may be placed on each slide, the four drops being in a staggered position.[24] In the case of the settling tower or horizontal sprayer, the first pair of spore suspension drops is of one fungus species and the second pair of another species. It is suggested that the drop should be approximately 0.05 mℓ in volume; on a plain glass slide the drops should spread to a diameter of about 10 mm, and on cellulose to a diameter of about 7.5 mm.[11,15]

Regarding the temperature for spore production in test cultures and for germination, it is well known that spores vary in their temperature requirements for germination,[36] and it has been shown that their resistance to a fungicide is greatest when the temperature is optimum.[13,37,38] In general, temperature ranges between 20 and 25°C are specified for spore production and germination.

It has also been reported that much of the day-to-day variation or replicate error may be due to the presence of a variable amount of water-soluble nutrients derived from the cultures when getting the spores.[12,18] These contaminating stimulants should be removed by obtaining the spores by a vacuum technique or they should be washed and centrifuged. Washed spores of some of the species may not germinate in distilled water; therefore, in such cases it is necessary to add a known quantity of stimulant like orange juice,[12,35] extracts from commercial dried potato-dextrose-agar,[18] or coenzyme R.[39] However, reports are on hand showing an increase in the resistance of the spores to chemicals by the use of stimulants[35,37] and, therefore, this may affect the comparisons of different species.

e. Observation for Germination

Spores may be examined 20 to 24 hr after the test begins. However, it has been observed that a fungus may respond to fungicides in a different order at a different time interval, though generally no further change may be expected after 50 hr.[13] Hence, observations at 6 and 24 hr are preferable for comparing various chemicals. One hundred potentially viable spores should be counted per concentration of each chemical and for each species. The spore is arbitrarily defined as germinated if the length of the germ tube exceeds half the smaller diameter of the spore.

f. Dosage Response or Toxicity Curve

The percentage of spores inhibited from germinating may be plotted against the dose or concentration of the chemical on the logarithmic probability paper and the best straight line drawn through the points giving greatest weight to those nearest the center.[40] However, sometimes the points may have a definite trend away from a straight line, either toward a curve concave upwards, convex upwards, or sigmoid,[11] or of a more irregular shape.[37] These departures from the normal straight lines should be substantiated by replication. In such cases[11] broken straight lines must be plotted. With the data plotted as straight lines it is possible to compare chemicals on the basis of concentration for equivalent inhibition of germination or control.[11,37,40]

The common unit for comparison is the LD_{50} or dose-inhibiting germination of 50% of the spores. In case the curves are convex or concave, they may be treated to the α-correction propounded by Parker-Rhodes[41] and McCallan.[42] This will straighten them out. Horsfall[43] has given further elaboration on this point.

The committee on standardization of fungicide[17] has advocated the use of a standard fungicide as a check on the reproducibility of technique, as a basis of comparison for new compounds and under certain conditions, as a means of adjusting day-to-day variations due to changes in the resistance level of the spores.[11,15] This committee has also described the method of preparing a standard laboratory bordeaux mixture.[44] The bordeaux mixture should be used for calculating the bordeaux coefficient. The bordeaux coefficient has been used to express a comparison of the fungicidal value of a candidate material with standard bordeaux.

To test the activity of a volatile compound, the dry spores can be exposed for different periods on filter paper or on a sintered glass disc inside a sealed petri plate or glass jar; their viability can then be examined by the regular spore germination technique.[45]

In order to meet the requirements and adapt to the facilities of various laboratories, several alternate procedures at certain instances have been presented along with the references to the original papers for detail study of the various techniques.

2. Test Tube Dilution

This technique is effective for preliminary slide germination tests of new chemicals as fungicides. It is simpler than the method described earlier, because it does not involve the use of elaborate apparatuses like a settling tower or a horizontal sprayer. It was with this idea that the Committee on the Standardization of the Fungicidal Tests came out with the test tube dilution method.[46] It is most precise for evaluating water-soluble materials, but only reasonably accurate for chemicals which give suspension with water. It is inaccurate for those chemicals which form aqueous suspensions with difficulty. This method consists of mixing toxicant and spore suspension together with an added nutrient and drops of resulting suspension pipetted on glass slides. Germinated spores were counted after 20 to 24 hr.

a. Preparation of Test Chemical Dilution

Weigh 0.5 g of the test chemical, dissolved or suspend in 50 mℓ of distilled water, and place in an Erlenmeyer flask. After thorough shaking, 2 mℓ of the solution or suspension is withdrawn and placed in a test tube without letting it run over the sides of the tubes. The contents of the tube would be of 1% (10,000 ppm). For preparing dilutions 5 mℓ solution or suspension is withdrawn from the stock (10,000 ppm) preparation. After discarding the remainder from the flask, it is rapidly rinsed with distilled water and 5 mℓ of aliquot is put back into it. Then, 45 mℓ of distilled water is added. Part of the samples (2 mℓ) is withdrawn from this 1000-ppm solution or suspension and placed in another tube. As mentioned above, the procedure is repeated until four to five test tube dilutions are obtained, giving 10,000, 1000, 100, 10, and 1 ppm of chemical. After obtaining the preliminary results from the above concentrations, a dose ratio of $\sqrt{10}$ should be prepared. The lowest dilution giving no germination would be the first dilution. For the second dilution place 16 mℓ of the first dilution and 34 mℓ of distilled water in a 50-mℓ graduated cylinder. After shaking, the contents may be poured in a 250-mℓ Erlenmeyer flask and withdraw 2-mℓ samples for test tubes. This procedure is repeated to get further dilutions, as desired.

b. Spore Germination Stimulants

In many cases, it is necessary to add spore stimulants to ensure a high and relatively stable percentage of germination in the control. Various stimulants have been tried, e.g., orange juice,[12] potassium citrate plus sucrose,[47] extracts from dried potato-dextrose-agar,[18] and coenzyme F.[38] Orange juice stimulant is prepared by filtering the juice of several good quality oranges through cheese cloth, followed by filter paper, and finally through a large Berkfield cylinder, type W, under vacuum. The filtrate is then diluted tenfold with distilled water and 10-mℓ portions placed in small corked vials which are stored at below-freezing temperatures. This constitutes 10% ultrafiltered orange juice. When needed for use, the contents of one vial are diluted to 100 mℓ, which gives the desired concentration for adding to the spore suspensions. The final concentration in which the spores are exposed for germination will be 0.1%. Details regarding other stimulants may be read from the original papers.

c. Adding Spores and Stimulants

Spore suspension is prepared according to the standard technique.[46] A concentration of 500,000 spores per milliliter should be obtained by means of a counting cell. For a volume of spore suspension, an equal volume of spore stimulant solution of a concentration ten times that finally desired is added. If no stimulant is added, the initial spore suspension concentration should be reduced to 25,000/mℓ. The mixture is stirred well by blowing through a 2-mℓ pipette graduated in 0.5-mℓ units; 0.5-mℓ samples of spore suspension stimulant mixture are then pipetted into each tube containing the 2 mℓ of the diluted chemical.

d. Placing Drops on Slides

Four or five glass slides are placed in each moist chamber in a horizontal position to the operator and two drops from each test tube of spores and chemical are pipetted side by side onto the left-hand side of a glass slide. If the chemicals are volatile and toxic and the range of dilution is wide, the chemical may be overrated because of the volatilization from concentrated drops on resulting condensation in dilute drops. This may be corrected in a second test in which a lower dose ratio and small dilution is used. When chambers are filled, the tops are placed in position, and germination counts are made 20 to 24 hr later. This technique should be used in conjunction with the standard slide germination method[17] in which relevant details of supporting techniques and necessary background information are given.

3. Agar Plate Germination

A modification of the slide germination technique has been suggested by Gattani.[48] It utilizes an agar surface instead of the conventional water drop, for determining spore germination. It has been found suitable for testing compounds insoluble in water, but soluble in liquids miscible with water. In this technique, dilutions of water-soluble fungicides are prepared in test tubes. For the fungicides which are insoluble in water, but their active ingredients are readily soluble in alcohol or other solvents miscible in water; 0.1 to 0.5 g of the chemical is dissolved in 2 to 5 mℓ of alcohol; and distilled water is then added to make the volume 100 mℓ. Dilutions of known concentrations are made from this stock solution. In flasks containing 99 mℓ of 1% sterilized water agar at 43°C, 1 mℓ of chemical dilution is added. After getting uniform distribution by vigorous shaking of flasks, 25-mℓ aliquots of agar are poured into petri plates and cooled.

Spores are obtained by the procedure suggested by Gattani.[48] Small quantities of spores are scraped with a sterile needle loop from the fungal culture grown in petri plates. This is then transferred gently to petri plates. The blunt end of a test tube, held at an angle of 45°, is pressed against the inoculum streaked on the plate. precaution being taken not to cut the agar. The large number of spores which at the beginning are clustered in the center will be so widely spaced that the individual spores may be examined under low power field. The number of spores is counted at four to five places in a petri plate. If the number of spores per low-power field is less than the number required, the process should be repeated by taking more inoculum on the needle loop.

The petri plates are kept at 21 to 23°C for 18 to 20 hr. The technique for the examination for spore germination is the same as given in the slide germination method.[17] However, the number of spores examined was much more than the number recommended by the American Phytopathological Society committee.[17] It is suggested that in concentrations where germination is less than 10%, 500 spores should be examined. When the percentage germination is more than 10, about 3000 spores need to be counted before deriving the percentage germination figures. Bordeaux mixture[44] or copper sulfate[49] may be used as the standard fungicide.

For determination of fungicidal activity of a chemical, the plates after incubation are tilted and washed with 5 mℓ of water. The washed water should be collected and centrifuged at 3000 rpm for 20 sec. After decantation another 5 mℓ of the spore wash from the second petri plate is added and the process is repeated with the third and fourth plate, to replenish the loss of spores in washings. After a final washing with distilled water the spores are centrifuged and decanted. The fungicidal action is arbitrarily recorded as having stopped with the final washing. Finally, the spores are streaked on water agar plates in which no fungicide has been incorporated and their germination counted.

Gattani's[48] main objection was that water drops containing spores tend to flatten or spread off the slide during the incubation period in a moist chamber. The loss of drops may result in marked differences in the germination test studies. This difficulty can be overcome by using cavity slides which would hold water drops well.[50] Gottlieb[51] compared the results obtained using agar plate and slide germination techniques and concluded that since different toxic values were obtained, it may be important to utilize more than one procedure in the primary evaluation of chemicals as fungicides.

4. Cell Volume Assay

This method was developed by Mandels and Darby,[52] based on the principles that spore cell volume increases[5] during germination and in the presence of a fungicide. The cell volume would be inhibited depending on the efficacy of the chemical. In this method, spores of *Myrothecium verrucaria* were obtained from cultures growing at 30°C on filter paper kept on agar in 250-mℓ Erlenmeyer flasks (Mandels[53]). The spores were removed from cultures by simply shaking gently with distilled water. They were washed twice and suspended in distilled water or buffer. By this method, a culture generally yields approximately 6×10^9 spores, having a centrifuged cell volume of 280 $\mu\ell$ and a dry weight of about 90 mg. This is adequate for 15 to 25 determinations.

For fungistatic properties of a chemical, 15 mℓ of a spore suspension in 0.05 M KH$_2$PO$_4$ + K$_2$HPO$_4$ buffer, pH 6.2, was added to 20 mℓ of sucrose and yeast extract in buffer containing the test chemical in the desired concentration in 125-mℓ flasks. The final concentration of sucrose and yeast extract was 1% each; final spore density was 1 to 2 $\mu\ell$/mℓ. The flasks are incubated at 30°C for 3 hr on a reciprocal shaker and after incubation, cell volume is determined on three 10-mℓ aliquots from each culture. Fungicidal tests are carried out by mixing 15 mℓ of spores suspension with 15 mℓ of the chemical. After overnight incubation on the shaker at 30°C, the suspensions are washed twice and resuspended in the sugar-yeast extract solution. Cell volumes are determined after 3 hr incubation on the shaker. So-called fungistatic-fungicidal tests are similar to the fungicidal tests except that the spores are not washed after incubation overnight, the sugar-yeast extract solution being added directly to the suspensions which still contained the chemical under test.

In the case of a viability test, samples should be removed from the fungicidal tests prior to washing and diluted to provide about 50 spores per milliliter. Petri plates having sucrose yeast extract agar are inoculated with 1 mℓ of the diluted suspensions. Colony counts should be made after a 2-day incubation. The test chemical should not be autoclaved. Sterile solutions are obtained by making up strong stock solutions and allowing them to stand for a short period before diluting them aseptically. Cell volumes are determined by centrifugation in triplicate for each treatment using standard hematocrit tubes of 10-mℓ capacity with the capillary part graduated to 0.10 mℓ in 0.005-mℓ intervals. Results are expressed in terms of changes in cell volume as percentage of check. The cell volume determinations =

$$\frac{\text{final volume in presence of inhibitor} - \text{initial volume}}{\text{final volume without inhibitor} - \text{initial volume}} \times 100$$

The major advantages of the technique are the quickness of the test (3 hr) and the elimination of aseptic precautions. Standard suspension can also be stored under refrigeration for at least 9 days without deterioration. Spores of other fungi were also found suitable for the utilization in the method.[52] The method is particularly adapted for fungistatic testing, although it can also be used for fungicidal properties.

5. Shaker Flask

The shaker flask method of culturing fungi was introduced by Kluyver and Perquin[54] and has also been used for spore germination studies.[55,56] Darby[57] studied the germination of spores in liquid culture in shaker flasks and observed this to be a rapid technique for evaluating fungicides by the inhibition of spore germination.

The shaker flasks should be fastened to a shallow tray and loaded with the components in the following order: buffer, substrate, chemical spore suspension. All the flasks are brought to the same final volume with distilled water. The several components are made up in concentrated form so as to provide the final desired concentration when mutually diluted with each other. Aseptic precautions are not essential. The loaded flasks should be loosely plugged and then transferred to a shaking machine in an incubator. After 4 hr of incubation, the cultures are removed and mounted in lactophenol plus cotton blue. It is essential to decide arbitrarily at what stage a given fungal spore may be considered as germinated, as the process of germination of fungus spores is successive over a period of time and comingles gradually in hyphal growth. Mandels and Darby[52] defined it as the stage at which the protruding germ tube is as long as its width. Percent germination is calculated from a count of several hundred spores.

Several components used by Darby[57] were buffer of 0.025 M potassium phosphate, pH 6.8; the substrate was 0.1% sucrose plus 0.1% yeast extract; the organism was *M. verrucaria* (Alb. and Schw.) Ditm. ex Fr. strain QM 460. The technique of propagating the fungus and preparing the spore suspension has been previously reported by Mandels and Darby.[52] The washed suspension was adjusted to provide a final concentration of 1 to 5 $\times$ 10^5 spores per milliliter. The flask was shaken at 225 rpm on a rotatory shaker at 30°C.

Fungi other than *M. verrucaria* have been successfully used by Darby.[57] He has recommended that the technique is suitable for routine studies of fungicides in view of the speed and ease with which it can be completed and better control of conditions such as pH, aeration, and nutrition afforded by shaker flasks in comparison with agar plate and slide germination methods.

B. Fungus Growth

Basically, the technique is to mix the test chemical with liquid agar medium, which is then poured into petri plates. Mycelial/spore discs are placed on the surface of agar medium. After incubating the plates for a certain period, the fungus growth is measured to assess the fungitoxicity of a test compound. The procedure was used for toximetric studies of wood preservatives.[58] Later, Palmiter and Keitt[59] adopted the method to an in vitro study of eradicant fungicides. Humphrey and Fleming[60] and Colley[61] have reviewed the methods and materials used in early toximetric studies of wood preservatives. Criticisms of the technique as applied to the testing of wood preservatives have been presented.[62-64] The main advantage of the technique is to use nonsporulating or several other fungi whose characteristics of spore production, germination, and size do not permit use of the spore germination method. The variations in techniques are multitudinous. Lee and Martin[65] described a technique in which spores, after exposure to a chemical, are transferred to growth medium and colony counts are the criteria for determining its activity. Carpenter[66] and Palmiter and Keitt[59] applied a fungus mycelial disc to the surface of agar, in which test chemicals are mixed thoroughly. Similarly, Bomar[67] applied hyphae growing on glass cover slips to treated agar. Forsberg[68] designed a method in which fungus-infested cotton thread is immersed in a chemical and transferred on the surface of agar. Sharvelle and Pelletier[69] reported a method in which paper discs are impregnated with a chemical, seeded with a test organism, and transferred to an agar plate. Other modifications are reported[70-76] which may be advantageous in specific cases. Gattani[48] recommended the use of pyridine-purified agar to reduce the chances of partial inactivation of chemicals which react with agar. Leben and Keitt,[77] Thornberry,[78] and Richardson[79] described inhibition zone techniques in which a test chemical is added to paper discs and these discs are placed on a seeded agar plate of the chemical is directly applied in holes in the agar. Efficacy of the chemical is determined by measuring the inhibition zone around the area. Torgeson[80] was of the opinion that differences in the solubility of test chemicals may affect their rate of diffusion in the agar medium and they may complicate the results.

The following techniques are widely accepted by researchers, for preliminary evaluation of the chemicals.

1. Poisoned Food: Method for Evaluating Volatility

The technique involves incorporation of the chemical into the nutrient medium and in this poisoned medium fungus is grown for a certain period. Both solid and liquid nutrient media have been used. Potato-dextrose-agar (PDA) medium is generally used as the solid medium

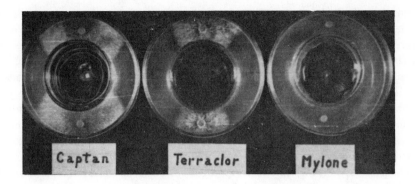

FIGURE 1.   Growth of *Rhizoctonia solani* after exposure to three fungicides for 6 days in double petri dish diffusion chambers: captan, not retarded; Terrachlor (PCNB), retarded; Mylone, none. (From Latham, A. J. and Linn, M. B., *Plant Dis. Rep.* 49, 398, 1965. With permission.)

to test the effect of the chemical on fungus growth. The required quantity of this medium is freshly prepared and autoclaved at 15 lb pressure for 15 min. The chemical is dissolved or suspended in sterilized water and a series of concentrations is prepared. To this medium, a requisite quantity of the chemical is added and thoroughly mixed by stirring. The medium is then poured onto petri plates. The test fungus is grown on PDA for a certain period at the optimum temperature required for its growth, which is variable depending on the fungus. A small disc (0.5 to 0.7 cm) of the fungus culture is cut with a sterile cork borer and transferred aseptically in an inverted position in the center of the petri plate containing the poisoned medium. Control plates should also be maintained in which the medium does not contain any chemical. The fungus growth is measured every 24 hr. The colony diameter, compared with check, is a criterion to measure fungitoxicity. Inhibition of growth in each concentration is calculated by the equation given by Vincent:[81]

$$I = 10 \ (C - T)/C$$

where I = inhibition, C = rate of growth in check treatment in mm/24 hr, and T = rate of growth in the treated lot in mm/24 hr.

By converting the percentage response (e.g., percent inhibition) to probits and plotting against logarithm of doses, a straight line (dose response curve) is obtained. The DR curves can be plotted on ordinary graph paper, semilog, or on log probability papers. In some cases, plotting of percent inhibition against logarithm of doses may yield a straight line. From these DR curves ED_{50} values are obtained which may be used to compare efficacy of various chemicals. Many times, fungitoxicity is measured by obtaining minimum inhibitory concentration (MIC), i.e., minimum concentration at which no growth is recorded[82] (see also Section 2.e).

Latham and Linn[73] described a method to evaluate volatility of fungicides. A 60 × 15-mm petri dish bottom is placed inside a 100 × 21-mm petri dish with cover and sterilized in an oven. The inside dish is centered, and 14 mℓ of sterile potato-dextrose-agar medium cooled to 47°C is poured in the annular space around the smaller petri dish (Figure 1). The test fungi, *Diplodia zeae*, *Rhizoctonia solani*, *Pyrenochaeta terrestris*, *Pythium debaryanum*, and *Sclerotium cepivorum*, were cultured on 13 mℓ of PDA in 100-mm petri dishes and used by Latham and Linn.[73] However, any test fungus can be used. Two inoculum plugs, 5 mm in diameter and 1.5 mm thick, are cut with a sterile cork borer from the periphery of the colony of each fungus and placed with the mycelium side down on the PDA on opposite sides in the annular space. One milliliter of a 1000-ppm suspension or solution of the chemical in distilled water is pipetted into a smaller petri dish bottom or well. The top edge of the bottom larger dish is thoroughly coated with a silicon lubricant which is oxidation resistant, insoluble in water, and nonfungitoxic. The cover is then pressed on to form an airtight chamber. The free air space in the chamber comes to approximately 120 cc. The chambers are incubated at 24 to 26°C. Measurement of the colony diameter is made at 24- to 48-hr intervals depending upon the fungus used. After 6 days, inoculum plugs showing no growth are transferred to fresh PDA and incubated further to determine viability. Jacks and Smith[83] used quart jars in which PDA medium was poured until the base was uniformly covered. A 4-mm disc from the soil fungi (*Verticillium dahliae*, *V. albo-atrum*, *Colletotrichum atramentarium*, *Fusarium oxysporum* var. *auriantiacum*, *F. lini*, *Corticium solani*, *Sclerotinia sclerotiorum*, *Phytophthora cryptogea*, *P. cactorum*, and *Pythium ultimum*) colony on PDA is placed in the center of the media. Inoculations are carried out on different days to allow for different growth rates of fungi, and in this manner colonies of similar size (1/8 to 1/4 of the jar diameter) are obtained. Jars are kept at room temperature (16 ± 4°C). Colony diameters are measured at regular intervals until hyphae reach the media margin. A colony not resuming growth after 14 days after exposure to fumigants is considered killed. The fumigants are measured as ppm of jar volume into glass dishes resting on the sealing discs. Sealing discs consisting of a thin metal disc lined with rubber give an airtight seal when fastened to jars with the aid of metal screw bands. Sealing discs and dishes are removed 48 hr after treatment, and the lids are then loosely replaced on jars.

Miller and Stoddard[84] placed fungicide samples in aluminum foil cups set on aluminum plate forms above PDA in 55-mm glass petri dishes. The fungi are seeded on the agar surface and dishes are kept covered. They evaluated volatile activity by rating mycelial growth, whereas Pryor and Walker[85] used an ordinary petri plate sealed with a double layer of grade "G" paraffin, or glass top fruit jars sealed by the usual rubber gaskets and wire clamps, for testing volatile compounds.

2. Inhibition Zone

The "paper-disc plate method"[86] for the assay of Streptomycin has been modified for application to the qualitative and quantitative evaluation of some fungicides by Thornberry.[78] During the same year, a technique based on the similar principles was reported by Leben and Keitt.[77]

a. Preparation of Assay Plates

Flat-bottomed, 90-mm Pyrex® petri plates are used. Fifteen milliliters of hot (80 to 90°C) agar medium are pipetted into leveled bottoms without any special aseptic precautions. A pipette with a large opening at its tip permits rapid filling of the plate when agar is hot. When this agar has solidified, 5 m𝓁 of the warm seeded agar is applied. The seeded agar is prepared by cooling the molten agar to 40°C and then adding the requisite amount of spore suspension to provide from 0.5 to 1×10^6 test fungus[50] or bacterial spores per milliliter of seeded agar. The plate is tilted to ensure even coverage before the agar solidifies. The plate is returned to the leveled surface for final leveling and solidification of the agar. The tops are then replaced and the paired plates are stacked in a refrigerator at 2 to 4°C upside down, to prevent condensation of moisture.

b. Preparation of the Test Samples

A series of dilutions of the test chemicals are prepared, which should be expressed as parts per million or micrograms per milliliter.

c. Setting up the Assay

Filter paper discs (12.7 mm diameter)[87] are speared individually with a needle and planted upon the chilled seeded agar of the plates according to the guidelines given below. A seven-disk guide, consisting of white paper with a 90-mm India-inked circle for plate alignment and seven evenly spaced 12.5-mm circles (one in the center and six in the outer position) for disc placement provides for zones of inhibitions up to about 30 mm in diameter. The liquid should be applied immediately after the pads are planted upon the agar. The volume of the test liquid varies according to the paper from which pads are prepared and should be slightly less than the amount necessary to saturate the disc. For a 12.7-mm disc, 0.09 mℓ applied with a 0.1-mℓ micropipette has proved to be satisfactory.

The plates were incubated at 28 to 30°C for 16 to 30 hr, depending upon the test fungus. To get the clear zones (rings) of inhibition, each disc is measured from the center of the disc.

d. Estimation of Toxicity

The relative toxicity of a candidate fungicide can be estimated from a standard curve of zones of inhibition produced from the known amounts of the reference fungicide. Chemicals with low diffusibility cannot be properly evaluated by this procedure.

Thornberry[78] suggested that the method may be used for the evaluation of fungistats, fungicides, bacteriostats, and bactericides, for general screening survey for toxic substances, and for certain toxicity studies. Test organisms that may be of particular concern can be substituted by certain modifications. For larger assays at one time, Lockwood et al.[88] recommended the use of a glass plate, instead of petri plates, 1/8 in. thick, kept in an aluminum pan (1 × 25 × 35 cm) and covered with an asbestos cover. After autoclaving, the agar is seeded and poured on the glass plate. Paper discs with fungicide are placed on the seeded agar.

e. Modified Paper Disc Method

The technique was given by Sharvelle and Pelletier[69] and Sharvelle[89] with the aim for use with fungi that do not sporulate readily in the culture. However, the method can also be used for sporulating fungi. In this method, the chemical and fungus are put on the same disc which is then placed on PDA medium. The colony diameter as compared to check is the criterion for fungitoxicity.

Filter paper bioassay discs, 12.7 mm in diameter, are sterilized, immersed in solutions or suspensions of test chemicals of known concentration, and then drained. The discs are seeded with a calibrated suspension of the test fungus (spore or mycelial) using a dropping pipette. Three impregnated seeded paper discs are then transferred aseptically to each of the two poured agar plates. A series of check discs is also maintained. After incubating the plates at the desired temperature, positive (+) or negative (−) growth for each of the chemicals tested for each concentration is recorded at 48 hr over a period of 7 to 10 days. Growth measurements from the periphery of treated and untreated discs provide dosage response data. From these observations, the MIC is recorded. The MIC is the dosage at which no growth occurs on the impregnated seeded disc over a period of 7 to 10 days. It serves as a basis for comparing the efficiency of various chemicals.

For testing volatile substances, the sterilized discs are first impregnated with the test fungus suspension and then transferred to the surface of sterilized glass plates. The plate is then placed in a gas-tight bottle of known capacity and liquid volatile chemicals are added to a pad in the bottle lid. Solid volatile materials in weighed amounts should be placed in open vials in the bottle. After the treatment period, the discs are transferred to the surface of the agar medium. Periodic observations of air dilution series permits determination of inhibition concentration.

C. Respiration

Nickerson[90] proposed that oxygen consumption be used for screening fungicides in the laboratory, and Mandels and Siu[91] described a "rapid, reliable manometric method" for determining fungistatic activity. This method was found useful in studying the effect of fungicides on oxygen consumption.[92-94] Horsfall's[43] conclusion about this technique is that "despite theoretical arguments in its favour, respirometry cannot be used for the primary screening of the chemical. It is a very valuable tool to aid in understanding the mechanism of action, but it seems inefficient as a technique for screening as the concentration of fungicide required to inhibit respiration is generally far more excess to that required to inhibit growth."

III. SOIL TREATMENT TESTS

After primary evaluation of hundreds or several thousands of chemicals for inherent fungistatic or fungicidal properties, the potential chemicals may be tested for practical fungitoxicity like seed treatment, soil treatment, or spray applications. Here, the discussion will be limited to soil treatment only.

Fungicides can be used to control soilborne diseases by eliminating the pathogen from the soil, by reducing the inoculum level, by fungistatic action which suppresses the growth of the pathogen in the soil, or by systemic action within the host plant. Until recently, standard laboratory techniques employing inhibition of spore germination or mycelial growth on nutrient media, etc. had been used in the primary screening of soil fungicides,[4] but these tests alone had little predictive value for selecting successful soil fungicides.[3] The lack of soil was probably the astounding factor. A successful soil fungicide must be able to act in a complex soil medium, in addition to being fungitoxic. Hence, soil is probably a necessary component of any laboratory screening procedure. Several techniques which are discussed below may be useful in determining the future of a compound in the field.

A. Soil-Fungus Methods

These methods involve the exposure of the test fungus to a chemical in soil contained in cups, jars, or soil column. After a certain period, the effect is decided by determining the viability of the fungus, counting of fungal spores by plating the soil samples or measuring the zone of inhibition around soil samples placed on agar plates seeded with fungus, or determining the viability of pathogens in infested tissue.

1. Methods for Primary Screening

Zentmyer[3] described a primary screening procedure for evaluating soil fungicides with *Phytophthora cinnamomi* as the test organism. However, any other test organism can also be used. Air-dried fine sandy loam (pH 7.0 to 7.5), soil is sieved through a 20-mesh sieve and autoclaved for 45 min at 15 lb of pressure per square inch in mason jars. One inch of soil is placed in a shell vial of 25 mℓ capacity (20 mm diameter $\times$ 85 mm deep). A disc 10 mm in diameter is cut from the outer margin of a PDA culture of the fungus to be tested and placed on the soil, and the disc is then covered with 1 in. of the sterilized soil. The total volume of the soil in the vial comes to about 17 mℓ (weight 20 g). The fungicide solution or suspension to be tested (5 mℓ) is then applied to the soil surface in the vial with a pipette. The vial is then incubated for 24 hr at 25°C. Suitable replications of vials are maintained. The vials are then emptied in a wire or perforated metal strainer and the soil is removed by washing with running water. The mycelial discs are picked out with sterile forceps, blotted on paper towels, and placed on a suitable medium in petri plates to determine viability of the fungus.

In tests of chemicals mixed dry with the soil, 60 g of sterilized soil is placed in 200-mℓ flasks. The measured quantity of the dry chemical is added to the flask and both are thoroughly mixed by shaking. One inch of the treated soil is placed in the bottom of the replicated vials. The inoculum disc is added and covered with 1 in. of the treated soil. Distilled water (5 mℓ) is applied to the surface of the soil. The vials are then incubated for 24 hr and the disc recovered as mentioned above. For obtaining low concentrations (1 to 5 ppm) the chemical is first added to the soil and thoroughly mixed to provide a higher dose (25 ppm), and this mixture is then diluted by adding soil until the desired concentration is obtained.

Results are recorded as presence or absence of growth of the fungus following exposure to the chemical. This technique primarily detects fungicidal chemicals. Fungistatic materials are not determined so readily, although an indication of fungistatic properties may be obtained by comparing the growth of mycelium from the disc from treated soil with that from the disc from untreated soil. Kendric and Middleton[4] and Domsch[95,96] further used this technique in the case of other plant pathogens. Corden and Young[97] have reported slight modification in the above procedure by wet mixing of the chemical rather than dry mixing to ensure better contact between chemicals and test fungi. Torgeson[98] described a primary test for soil fungicides in which artificially infested soil in small (4-oz) cups is drenched with the test chemical and the efficacy of the chemical is determined by rating the mycelial growth on the soil surface. Four test fungi, *Fusarium oxysporum* f.sp. *cubense*, *Pythium* sp., *Rhizoctonia solani*, and *Sclerotium rolfsii*, were used. Results obtained appear to correlate well with those obtained in greenhouse evaluation tests.[99]

Lingappa and Lockwood[100] described a technique in which spore suspension of a test fungus is applied to the surface of treated, compacted soil, and after incubation a dye solution is placed on the soil surface to kill and stain the spores. The spores are recovered on a film applied over the soil surface and examined for spore germination. A simple technique was designed by Chinn and Ledingham[101] for testing the fungicidal effect of chemicals on fungal spores in soil. Fungicides are mixed dry with soil or used as fumigants. Spores of test organisms (*Helminthosporium sativum*, *Fusarium culmorum*, *Alternaria tenuis*) were cultured on wheat straw, exposed 4 days to treated soil, recovered, and tested for viability by a modification of the floatation-viability count method.[102] Thayer and Wehlberg[103] successfully used bean stem sections to trap *Pythium* and *Rhizoctonia* for determining the toxicity of chemicals against these fungi. A rapid method was described by Rodriguez-Kabana et al.[104] in which autoclaved oat kernals colonized with *S. rolfsii* are placed equidistantly in soil plates containing moist, unsterilized sandy loam. The infested plates are then treated and incubated at 30°C. Data on the mycelial development and sclerotia formation are recorded. This method has been adopted for other pathogens, also (*R. solani*, *Rhizopus stolonifer*, and *Cylindrocladium crotalariae*). Corden and Young[97] described a method for artificially infesting nonsterilized field soil with *F. oxysporum* f. *cubense* without adding culture substrates to the soil. The soil dilution technique is used to evaluate the effectiveness of the chemicals. Segments of sugar beet, carrot, etc. have been successfully used to trap *S. rolfsii* from soil.[105] Several techniques have been suggested for the placement of inoculum mycelium and/or spores which can be added to soil as part of agar cylinders, or as discs incorporated as membrane filters or nylon membrane.[106,107]

2. Methods for Evaluating Volatility

Richardson and Munnecke[108] described a method in which plate cultures of test fungi are exposed at a fixed distance above treated soil in closed containers. The relative toxicity of vapors emanating from the applied chemicals is assessed on the basis of the degree of inhibition of radial growth.

Replicated 100-g lots of treated and control soil are dispensed in 16-oz-widemouth jars (Figure 2). The moisture content is adjusted to the required level by adding 15 to 25 mℓ of water, depending on the soil type. A petri plate containing a layer of PDA and a 6-mm inoculum disc, cut from the margin of an active plate culture of the test fungus (*P. irregularae*,

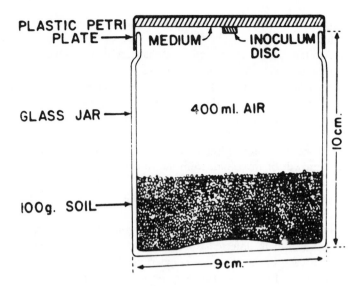

FIGURE 2.   Apparatus used to determine vapor toxicity of chemicals applied to soil. (From Richardson, L. T. and Munnecke, D. E., *Phytopathology*, 54, 836, 1964. With permission.)

R. solani, and *T. viride*), is inverted over the mouth of each jar and held in place by strips of cellophane tape. Control culture grows at the same rate as regular plate cultures so that the oxygen supply appears to be adequate. The jar cultures are incubated at room temperature until the colonies in the check approach the margins of the plates. The degree of growth retardation in each case is calculated from the mean difference in colony diameter between treated and check cultures as percentage of the latter. Dosage response data are plotted on log probability paper. To distinguish fungistatic from fungicidal effects inoculum discs that failed to develop are transferred to a fresh medium in regular culture plates and incubated further. Residual effects, indicating sorption of a toxicant by the medium, can be demonstrated by placing a fresh inoculum disc on an exposed culture plate and covering with a regular petri lid during incubation. To determine the rate and duration of toxic vapor production, fresh culture plates are plated over jars of treated soil periodically. Corden and Young[97] placed agar inoculum (*Fusarium* sp.) discs around the base of a beaker of fungicide-treated soil, then transferred the disc to culture plates. Since only presence or absence of growth can be indicated by this method, the results are not quantitative and fungistasis is not revealed.

Maurer et al.[109] used soil microbiological sampling tubes (SMST) to evaluate the efficacy of fumigants applied to soil in sealed mason jars (Figure 3). Nonsterilized loam soil (300 g [odb]) infested with *F. roseum* f.sp. *cerealis* is placed in pint mason jars. Fumigants at various concentrations are applied to cotton pads placed on the top of the soil. Lids with a hole (so that SMSTs can be inserted later) are placed on the jars immediately after treatment. The holes are sealed with keroseal electric tape and jars are incubated at 25°C. After 72 hr the tape is removed and sterile cotton plugs are placed in these holes. Aeration proceeds for 10 days. At the end of this period, SMSTs (12 holes/tube) are inserted into the soil, through holes in the lid. Three days later, isolates from tubes are transferred to PDA containing 500 ppm pentachloronitrobenzene (PCNB) for identification.

Another simple and quick technique for the determination of the fungicidal effect of vapors from chemicals applied to soil is given by Oserkowsky.[110] In this method, a small vial is placed upright in a stender dish and melted agar is poured into the dish. After sterilization and cooling, the vial is held in place by the hardened agar. The volatile toxicant to be tested

FIGURE 3. Soil in mason jar with a soil microbiological sampling tube inserted through the lid. (From Maurer, C. L., Baker, R., Phillips, D. J. and Danielson, L., *Phytopathology,*, 52, 957, 1962. With permission.)

is placed in the vial and a small block of agar supporting a fresh mycelial growth of the fungus (*S. rolfsii*) is transferred to the surface of agar in the stender dish. After incubating the stender dish at 29 to 30°C for 2 to 5 days, the mycelial growth in the stender dish is compared with that of check cultures grown in stender dishes but without chemicals. If no growth is made by the mycelium subjected to the fungicidal vapors, it is transferred to fresh agar medium free of fungicidal vapors and incubated at 29 to 30°C for 5 to 22 days. If no growth takes place fungicidal vapors are considered as lethal. If sclerotia is tested, the same are placed on the bottom of a weighing glass equipped with ground glass cover and containing a small vial of the substance to be tested. The whole set is incubated at 25°C for 3 days. The sclerotia are then removed from the weighing glass, plated on agar media, and incubated at 29 to 30°C. Several other methods for evaluating the fungitoxicity of volatile compounds have been reported.[85-87,111] However, the vapor phase activity of fungicides in soil could not be determined by either of these methods. Hence, they have been discussed elsewhere.

3. Methods for Evaluating Drenchability/Movement

Generally, columns filled with soil are utilized for testing drenchability of soil fungicides. Fungus discs are kept at different heights in the column and the chemical solution or suspension is applied at the top. The fungus discs are recovered and tested for viability. There are various modifications of the technique.

Latham and Linn[5] used Lucite® tubing with an inside diameter of 4.5 cm and a thickness of 6.5 mm, which was cut and lap jointed in a lathe to form four 5-cm sections and one

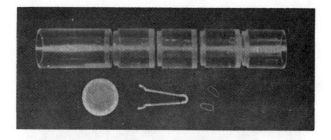

FIGURE 4. Disassembled soil column cylinders used for testing soil fungicides with base plate, clip, and elastic bands. (From Latham, A. J. and Linn, M. B., *Phytopathology*, 58, 460, 1968. With permission.)

FIGURE 5. Placement of mycelia-agar discs in the soil column. (From Latham, A. J. and Linn, M. B., *Phytopathology*, 58, 460, 1968. With permission.)

10-cm basal section. When united, the sections make a tube 30 cm long. Scotch® electrical tape is placed around each joint to form a water-tight seal. A fine-mesh Saran screen cemented to a circular celluloid ring is held on the bottom of the tube with plastic clips reinforced with rubber bands (Figure 4). The tube is then filled with 475 g of soil at pH around 7.0 that has been passed through a 3-mm mesh screen and autoclaved in shallow glass trays for 5 min at 121°C. The soil column is tamped tightly, 128 mℓ of distilled water is added, and the tube is covered and allowed to stand for 24 hr to bring the soil to field-capacity moisture content. The columns are then opened and mycelial agar discs 5 mm in diameter and 2 mm thick are placed in a circle at the 2.5-, 7.5-, 12.5-, and 17.5-cm soil level in the tube (Figure 5, lower). When applying chemicals, a dissecting needle is run around the periphery of the soil during pouring to puddle thin areas and prevent channeling of the chemical suspension of solution down the side of the tube. The discs are recovered after 24 hr by collecting a small amount of soil above and below each disc level onto a 1000-μm sieve and washing the material for 5 min in cold running water to remove excess soil. The viability of the mycelia is determined on PDA medium plus Streptomycin and vancomycin (PDASV).[112] The minimum fungicide concentration required to inhibit mycelial growth is determined after 7 days of incubation. If sclerotia are to be tested, 25 sclerotia are placed in a 2-cm square, 15-denier nylon envelope and positioned at each of the four levels in other columns. Soil columns with sclerotia are prepared identically. Chemicals are applied as described earlier. The sclerotia are removed from the soil after 48 hr of exposure, soaked for 3 min in 10% sodium hypochlorite, rinsed three times in sterile deionized water, and plated on PDASV medium. Germination is recorded after 12 days at 21 to 24°C.

Ezekiel[113] described the soil chamber method for testing penetration of fungicide through soil. Jars are filled with sieved soil, and chemicals are placed on the surface of the soil after the inoculum had been inserted in the jars. The quantity of material added is calculated as parts per million of the air-dried weight of soil in the jar. Liquid materials are poured rapidly on the soil surface; solid materials are placed in piles on the surface. In filling a jar, two masses of inoculum are placed on the bottom at opposite sides. Half of the soil (by weight) for the jar is added and compacted to the usual density. Two more masses of inoculum are then placed on this soil and next to opposite walls of the jar. A major part of the soil is added and compacted, the final two masses of inoculum inserted, and finally the remaining soil added and compacted. After this setup is completed, the material to be tested is placed on the surface and the jar lid sealed using the rubber ring. The upper inocula are at an average depth of about 15 mm below the surface of the soil, the middle inocula at about 75 mm, and the bottom inocula at about 135 mm. The various chemicals could prevent growth from the depth mentioned. Growth of the fungus (*Phymatotrichum omnivorum*) is observed through the glass walls of the container. Direct observation is possible, because the characteristic strands of the fungus grow to a greater extent along the glass-soil surface than into the soil. Ezekiel[113] has clearly mentioned that penetration observed through the moist soil materials in the jar would not prove that the fungicide would penetrate soil in the field; lack of penetration would strongly indicate the improbability of such penetration through soil under natural conditions.

Evans[114] utilized soil recolonization tubes (Figure 6) to investigate movement of fumigants through moist soil. Soil recolonization tubes are filled with soil at various moisture contents. The moisture holding capacity (MHC) of soil is maintained by adding water through side arms. The tubes are then allowed to stand overnight to permit distribution of the added water through the soil. The chemical is added to the center side arm of each tube and the apparatus is sealed by rubber bands for 24 hr at room temperature. After that the bands are removed and replaced by "oxoid" metal caps. After a further 24 hr, samples of soil are taken from the seven side arms of each tube for the preparation of soil plates and fungal colonies are counted after 5 days of incubation. The soil recolonization tubes consist of an 18-in. length of Pyrex® glass tubing of 1 in. interval diameter, to which seven side arms 1 in. long and 0.5 in. interval diameter are fused at 2-in. intervals.

An assay with artificially infested soil is designed[97] to determine the drenchability of fungitoxic chemicals into the soil. Frozen juice concentrate cans, 5 cm in diameter and 9.5 cm tall and open at one end with 1/8-in. holes covered by one mat of glass wool in the other end, are filled with air-dry *Fusarium*-infested soil (200 g) making a column of soil 7 cm high. Chemicals are drenched on the soil surface and soils are incubated for 48 hr at about 22°C. A sample for dilution is taken from the center of the cans 5 to 6 cm below the surface with a soil sampler that cut a core 22 mm in diameter. A 3-g sample on the air-dry basis is weighed from the 5- to 6-cm soil disc and the *Fusarium* population is determined by the dilution plate technique.

Several other assays have been devised to measure the penetration of soil by fungicides drenched on the soil surface. The effectiveness of chemical treatments has been evaluated by determination of the viability of a test fungus on a nutrient agar disc,[3,115,116] determining the viability of spores and mycelia on infected paper discs,[117] determining the viability of *Phytophthora parasitica* in roots of infected citrus seedlings that had grown in the soil columns at various levels below the surface,[118] counting the number of fungal colonies from small samples of artificially infested soil collected from the columns at various depths,[115,119] assaying the leachate for presence of the fungicide,[120] and measuring the zones of inhibition

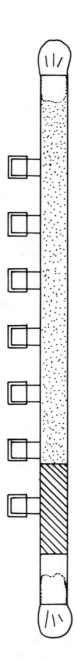

FIGURE 6.   The soil recolonization tube.
Key: hatched, nonsterile inoculum; stippled,
sterile soil. (From Evans, E., *Trans. Br. Mycol. Soc.*, 38, 335, 1955. With permission.)

around soil samples taken at various depths from the soil columns and placed on nutrient
agar inoculated with fungal spores.[121]

4. Method for Evaluating Persistence

A very simple and quick technique for qualitative/quantitative determination of fungicides
in soil is described by Munnecke[122] and subsequently modified by Sinha[123] and Sinha et
al.[124] In this technique plugs of soil treated with a fungicide are placed on PDA seeded
previously with spores of *Myrothecium verrucaria* and are held at 70 to 75°F for 48 hr. The

diffusion of the fungicide into the agar and the subsequent inhibition of growth of the test fungus result in a clear zone around the soil plug. The diameter of the clear areas is measured after 48 hr and utilized as an index of fungicidal effect.

a. Test Organism

Spores of *M. verrucaria* are usually used in the method. However, any other sporulating fungi can also be used depending on its sensitivity to the fungicide. An 8-day-old culture grown on PDA of the test fungus (*M. verrucaria*) is flooded with a small quantity of sterile, distilled water and rubbed gently with a blunt transfer needle to liberate spores. The number of spores is counted with the help of a hemocytometer. Four counts are made using the fourth drop from a pipette.

In order to make the technique easier, Sinha[123] standardized the concentration of spores using a Systronic Photometric Colorimeter. The spore suspension is read in a colorimeter against blue filter (420 μm) and it was standardized that spore concentration obtained at 60% transmission gave a clear inhibition zone. A required amount of spore suspension plus sterilized PDA were poured into the petri plate, and after solidification the soil plug with a fungicide was kept in the center of the plate and the inhibition zone measured.

b. Calibration of Soil Fungicide Bioassay

Test fungicide is added to soil having 45% moisture level to get several concentrations. The fungicide is thoroughly mixed with the test soil, divided into four portions, and cylindrical plugs are prepared from each portion. The plugs are molded in small pieces of plastic tubing placed over PDA seeded earlier with spores (2500 spores per milliliter) of *M. verrucaria*. After 48 hr of incubation at 25 ± 1°C, a clear zone of inhibition developed in the plate where the fungicide had diffused from the plug into agar, while the rest of the surface was covered with the mycelial growth of *M. verrucaria*. The diameter of each zone is measured in millimeters. Dosage response curves are plotted from the mean diameter of 12 plugs for the logarithm of each concentration of the fungicide in soil.

c. Determination of Persistence in Soil

Soil plugs (45% moisture content) having different concentrations of fungicide are prepared as mentioned earlier, and each treatment is suitably replicated. The amount of fungicide, at different time intervals, is obtained by plotting the mean diameter of replicate samples on the standard curve. If the technique is used for qualitative rather than quantitative purposes, it is not necessary to plot the standard curve.

5. Methods for Evaluating Penetration of Fungicides into Plant Tissues

Corden and Young[97] developed an assay method to determine penetration of plant tissues by fungicides using stems from tomato plants infected by *F. oxysporum* f. *lycopersici*. The stems are cut into 1/8-in. sections and ten sections are buried in each of the three replicate soil samples of 100 g of mix treated with the test chemical. After 24 hr, the stem sections are washed from the soil and placed on PDA. Viability of the fungus is determined 10 days after planting.

Another technique under the soil-fungus group is given by Domsch[125] in which oxygen consumption by treated soil has been utilized to study the effect of soil fungicides on the respiration of soil microorganisms. Rushdi and Jeffers[126] have used a technique for measuring linear growth of fungi in soil and demonstrated that adsorption of fungicides by soil is an important factor controlling activity.

B. Soil-Fungus-Host Methods

The techniques for testing of soil fungicide treatment in a soil-fungus-host system involve the use of test chemicals in infested soil planted with susceptible hosts. Disease control of healthy plants is calculated in infested soil as compared to the noninfested soil or to a standard fungicide treatment.

The evaluation of chemicals in the greenhouse for the control of soilborne plant pathogens is an important intermediate step in the development of a new fungicide, falling logically between the initial testing of chemicals in the laboratory and the eventual testing of promising ones under actual field conditions. In vitro laboratory tests usually determine which chemicals tested are sufficiently lethal or inhibitory to certain plant pathogens to warrant their further exploration. These materials are then tested under controlled greenhouse conditions to select those which appear to satisfy, to a reasonable degree, certain principal requirements of a desirable chemical. They are then considered to qualify for more conclusive evaluation for disease control in the field.

Soil-fungus-host techniques of evaluation of fungicides for control of soilborne plant diseases are not confined to any fixed, standard technique that might be used universally. Considering all the variables including chemicals, pathogens, host plants, environment control, and facilities, there is a wide range for adaptation of diversified techniques to the practical usefulness in determining how promising a chemical might be as a fungicide.

Arndt[127] reported a sand nutrient culture method utilizing aluminum dishes in which the test chemical is applied only to the sand immediately around the cotton seed. The inoculum (*Rhizoctonia solani*) is placed in the center of the dish and the number of healthy seedlings remaining after 12 days determines chemical effectiveness. Similar techniques using treated media in dishes, flasks, and pots have been reported.[128,129] The details of the techniques as reported by Darrag and Sinclair[129] are given below. The methods were aimed to evaluate chemotherapeutic activity of fungicides against cotton seedlings.

Greenhouse method — Large dishes (17.5 × 6.5 cm) are sterilized with mercuric chloride (1:1000) and thoroughly rinsed with sterile, distilled water. Approximately 1500 mℓ of autoclaved vermiculite (Terralite® brand) is placed in each dish. The chemical is mixed with potato dextrose broth (PDB) at a desired concentration and added to the vermiculite except for check dishes, which contain only PDB. A polyethylene disc cut slightly larger than the dish diameter and with 20 equidistant holes punched out with a sterile cork borer is placed on the top of the vermiculite. Surface-sterilized seed is sown by one of three variations: (1) a single seed is placed in each hole and covered with a thin layer of sterilized vermiculite at the time of planting, (2) single seeds are planted in each hole and covered with vermiculite after most of them are germinated, or (3) a single, germinated (4 days in sterilized vermiculite) seed is transplanted into each hole and then covered. Dishes are inoculated with 15 agar discs of the test fungus placed between the seeds before covering with vermiculite.

Flask method — Erlenmeyer flasks (125 or 250 mℓ) are plugged with cotton and autoclaved. Soil or vermiculite is added to some flasks before autoclaving to determine the effects on seedlings when compared with no additive. Fungicide solutions are added after flasks are cooled. A stock solution of the chemical at various concentrations is prepared. The neck of each flask is covered with a piece of plastic wrap after adding the required amount of chemical solution. A cuplike depression is made in the plastic wrap, which is held in place with a rubber band. The plastic wrap provides support for cotton seedlings above the fungicide solution and keeps inoculum separated from the fungicide solution. A single germinated cotton seed (4 days) with the seed coat removed is placed into each of two or four holes punched into the plastic, so that its roots are immersed in the fungicide solution. An agar disc with *R. solani* is placed adjacent to the seedling hypocotyls. The depression is filled with sterile vermiculite and moistened with sterile distilled water. Flasks are kept at room temperature (22 to 27°C) and the surface of the vermiculite is kept moist to ensure good conditions for growth of the fungus. The number of healthy seedlings is recorded each day for the duration of each experiment.

Greenhouse method — Autoclaved 6-in. clay pots are filled with nonautoclaved field soil to within about 4 to 5 cm of the top (approximately 1.5 kg soil). Soil is either treated or non-treated by mixing the soil with 200 mℓ of chemical solution per pot at desired concentrations. A disc of polyethylene cut slightly larger than the diameter of the pot has 20 equidistant holes punched out with a sterile cork borer and is placed on top of the soil. Surface-sterilized seeds are sown by one of three methods as described for the glass dish method. Infested pots have 15 mℓ of mycelial suspension of *R. solani* added to the layer of soil above the polyethylene disc. The disc will keep treated soil separated from infested soil. The preparation of inoculum for soil infestation is described by Sinclair.[130] The number of germinated seeds is recorded after 1 week and the number of healthy seedlings after 4 weeks for all experiments.

There are several other published methods for determining efficacy of materials for the control of organisms causing diseases of seeds and seedlings. Flats of *Rhizoctonia*-infested soil treated with a chemical followed by sowing of cotton seed have been successfully utilized by Sinclair[130,131] and Brinkerhoff et al.[132] Elsaid and Sinclair[133] also used to simulate field conditions by placing infested soil on the edge of treated soil and determining the progress of disease development. In 1960 Reinhart[134] described a paper cup method in which treated soil infested with *F. oxysporum* is planted with cucumber seed. The percent of control is expressed from the number of diseased plants in the treated and nontreated soil. Ranney and Bird[135] demonstrated the role of soil temperature in determining effectiveness of a compound. Several greenhouse tests have been developed for fungicide evaluation against many plant pathogens, including *Pythium* damping-off,[136,137] tobacco black shank (*Phytophthora parasitica* var. *nicotianae*), tomato wilt (*F. oxysporum* f. *solani*), southern blight of peas or peanuts (*Sclerotium rolfsii*),[98] *Fusarium* root rot of beans,[138] and melon wilt (*F. oxysporum* f. *melonis*).[139]

A detailed procedure has been reported by Colhoun[140] for the determination of fungicidal efficacy in the control of cabbage club root (*Plasmodiophora brassicae*). The required amount of a chemical in powder form is thoroughly incorporated with the air-dry soil. The soil is uncontaminated with *P. brassicae*; it should be inoculated with the appropriate spore load. Water is then added to the naturally or artificially contaminated soil to provide a standard moisture content of 70% of the maximum WHC. The soil is again thoroughly mixed and placed in water-tight tins, which serve as pots. Throughout the work, cabbage seedlings raised in sterilized compost are transplanted into the soil which is untreated or treated with a fungicide. Twenty seedlings are planted in each pot. In limed soil a delay of 5 days occurs between treatment and transplanting, but there is little delay when unlimed soils are used. The soil is maintained throughout each experiment at a relatively constant moisture content equal to 70% of the maximum WHC by weighing the pots on alternate days and replacing the water lost. At the conclusion of the experiment a value of 10 is given to pots in which plants are grown best, the others being assessed at between 10 and 0 according to the growth made.

Colhoun[141] recommended that fungicides to be tested should be subjected to a preliminary pot test and those which are considered satisfactory should then be subjected to a more detailed test.

- Preliminary pot test: this test should be carried out at a spore load of 10^5. Each fungicide should be tested in acid soil of pH 5.0 to 6.3 as well as in soil treated with Ca (OH)$_2$ to adjust the reaction to pH 7 to 7.2. Two replicates, each pot containing 20 plants, should be included.
- Detailed pot test: in this test spore loads of 10^3, 10^5, and between 10^7 and 2.5×10^7 should be employed. Each fungicide should be tested at each spore load in acid soil (pH 5 to 6.3) and in soil of pH 7.0 to 7.2, the reaction having been adjusted by the addition of Ca(OH)$_2$. The number of replicates may vary between two and five.

The following methods should be adopted to carry out the preliminary and detailed pot tests:

1. Fungicides in powder form should be applied to air-dried soil (light loam) which has already been treated with $Ca(OH)_2$, if the soil reaction required adjustment. Fungicides in solution should be added after the spore suspension has been incorporated with the soil. Very thorough mixing by hand should be given to the soil after the addition of fungicides, as well as after adding the spore suspension.
2. The spores used for inoculating soil should not have been stored for longer than 2 months. When possible, fresh spores from diseased roots should be employed.
3. The soil moisture should be maintained at 70% of the maximum WHC by weighing the pots on alternate days and replacing the water lost.
4. Tests should be made at a mean air temperature of 23°C, but those in an acid soil may also be made at 18°C.
5. A period of 4 to 5 weeks is required, but the plants should continue to be grown until there are well-developed clubs on the roots of those in soil not treated with any fungicides.

To assess the efficiency of fungicides it is suggested that the value obtained for the percentage number of diseased plants and for the disease index should be used to compare the efficacy of each fungicide with that of a standard treatment. Treatment with 200 mℓ of a 0.1% solution of $HgCl_2$ per 3000 g of oven-dry soil should be accepted as the standard.

Numerous reports are on hand regarding testing of the basic aspects of the environmental factors which influence the development of major soilborne diseases. It makes it possible to design a soil-fungus-host test system for greenhouse evaluation of chemicals for those diseases whose published techniques are not available. Various factors such as soil temperature, soil moisture, soil type, residual effectiveness, methods of application, formulation, inoculum potential, etc., which may have an important bearing on how well a potential soil fungicide will perform under field conditions, can be determined by suitable modifications of standard greenhouse evaluation techniques.

IV. CORRELATION BETWEEN BASIC, SOIL TREATMENT, AND FIELD TESTS

The primary purpose of laboratory test techniques, basic and soil treatment, apart from their utility in studies on the nature of fungicidal action, is to predict field performance. It is very important to see how the laboratory techniques correlate with one another with the intermediate greenhouse techniques and, above all, with field tests. Limited published reports on the subject have been well reviewed by McCallan[142] and Domsch.[143] Zentmyer[3] and Kendrick and Middleton[4] pointed out the inadequacy of slide germination/agar plate evaluation of soil fungicides. Klomparens and Vaughn[144] demonstrated good correlation between laboratory performance of fungicides tested on malt agar and field control of several turf pathogens. The authors agree with the views of Domsch[143] that wherever large numbers of chemicals are being run through routine tests, this has to be done on an economical scale and with uncomplicated techniques. In these cases screening tests on artificial media are to be recommended. However, primary screening procedures using soil have been described giving higher correlation between laboratory and greenhouse or field performance.[3,98]

Since fungitoxicity is related to the sensitivity of the test organisms, it is suggested that as many fungi as possible should be included in screening tests of this preliminary type.[145] A fungus can easily be killed in laboratory tests on artificial media[83] or in soil tubes, closed containers, or pots,[145,148] but it seems to be unrealistic to transfer this idea of total eradication to soil conditions. At reasonable dosage levels, a high proportion of most of the single organisms can be eliminated from the total population of pathogenic soilborne fungi. However, resting structures remain viable to a certain extent, and to kill the remaining part an unduly high concentration should be applied. Basic and soil treatment tests will continue as several chemicals are being developed, but the final evaluation must be made under a variety of environments in the field before the true potential of a chemical can be realized.

REFERENCES

1. **Prevost, B.,** Memoine sur la cause immediate de la carie en charbon des blés et de plusieurs autres maladies des plantes, et sur les préservatigs de la carie Bernard, Paris, *Phytopathol. Classics,* 6, 1, 1930.
2. **Block, S. S.,** Getting the most from fungicide tests, *J. Agric. Food Chem.,* 7, 18, 1959.
3. **Zentmyer, G. A.,** A laboratory method for testing soil fungicides, with *Phytophthora cinnamomi* as test organism, *Phytopathology,* 45, 398, 1955.
4. **Kendrick, J. B., Jr. and Middleton, J. T.,** The efficacy of certain chemicals as fungicides for a variety of fruit, root and vascular pathogens, *Plant Dis. Rep.* 38, 350, 1954.
5. **Latham, A. J. and Linn, M. B.,** A comparison of soil column and petri dish techniques for the evaluation of soil fungitoxicants, *Phytopathology,* 58, 460, 1968.
6. **Carleton, M. A.,** Studies in the biology of the Uredineae. I. Notes on germination, *Bot. Gaz.,* 18, 447, 1893.
7. **Swingle, W. T.,** Bordeaux mixture: its chemistry, physical properties, and toxic effects on fungi and algae, *U.S. Dept. Agric. Div. Vegetable Physiol. Pathol. Bull.,* p.9, 1896.
8. **Reddick, D. and Wallace, E.,** On a laboratory method of determining the fungicidal value of a spray mixture or solution, *Science,* 31, 798, 1910.
9. **Montgomery, H. B. S. and Moore, M. H.,** A laboratory method for testing the toxicity of protective fungicides, *J. Pomol. Hortic. Sci.,* 15, 253, 1938.
10. **Martin, H.,** The significance of the bioassay in studies of fungicidal action, *Ann. Appl. Biol.,* 29, 326, 1942.
11. **McCallan, S. E. A., Wellman, R. H., and Wilcoxon, F.,** An analysis of factors causing variation in spore germination tests of fungicides, III. Slope of toxicity, curves, replicate tests, and fungi, *Contrib. Boyce Thompson Inst.,* 12, 49, 1941.
12. **McCallan, S. E. A. and Wilcoxon, F.,** An analysis of factors causing variation in spore germination tests of fungicides. I. Methods of obtaining spores, *Contrib. Boyce Thompson Inst.,* 11, 5, 1939.
13. **Wellman, R. H. and McCallan, S. E. A.,** An analysis of factors causing variation in spore germination tests of fungicides. IV. Time and temperature, *Contrib. Boyce Thompson Inst.,* 12, 431, 1942.
14. **Horsfall, J. G.,** Biological assay of protective fungicides, *Chron. Bot.,* 6, 292, 1941.
15. **Horsfall, J. G., Heuberger, J. W., Sharvelle, E. G., and Hamilton, J. M.,** A design for laboratory assay of fungicides, *Phytopathology,* 30, 515, 1940.
16. **Heuberger, J. W. and Turner, N.,** A laboratory apparatus for studying settling rate and fractionation of dusts, *Phytopathology,* 32, 166, 1942.
17. American Phytopathological Society, Committee on Standardization of Fungicidal Tests, The slide-germination method of evaluating protectant fungicides, *Phytopathology,* 33, 627, 1943.
18. **Peterson, P. D.,** The spore germination method of evaluating fungicides, *Phytopathology,* 31, 1108, 1941.
19. **Evans, A. C. and Martin, H.,** The incorporation of direct with protective insecticides and fungicides. I. The laboratory evaluation of water-soluble wetting agents as constituents of combined washes, *J. Pomol. Hortic. Sci.,* 13, 261, 1935.
20. **Heuberger, J. W., Marsh, R. W., and Martin, H.,** Studies upon the copper fungicides. IV. The fungicidal value of the copper oxides, *Ann. Appl. Biol.,* 24, 867, 1937.
21. **Heuberger, J. W. and Horsfall, J. W.,** Relation of particle size and color to fungicidal and protective value of cuprous oxide, *Phytopathology,* 29, 303, 1939.
22. **Marsh, R. W.,** Notes on a technique for the laboratory evaluation of protective fungicides, *Br. Mycol. Soc. Trans.,* 20, 304, 1936.
23. **Marsh, R. W.,** Some applications of laboratory biological tests to the evaluation of fungicides, *Ann. Appl. Biol.,* 25, 583, 1938.
24. **McCallan, S. E. A.** Studies on fungicides. II. Testing protective fungicides in the laboratory, *N.Y. (Cornell) Agric. Exp. Stn. Mem.,* 128, 8, 1930.

25. **Shafer, W. B.,** Fungitoxicity investigations using a novel spore germination technique, *Phytopathology,* 42, 519, 1952.
26. **Spencer, D. M.,** Polystyrene microbeakers for spore germination test, *Plant Pathol.* 11, 41, 1962.
27. **Tamura, H.,** Study on the evaluating method of organic fungicides on the "petri dish method" as a new method of bioassay in laboratory, *Bull. Natl. Inst. Agric. Sci.,* 4, 95, 1954.
28. **McIntosh, A. H.,** Graphical and other short statistical methods for "all-or-none" bioassay tests, *J. Sci. Food Agric.,* 12, 312, 1961.
29. **McCallan, S. E. A. and Wilcoxon, F.,** An analysis of factors causing variation in spore germination tests of fungicides. II. Methods of spraying, *Contrib. Boyce Thompson Inst.,* 11, 309, 1940.
30. **Blumer, S. and Kundert, J.,** Methoden der biologischen Laborpriifung von kupferpraparaten, *Phytopathol. Z.,* 17, 161, 1950.
31. **Ciferri, R. and Baldacci, E.,** Le cause d'errore del metodo di controllo *"in vitro"* e descrizione del" banco di nebulizzazione" Palagi, construzione Terzano, *Atti Inst. Bot. Univ. Pavia,* 5, (1), 106, 1943.
32. **Ten Houten, J. G. and Kraak, M.,** A vertical spraying apparatus for the laboratory evaluation of all types of liquid pest control materials, *Ann. Appl. Biol.,* 36, 394, 1949.
33. **Rabe, W.,** Studies on the action of wettable sulphur against conidia of *Venturia inaequalis* (Cke.) Wint, *Hoefchen Briege* (English ed.), 9, 1, 1956.
34. **Frick, E. L.,** Methods of reducing variability in the results of glass slide spore-germination assays of fungitoxicity, *Ann. Appl. Biol.,* 54, 349, 1964.
35. **McCallan, S. E. A. and Wellmam, R. H.,** Fungicidal versus fungistatic, *Contrib. Boyce Thompson Inst.,* 12, 451, 1942.
36. **Doran, W. L.,** Effect of external and internal factors on the germination of fungus spores, *Bull. Torr. Bot. Club,* 49, 313, 1922.
37. **Dimond, A. E., Horsfall, J. G., Heuberzer, J. W., and Stoddard, E. M.,** Role of the dosage-response curve in the evaluation of fungicides, *Conn. Agric. Exp. Stn. Bull.,* 451, 635, 1941.
38. **McClellan, W. D.,** Temperature as it affects spore germination in the presence of copper and sulphur, *Phytopathology,* 32, 394, 1942.
39. **Goldsworthy, M. C. and Green, E. L.,** Effect of low concentrations of copper on germination and growth of conidia of *Sclerotinia fructicola* and *Glomerella cingulata, J. Agric. Res. (U.S.),* 56, 489, 1938.
40. **Wilcoxon, F. and McCallan, S. E. A.,** Theoretical principles underlaying laboratory toxicity tests of fungicides, *Contrib. Boyce Thompson Inst.,* 10, 329, 1939.
41. **Parker-Rhodes, A. F.,** Studies on the mechanism of fungicidal action. II. Elements of the theory of variability, *Ann. Appl. Biol.,* 29, 126, 1942.
42. **McCallan, S. E. A.,** Characteristic curve for the action of copper sulfate on the germination of spores of *Sclerotinia fructicola* and *Alternaria oleracea, Contrib. Boyce Thompson Inst.,* 15, 77, 1948.
43. **Horsfall, J. G.,** *Principles of Fungicidal Action,* Chronica Botanica, Waltham, Mass., 1956, 279.
44. American Phytopathological Society, Committee on Standardization of Fungicidal Tests, Standard laboratory Bordeaux mixture, *Phytopathology,* 33, 633, 1943.
45. **Sampson, R. E. and Ludwig, R. A.,** Laboratory studies on the evaluation and activity of antifungal fumigants, *Can. J. Bot.,* 34, 37, 1956.
46. American Phytopathological Society, Committee on Standardization of Fungicidal Tests, Test tube dilution technique for use with the slide-germination method of evaluating protectant fungicides, *Phytopathology,* 37, 354, 1947.
47. **Miller, H. J.,** The use of *Venturia inaaquelis* and *Sclerotinia fructicola* with pure chemical stimulants in slide-germination tests of fungicides, *Phytopathology,* 34, 1009, 1944.
48. **Gattani, M. L.,** The agar plate spore germination method for testing fungicides, *Phytopathology,* 44, 113, 1954.
49. **Young, H. C. and Cooper, W. H.,** A method for determining the fungicidal coefficient of lime sulphur and other common fungicides. *Mich. Acad. Sci. Rep.* 19, 221, 1917.
50. **Nene, Y. L. and Thapliyal, P. N.,** *Fungicides in Plant Disease Control,* Oxford and IBH, New Delhi, 1979, 507,
51. **Gottlieb, D.,** A comparison of agar plate and test tube-dilution methods for the preliminary evaluation of fungicides, *Phytopathology,* 35, 485, 1945.
52. **Mandels, G. R. and Darby, R. T.,** A rapid cell volume assay for fungitoxicity using fungus spores, *J. Bacteriol.,* 65, 16, 1953.
53. **Mandels, G. R.,** The invertase of *Myrothecium verrucaria* spores, *Am. J. Bot.* 38, 213, 1951.
54. **Kluyver, A. J. and Perquin, L. H. C.,** Zur Methodik der Schimmelstoffwechse lunter-suchungen, *Biochem. Z.,* 266, 68, 1933.
55. **Davies, O. L., Duckworth, R. B., and Harris, G. C. M.,** A method for estimating percentage germination of fungal spores, *Nature (London),* 161, 642, 1948.

56. **Mandels, G. R. and Norton, A. B.**, Studies on the physiology of spores of the cellulytic fungus *Myrothecium verrucaria*, Quarter Master General Laboratories Research Report, Microbiology Series No. 11, 1948.

57. **Darby, R. T.**, Fungicide assay by spore germination in schaker flasks, *Appl. Microbiol.*, 8, 146, 1960.

58. **Schmitz, H.**, A suggested toximetric method for wood preservatives, *Ind. Eng. Chem. Anal. Ed.*, 2, 361, 1930.

59. **Palmiter, D. H. and Keitt, G. W.**, The toxicity of copper-lime-arsenic mixtures to certain phytopathogenic fungi grown on malt agar plates, *J. Agric. Res.*, 55, 439, 1937.

60. **Humphrey, C. J. and Fleming, R. M.**, The toxicity to fungi of various oils and salts; particularly those used in wood preservation, *U.S. Dept. Agric. Bull.*, p. 227, 1915.

61. **Colley, R. H.**, The evaluation of wood preservations, *Bell Teleph. Syst. Tech. Publ. Monogr.*, 2118, 1, 1952.

62. **Bateman, E.**, The effect of concentration on the toxicity of chemicals to living organisms, *U.S. Dept. Agric. Tech. Bull*, 346, 1, 1933.

63. **Findlay, W. P. K.**, Laboratory methods of testing wood preservatives, *Ann. Appl. Biol.*, 19, 271, 1932.

64. **Richards, C. A.**, Methods of testing the relative toxicity of wood preservatives, *Proc. Am. Wood Preserv. Assoc.*, 19, 127, 135, 1923.

65. **Lee, H. A. and Martin, J. P.**, A method for testing *in vitro* the toxicity of dust fungicides to fungus spores, *Phytopathology*, 17, 315, 1927.

66. **Carpenter, J. B.**, A toximetric study of some eradicant fungicides, *Phytopathology*, 32, 845, 1942.

67. **Bomer, M.**, Estimation of efficiency of fungitoxic compounds according to the inhibition of mycelial growth, *Folia Microbiol.*, 7, 185, 1962.

68. **Forsberg, J. L.**, A new method of evaluating fungicides, *Phytopathology*, 39, 172, 1949.

69. **Sharvelle, E. G. and Pelletier, E. N.**, Modified paper disc method for laboratory fungicidal bioassay, *Phytopathology*, 46, 36, 1956.

70. **Eastburg, P. H., McCaskey, B. L., and Thomas W. D.**, A mist method for the evaluation of fungicides, *Phytopathology*, 47, 519, 1957.

71. **Grosser, A. and Friedrich, W.**, Eine Testmethode zur Bestimmung von Pilzhemmstoffen, *A. Naturforsch.*, 26, 425, 1947.

72. **Henry, B. W. and Wagner, E. C.**, A rapid method of testing the effects of fungicides on fungi in culture, *Phytopathology*, 30, 1047, 1940.

73. **Latham, A. J. and Linn, M. B.**, An evaluation of certain fungicides for volatility, toxicity and specificity using a double petri dish diffusion chamber, *Plant Dis. Rep.*, 49, 398, 1965.

74. **Manten, A., Klopping, H. L., and Vander Kerk, G. J. M.**, Investigation on organic fungicides. II. A new method of evaluating antifungal substances in the laboratory, *Antonie van Leeuwenhoek J. Microbiol. Serol.*, 16, 282, 1950.

75. **Mason, C. L. and Powell, D.**, A *Phythium* plate method for evaluating fungicides, *Phytopathology*, 37, 527, 1947.

76. **Teschner, G.**, Einfache laboratoriumsteste als Beitrag zur fungiziden Muttelprufung, *Nachrichtenbl. Dtsch. Pflanzenschutzdienst (Brunsurck)*, 7, 170, 1955.

77. **Leben, C. and Keitt, G. W.**, A bioassay for tetramethyl-thiuram-disulfide, *Phytopathology*, 40, 950, 1950.

78. **Thornberry, H. H.**, A paper disc plate method for quantitative evaluation of fungicides and bactericides, *Phytopathology*, 40, 419, 1950.

79. **Richardson, L. T.**, Bioassay by the paper disk plate method, *Proc. Can. Phytopathol. Soc.*, 20, 21, 1953.

80. **Torgeson, D. C.**, Determination and measurement of fungitoxicity, in *Fungicides, an Advanced Treatise*, Academic Press, New York, 1967, 93.

81. **Vincent, J. M.**, Distortion of fungal hyphae in the presence of certain inhibitors, *Nature (London)*, 159, 850, 1927.

82. **Grover, R. K. and Moore, J. D.**, Toximetric studies of fungicides against brown rot organisms, *Sclerotinia fructicola* and *S. laxa*, *Phytopathology*, 52, 876, 1962.

83. **Jacks, H. and Smith, H. C.**, Soil disinfection. XII. Effect of fumigants on growth of soil fungi in culture, *N. Z. J. Sci. Technol.*, 6, 69, 1952.

84. **Miller, P. M. and Stoddard, E. M.**, Importance of fungicide volatility in controlling soil fungi, *Phytopathology*, 47, 24, 1957.

85. **Pryor, D. E. and Walker, J. C.**, A method for testing the toxicity of volatile compounds, *Phytopathology*, 29, 641, 1939.

86. **Loo, Y. H., Skell, P. S., Thornberry, H. H., Ehrlich, J., McGuire, J. M., Savage, G. M., and Sylvester, J. C.**, Assay of streptomycin by the paper-disc plate method, *J. Bacteriol.*, 50, 701, 1945.

87. **Vincent, J. G. and Vincent, H. W.**, Filter paper disc modification of the Oxford cup penicillin determination, *Proc. Soc. Exp. Biol. Med.*, 55, 162, 1944.

88. **Lockwood, J. L., Leben, C., and Keitt, G. W.,** A culture plate for agar diffusion assay, *Phytopathology,* 42, 447, 1952.

89. **Sharvelle, E. G.,** *The nature and Uses of Modern Fungicides,* Burgess Publishing, Minn., 1960, 308.

90. **Nickerson, W. J.,** Inhibition of fungus respiration: a metabolic bioassay method., *Science,* 103, 484, 1946.

91. **Mandels, G. R. and Siu, R. G. H.,** Rapid assay for growth: determination of microbiological susceptibility and fungistatic activity, *J. Bacteriol.,* 60, 249, 1950.

92. **Torgeson, D. C.,** Effect of fungicides on the respiration of three species of soil fungi, *Phytopathology,* 53, 891, 1963.

93. **McCallan, S. E. A. and Miller, L. P.,** Effect of fungicides on oxygen consumption and viability of mycelial pellets, *Contrib. Boyce Thompson Inst.,* 18, 484, 1957.

94. **Walker, A. T.,** Germination and respiration responses of *Myrothecium verrucaria* to organic fungicides, *Iowa State Coll. J. Sci.,* 30, 229, 1955.

95. **Domsch, K. H.,** Die wirkung von Bodenfungiciden. I. Wirkstoffspektrum, *Z. Pflanzenkr. Pflanzenschutz,* 65, 385, 1958.

96. **Domsch, K. H.,** Prufgang fur *Thielaviopsis* — und *Fusarium* — aktive Wirkstoffe, *Z. Pflanzenkr. Pflanzenschutz,* 69, 1, 1962.

97. **Corden, M. E. and Young, R. A.,** Evaluation of eradicant soil fungicides in the laboratory, *Phytopathology,* 52, 503, 1962.

98. **Torgeson, D. C.,** Laboratory and greenhouse evaluation of compounds for soil fungicidal activity, in Proc. 6th Pacific Coast Research Conf. on Soil Fungi, Horner, C. E., Ed. 24, 25, Mimeo, 1959.

99. **Domsch, K. H.,** Die Prufung von Bodenfungiciden. I. Pilz-Substrat-Fungicid-Kombinationen, *Plant Soil,* 10, 114, 1958.

100. **Lingappa, B. T. and Lockwood, J. L.,** Direct assay of soils for fungistasis, *Phytopathology,* 53, 529, 1963.

101. **Chinn, S. H. F. and Ledingham, R. J.,** A laboratory method for testing the fungicidal effect of chemicals on fungal spores in soil, *Phytopathology,* 52, 1041, 1962.

102. **Chinn, S. H. F., Ledingham, R. J., and Sallans, B. J.,** Population and viability studies of *Helminthosporium sativum* in field soils, *Can. J. Bot.,* 38, 533, 1960.

103. **Thayer, P. L. and Wehlberg, C.,** A method for evaluating soil fungicides, *Phytopathology,* 47, 535, 1963.

104. **Rodriguez-Kabana, R., Backman, P. A., and McLeod, C.,** A soil plate method for rapid screening of pesticides against *Sclerotium rolfsii, Plant Dis. Rep.,* 59, 439, 1975.

105. **Zehara, A. and Shacked, P.,** A baiting method for estimating the saprophytic activity of *Sclerotium rolfsii* in soil, *Phytopathology,* 58, 410, 1968.

106. **Adams, P. B.,** A buried membrane filter method for studying behaviour of soil fungi, *Phytopathology,* 57, 602, 1967.

107. **Nesheim, O. N. and Linn, M. B.,** Nylon mesh discs useful in the transfer of fungi and the evaluation of soil fungitoxicants, *Phytopathology,* 60, 395, 1970.

108. **Richardson, L. T. and Munnecke, D. E.,** A bioassay for volatile toxicants from fungicides in soil, *Phytopathology,* 54, 836, 1964.

109. **Maurer, C. L., Baker, R., Phillips, D. J., and Danielson, L.,** Evaluation of applied soil fumigants with the soil microbiological sampling tube, *Phytopathology,* 52, 957, 1962.

110. **Oserkowsky, J.,** Fungicidal effect on *Sclerotium rolfsi* of some compounds in aqueous solution and in the gaseous state, *Phytopathology,* 24, 815, 1934.

111. **Kendrick, J. B., Jr.,** Comparative fungitoxicity of some mono and dialkylsubstituted dethiocarbamate vapors and solutions, *Phytopathology,* 50, 641, 1960.

112. **Latham, A. J. and Linn, M. B.,** A non-fungistatic bacterio-static medium containing streptomycin and vancomycin, *Plant Dis. Rep.* 45, 866, 1961.

113. **Ezekiel, W. N.,** Evaluation of some soil fungicides by laboratory tests with *Phymatotrichum omnivorum, J. Agric. Res.,* 56, 553, 1938.

114. **Evans, E.,** Survival and recolonization by fungi in soil treated with formalin or carbon disulphide, *Trans. Br. Mycol. Soc.,* 38, 335, 1955.

115. **Cetas, R. C. and Whidden R.,** Evaluation of soil fungicides against *Fusarium solani* isolated from feeder roots of citrus trees, *Plant Dis. Rep.,* 44, 465, 1960.

116. **Newhall, A. G.,** An improved method of screening potential soil fungicides against *Fusarium oxysporum* f. *cubense, Plant Dis. Rep.,* 42, 677, 1958.

117. **Thomas, W. D., Jr.,** A soil column apparatus for the study of soil fungicides, *Phytopathology,* 47, 535, 1957.

118. **Klotz, L. J., Dewolfe, T. A., and Baines, R. C.,** Laboratory method for testing effectiveness of soil disinfestants, *Plant Dis. Rep.,* 43, 1174, 1959.

119. **Corden, M. E. and Young, R. A.,** The fungicidal activity and sorption of nabam in soil, *Phytopathology,* 50 (Abstr.), 83, 1960.

120. **Pote, H. L. and Thomas, W. O.,** An apparatus for testing fungicides in a soil column, *J. Colo. Wyo. Acad. Sci.,* 4, 49, 1954.
121. **Munnecke, D. E.,** Biological assay technique for studying fungicide drenches in soil, *Phytopathology,* 44 (Abstr.), 499, 1954.
122. **Munnecke, D. E.,** A biological assay of nonvolatile, diffusible fungicides in soil, *Phytopathology,* 48, 61, 1958.
123. **Sinha, A. P.,** Effect of Pesticides on Soil Microflora and Their Related Physiological Activities with Special Reference to Sugarbeet Crop, Ph. D. thesis, Kanpur University, Kanpur, India, 1976, 250.
124. **Sinha, A. P., Agnihotri, V. P., and Singh, K.,** Persistence of carbendazim in soil and its effect on rhizosphere fungi in sugarbeet seedlings, *Indian Phytopathol.,* 30, 21, 1980.
125. **Domsch, K. H.,** Der Einfluss von fungiziden Wirkstoffen auf die Bodenatmung, *Phytopathol. Z.,* 49, 291, 1964.
126. **Rushdi, M. and Jeffers, W. F.,** Effect of some soil factors on efficiency of fungicides in controlling *Rhizoctonia solani, Phytopathology,* 46, 88, 1956.
127. **Arndt, C. H.,** Evaluation of fungicides as protectants of cotton seedlings from infection by *Rhizoctonia solani, Plant Dis Rep.,* 37, 397, 1953.
128. **Mukhopadhyay, A. N. and Thakur, R. P.,** Control of *Sclerotium* root rot of sugar beet with systemic fungicides, *Plant Dis. Rep.,* 55, 630, 1971.
129. **Darrag, I. E. A. and Sinclair, J. B.,** Technique to evaluate chemotherapeutic activity of certain fungicides against *R. solani* in cotton seedlings, *Plant Dis. Rep.,* 52, 399, 1968.
130. **Sinclair, J. B.,** Greenhouse screening of certain fungicides for control of *Rhizoctonia* damping-off of cotton seedlings, *Plant Dis. Rep.,* 42, 1084, 1958.
131. **Sinclair, J. B.,** Laboratory and green house screening of various fungicides for control of *Rhizoctonia* damping-off of cotton seedlings, *Plant Dis. Rep.,* 41, 1045, 1957.
132. **Brinkerhoff, L. A., Brodie, B. B., and Kortsen, R. A.,** Cotton seedling tests with chemicals used as protectants against *Rhizoctonia solani* in the green house, *Plant Dis. Rep.,* 38, 476, 1954.
133. **Elsaid, H. M. and Sinclair, J. B.,** A new green house technique for evaluating fungicides for control of cotton sore-skin, *Plant Dis. Rep.,* 46, 852, 1962.
134. **Reinhart, J. H.,** A method of evaluating fungicides in the soil under controlled conditions, *Plant Dis. Rep.,* 44, 648, 1960.
135. **Ranney, C. D. and Bird, L. S.,** Green house evaluation of in-the-furrow fungicides at two temperatures as a control measure for cotton seedling necrosis, *Plant Dis. Rep.,* 40, 1032, 1956.
136. **Domsch, K. H.,** Die Prufung von Bodenfungiciden. II. Pilz Boden-Wint-Fungicid-Kombination en, *Plant Soil,* 10, 132, 1958.
137. **Torgeson, D. C., Hensley, W. H., and Lambrech, J. A.,** N-Phenylmaleimiden and related compounds as soil fungicides, *Contrib. Boyce Thompson Inst.,* 22, 67, 1963.
138. **Davison, A. D. and Vaughn, J. R.,** Effect of several antibiotics and other organic chemicals on isolates of fungi which cause bean root rot, *Plant Dis. Rep.,* 41, 432, 1957.
139. **Rouxel, F., Mention, M., Cassini, R., and Bouhot, D.,** A test design for the identification of soil fungicides effective against *Fusarium oxysporum* cause of wilt, *Phytiatr. Phytopharm.,* 20, 233, 1971.
140. **Colhoun, J.,** Biological technique for the evaluation of fungicides. III. The evolution of a technique for the evaluation of soil fungicide for the control of club-rot disease of Brassicae, *Ann. Appl. Biol.,* 41, 290, 1954.
141. **Colhoun, J.,** A study of the epidemiology of club-root disease of Brassicae, *Ann. Appl. Biol.* 40, 262, 1953.
142. **McCallan, S. E. A.,** Evaluation of fungicides in the laboratory, in *Plant Pathology, Problems and Progress, 1908-1958,* Holton, C. S., Fisher, G. W., Fulton, R. W., Hart, H., and McCallan, S. E. A., Eds., University of Wisconsin Press, Madison, 1959, 248.
143. **Domsch, K. H.,** Soil fungicides, *Annu. Rev. Phytopathol.,* 2, 293, 1964.
144. **Klomparens, W. and Vaughn, J. R.,** The correlation of laboratory screening of turf fungicides with field studies, *Mich. Agric. Exp. Stn. Q. Bull,* 34, 425, 1952.
145. **Kendrick, J. B., Jr., and Zentmyer, G. A.,** Laboratory evaluation of chemicals as potential soil fungicides, *Phytopathology,* 47, 20, 1957.
146. **Schmitt, C. G.,** Comparison of volatile soil fungicides, *Phytopathology,* 39, 21, 1949.
147. **Wilhelm, S. and Ferguson, J.,** Soil fumigation against *Verticillium alboatrum, Phytopathology,* 43, 593, 1953.
148. **Zentmyer, G. A. and Kendrick, J. B.,** Fungicidal action of volatile soil fumigants, *Phytopathology,* 39, 864, 1949.

INDEX